The Cell as A Machine

This unique introductory text explains cell functions using the engineering principles of robust devices. Adopting a process-based approach to understanding cell and tissue biology, it describes the molecular and mechanical features that enable the cell to be robust in operating its various components, and explores the ways in which molecular modules respond to environmental signals to execute complex functions. The design and operation of a variety of complex functions are covered, including engineering lipid bilayers to provide fluid boundaries and mechanical controls, adjusting cell shape and forces with dynamic filament networks, and DNA packaging for information retrieval and propagation. Numerous problems, case studies and application examples help readers connect theory with practice, and solutions for instructors and videos of lectures accompany the book online. Assuming only basic mathematical knowledge, this is an invaluable resource for graduate and senior undergraduate students taking single-semester courses in cell mechanics, biophysics and cell biology.

Michael Sheetz is the Founding Director of the Mechanobiology Institute and a distinguished professor in the Department of Biological Sciences both at the National University of Singapore.

Hanry Yu is a professor in both the Department of Physiology and the Mechanobiology Institute at the National University of Singapore. He is also the Director of the National University Health System's research facilities in microscopy and cytometry.

CAMBRIDGE TEXTS IN BIOMEDICAL ENGINEERING

Series Editors
W. Mark Saltzman, Yale University
Shu Chien, University of California, San Diego

Series Advisors
Jerry Collins, Alabama A & M University
Robert Malkin, Duke University
Kathy Ferrara, University of California, Davis
Nicholas Peppas, University of Texas, Austin
Roger Kamm, Massachusetts Institute of Technology
Masaaki Sato, Tohoku University, Japan
Christine Schmidt, University of Florida
George Truskey, Duke University
Douglas Lauffenburger, Massachusetts Institute of Technology

Cambridge Texts in Biomedical Engineering provide a forum for high-quality textbooks targeted at undergraduate and graduate courses in biomedical engineering. They cover a broad range of biomedical engineering topics from introductory texts to advanced topics, including biomechanics, physiology, biomedical instrumentation, imaging, signals and systems, cell engineering and bioinformatics, as well as other relevant subjects, with a blending of theory and practice. While aiming primarily at biomedical engineering students, this series is also suitable for courses in broader disciplines in engineering, the life sciences and medicine.

The Cell as A Machine

Michael Sheetz

National University of Singapore

Hanry Yu

National University of Singapore

CAMBRIDGE
UNIVERSITY PRESS

University Printing House, Cambridge CB2 8BS, United Kingdom

One Liberty Plaza, 20th Floor, New York, NY 10006, USA

477 Williamstown Road, Port Melbourne, VIC 3207, Australia

314-321, 3rd Floor, Plot 3, Splendor Forum, Jasola District Centre, New Delhi - 110025, India

79 Anson Road, #06-04/06, Singapore 079906

Cambridge University Press is part of the University of Cambridge.

It furthers the University's mission by disseminating knowledge in the pursuit of education, learning and research at the highest international levels of excellence.

www.cambridge.org
Information on this title: www.cambridge.org/9781107052734
DOI: 10.1017/9781107280809

First published 2018

A catalogue record for this publication is available from the British Library

Library of Congress Cataloging in Publication data
Names: Sheetz, Michael P., author. | Yu, Hanry, author.
Title: The cell as a machine / Michael Sheetz, Columbia University, New York, Hanry Yu,
    National University of Singapore.
Description: Cambridge, United Kingdom ; New York, NY : Cambridge University Press, 2018. |
    Series: Cambridge texts in biomedical engineering | Includes bibliographical references and index.
Identifiers: LCCN 2017046256 | ISBN 9781107052734 (hardback)
Subjects: LCSH: Cells–Mechanical properties. | Cell physiology.
Classification: LCC QH645.5 .S54 2018 | DDC 571.6–dc23 LC record available at
    https://lccn.loc.gov/2017046256

ISBN 978-1-107-05273-4 Hardback

Additional resources available at www.cambridge.org/sheetz

Contents

Preface and acknowledgments

The motivation for this book is to provide an approach for understanding and integrating the complex functions of cells that shape tissues and drive growth and differentiation. Correlative information in the catalogs of DNA sequences, mRNA levels, interactomes, and biochemical reactions involved in complex cellular functions provides an insufficient understanding. Having the parts list of a complex machine does not enable one to assemble that machine, much less to understand how it works. In contrast, a bioengineering approach that focuses on the physical aspects of cells and tissues provides only a part of the function. A detailed engineering diagram of the steps involved and how the physical parameters relate to the biochemical functions is needed to understand most cell functions. For example, the biochemical reactions involved in the uncontrolled cell proliferation of cancer cells are known, as are the physical behaviors of tumor cells, but how physical signals from the cell environment are integrated with the biochemistry of growth is not known. In normal cells, physical feedback from matrix rigidity and morphology controls normal cell growth, whereas in cancer cells, those same parameters are misread, resulting in uncontrolled growth. By considering the reverse engineering of complex cell functions as a problem similar to reverse engineering a complex automobile, it should be possible to understand cellular functions.

The premise of this book is that cells are small, self-replicating machines that exploit fundamental physical and physical–chemical principles of mesoscale objects to pass on the DNA of the organism that they form. From the revolution in molecular biology, we know the sequences of many genomes, as well as the sequences of the mRNAs in many specific cell types, which identify the proteins that are present. Further, we know many of the mutations that correlate with specific genetic diseases or altered phenotypes. Thus, we possess a cellular parts list. Fortunately, there are now ways to measure subcellular forces, protein positions at the nanometer level, and protein dynamics. By coupling those tools with real-time observations of cell functions, it will be possible to determine the steps in very complex functions.

Our approach is to model the cell as a complex machine that has been selected for robustness over many millions of years. Thus, the functions in cells follow the general rules for robust devices, and can be described by an engineering diagram that details the steps in the function, the potential branch points, and the various outcomes. At this time, there is detailed knowledge of only a few functions that enable this approach, but the paradigm provides a path to a detailed understanding of how cells actually work. This book describes how such an approach can help with understanding the major functions of a typical mammalian cell such as a fibroblast.

There are a number of basic tools that one needs to understand how cell functions are performed. For example, all intracellular processes are diffusively driven and are therefore stochastic. In contrast, we are used to thinking about macroscopic machines that use momentum and are naturally deterministic. Noisy, stochastic processes can be averaged to give a deterministic outcome, but we don't understand how cells with the same genetic code can produce nearly identical twins using those noisy processes.

To replicate themselves, cells rely upon a number of basic functions that can be broken down into component functional steps. These steps, in turn, are typically driven by multi-protein complexes (cofactors and environmental factors such as lipid surfaces or membranes often play critical roles). Each of the functional processes will be considered as a system that is engineered for robustness and efficiency. From such a systems-engineering viewpoint, physical and physical–chemical parameters are critical in determining the throughput, efficiency, and fidelity of the product of a given function. Because we have an incomplete understanding of many cell functions, this description will also be incomplete. Our emphasis is on the principles that govern these functions and on providing senior undergraduate and beginning graduate students with the tools needed to address the general problem of understanding cellular functions in other systems.

For example, over 70+ known proteins participate in clathrin-dependent endocytosis. This process bends the plasma membrane inward, ultimately to form an intracellular vesicle. Knowing the parts involved, and roughly where they are in the complex, we can start to build a detailed description of the process at the nanometer level. It starts with simple steps from initial recruitment of the components to be taken into the vesicle, to the inward bending of the membrane, to forming a deep invagination of the membrane, and, finally, to pinching off a vesicle from the plasma membrane. The enzymatic activities of the proteins must then be linked with the specific sub-steps in these larger steps. In the extreme, the goal is to produce a detailed description of the steps and the roles of each of the 70+ proteins in those steps. The functions provide a simple way to organize our knowledge of cellular physiology. Further, the characterization of the links between the different functions will enable the control processes to be understood.

Although we are far from having an atlas that describes all cellular functions, we are at a point where we can imagine building such an atlas for all 300+ cell types in the human body. We suggest that the vast amount of information of biological cells can be best linked with the functions that are involved and their regulatory pathways. In addition to knowing how to treat diseases and engineer organisms better, having a better understanding of subcellular functions will enable us to exploit nanodevices, such as computers and nanomachines. As they approach the scale of functional complexes in cells, they will also be governed by the principles that govern mesoscale processes. From a complete understanding of the several hundred major cellular functions that are commonly used by mammalian cells, it should be possible to develop a repair manual for the human organism. This would have many benefits for health and welfare.

For the past few years, both basic biology and translational biomedical engineering research communities have paid increasing attention to integrating biochemical reactions in relevant physical contexts, such as the hardened tissue in wound healing, to causally determine the cellular and tissue functions. Many of these are described with biochemical and physical terms in this book. This emerging research approach of understanding and controlling both the physical and biochemical aspects of cellular and tissue functions is what we called the mechanobiology approach, which systematically describes concepts, terminology, technologies, and methodology to develop detailed descriptions of biological functions.

Finally, I would like to acknowledge the support of the many people who made this book possible. First, there were many students and colleagues who helped to hone this message and provide relevant examples. Second, my wife, Dr. Linda Kenney, has patiently supported this effort, which has taken much longer than either of us had expected. My co-author, Hanry Yu, has provided a lot of encouragement and has really helped to make this more relevant to the engineering community. Our illustrator, Dr. Cindy Zhang, has done an excellent job in enlivening important concepts. Steven Wolf and Stuart McLaughlin have kindly proofread the book. This would not have been possible without the support of the Mechanobiology Institute at the National University of Singapore (supported by an RCE grant from the Singapore National Research Foundation and Ministry of Education). Early concepts for this book were formed in a Nanomedicine Center funded by National Institutes of Health, USA.

The full figure caption information can be found online here: www.cambridge .org/sheetz

PART I

Principles of Complex Functions in Robust Machines

1 Robust Self-replicating Machines Shaped by Evolution

Biological cells are wonderfully robust self-replicating machines that selfishly propagate their DNA until environmental factors such as nutrients limit them. Cells have mastered the use of nanoscale devices in a water environment to fabricate their progeny and perform many critical functions for survival. To understand how cellular machines work, we need to understand the identity and structure of molecular components as well as the network of interactions between these molecules. We also need to study the complex cell functions in which an orchestra of molecules interacts transiently in space and time to cause an observable cellular behavior. Such an integrated systems engineering view of the cell is currently far from our reach; however, there are sufficient tools and techniques in place to provide us with many useful insights into the complex functions of the wonderful machines that are cells.

In this book, we will mainly discuss eukaryotic cells and where possible the details of mouse embryo fibroblasts (3T3 cells) and a human cancer line (HeLa). Because there are over 300 different cell types with important specializations, we have chosen a relatively general cell type to describe what cells do. Many complex functions are missing in fibroblasts and there will be some necessary extrapolations at certain points. 3T3 and HeLa cells do not form multicellular tissues; cell–cell interactions and the higher-order functions of tissues will be discussed in later chapters. The level of integration and complexity needed for the control of multicellular complex functions is at least an order of magnitude greater than for single cell complex functions. We first focus on dissecting the individual complex functions in single cells, because it is then possible to analyze multicellular complex functions in a modular fashion.

In this chapter, we will define the biological cell that we wish to understand and discuss how evolution has made it more robust over the last two billion years (Huxley, 1958). A cutaway diagram of a cell in suspension (Figure 1.1) provides an illustration of the basic organelle compartments that are found in mammalian cells. Compartmentalization of functions is critical for a complex machine like a cell. Each compartment performs some basic function(s) and communicates bidirectionally with the rest of the cell. Often compartments, like the cell itself,

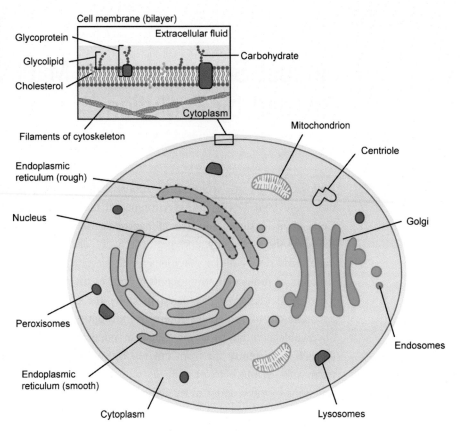

are bounded by lipid bilayers, whose hydrophobic cores effectively prevent open movements of ions and charged materials into or out of the compartments. The nucleus contains the DNA that encodes the plan for the cell that is read out by RNA polymerases. The mRNAs produced in the nucleus are transported through nuclear pores to the cytoplasm where they can be translated into protein by ribosomes. Proteins destined for extracellular secretion or the outer surface of the plasma membrane (the boundary of the cell) are translated by ribosomes on the rough endoplasmic reticulum. Carbohydrates are added to the proteins (making them glycoproteins) in the endoplasmic reticulum and the Golgi before movement to the plasma membrane in secretory vesicles (not shown). The plasma membrane barrier is primarily formed by a lipid bilayer with cholesterol and embedded glycoproteins. Actin filaments support the plasma membrane mechanically and attach it to the cell cytoskeleton that structures the cytoplasm. In cell mitosis, the centrioles organize the microtubule arrays that carry the chromosomes to the daughter cells and specify the location of the cleavage furrow for forming the two daughter cells. Invaginations of the plasma membrane form endocytic vesicles that pass inactive proteins on to the lysosomes where they are degraded. ATP to power the cell comes primarily from the mitochondria. This is only a partial list of the cellular compartments, as many other cell functions are performed by localized complexes that could be considered as compartments even when they aren't surrounded by a membrane bilayer. This machine is able to grow and divide once a day.

1.1 The Cell is a Self-contained and Self-replicating Machine

A defining aspect of cellular life is the ability of a cell to grow and produce two daughter cells, i.e. to replicate itself. Thus, our working definition of a eukaryotic cell is a self-contained unit that can replicate itself in an organism. With proper nutrients and a growth-stimulating environment, a cell can make new proteins, nucleic acids, carbohydrates and lipids in a controlled way so that two daughter cells can be formed after 24 hours. This is not a continuous process and the cell goes through at least three distinct phases of the cell cycle: growth phase one, synthesis of DNA, and growth phase two. Further, the synthesis of at least proteins and ribonucleic acids occurs in bursts followed by periods of inactivity. As in any complex process, a series of steps needs to be completed in the proper sequence to replicate the original cell. Because the cell has many functional compartments, parallel processing can occur, but there will be check points where all of the necessary steps need to be completed before the cell can transition to the next phase in the cycle of self-replication.

An important aspect of the self-replication process is that each cell of an organism contains the information needed to form that organism in its nucleus. Thus, the cell replication also replicates the genetic material needed to direct the continued propagation of the organism. Recent studies have demonstrated that an adult cell can be used to create a new organism either through transplantation of an adult cell nucleus into an egg or by formation of induced pluripotent stem cells. With the exponential growth of the number of cells, the mass of the cells can rapidly exceeds the supply of nutrients required for further growth. Controls on growth are also needed to regulate the size of the organism. Thus, the self-replication aspect of cells is limited in most organisms and even bacteria have phases with limited growth. The organism and not the individual cell is critical for the propagation of the DNA, and exponential growth of organisms is limited only by resources.

The notion of self-replicating machines is not new, and several have considered the theoretical aspects of what is needed in a self-replication machine. With the recent development of rapid-prototyping printers that can build three-dimensional structures, it is conceivable that a machine could be developed to replicate itself with a few raw materials. The critical issue is how to encode the information needed to direct the self-replication process in a form that the machine could reproduce. This aspect of bacterial cells has been addressed in the field of synthetic biology, Gibson and colleagues have synthesized the DNA of a bacterium and succeeded in putting that DNA into the cytoplasm of another bacterium. The bacterium grew with the synthesized DNA and the previous genome was degraded, making a synthesized bacterium (Gibson et al., 2008). Thus, it is possible to reprogram a bacterial self-replicating machine to produce a new bacterium with a totally defined set of genes. Understanding the problems that must be overcome to enable an artificial organism to replicate itself will help us further understand cellular life, which will in turn serve as a foundation for the engineering of drugs and biologics to treat diseases and facilitate tissue regeneration for biomedical applications.

1.2   Cell Functions have Evolved According to Darwinian Selection

There is geological evidence of cellular life in fossilized remains of organisms over the past 2 billion years on earth. Whether the first cells formed directly from some primordial soup, were created by God or came to Earth from some other planet is not considered here. Instead, we are concerned about how the cell functions observed today were selectively preserved through an evolutionary process known as Darwinian selection. As Julian Huxley wrote in the introduction to the 150th Anniversary edition of Darwin's *On the Origin of the Species*, "Natural selection was seen, not as involving the sharp alternative of life or death, but as the result of the differential survival of variants; and it was established that even slight advantages, of one-half of one percent or less, could have important evolutionary effects." Thus, over such a long period, countless individual cells have lived and the process of evolution has modified the cells in any given organism for robustness (see textbox for analogy to selective evolution of automobiles). Basically, mutations in the genome will modify the organism's ability to survive to produce offspring. As the environment changes over time, life vs. extinction should be considered in a stochastic context where greater probability of survival is key and small advantages over large populations and many generations will dominate. The critical message is that we only find the successes alive. No moral message is intended to be drawn from the fact that some survived and others did not; rather, it is important to understand how optimization of certain functions increased survivability.

Selection in the Marketplace and Relation to Performance

Darwinian selection is analogous in many ways to the process of evolution in the commercial marketplace. Since the first automobile was developed in the late nineteenth century, billions of vehicles have been produced with increasing sophistication. Refinements such as fuel injection, seatbelts and even cup-holders have been incrementally introduced into most modern vehicles according to buyers' preferences. Features that resulted from the fashion-of-the-day or whimsical fads were not often retained in later models. However, features that improve performance, efficiency and comfort are selectively preserved, and stand the test of time. Small differences in the perceived advantage of features are often sufficient for those features to be incorporated into almost all vehicles. For example, cup-holders are a small convenience making it marginally more likely that a person will buy a car with them than without. However, virtually all cars sold in the USA have cup-holders.

This analogy can help us to understand some of the difficulties faced by scientists who commonly try to understand how cells work by removing or mutating proteins in those cells. They assume that the proteins have functional advantages for the cell and altering the proteins will alter the relevant function. However, many proteins have been deleted from cells or organisms without any obvious change in function. If those proteins confer a small selective advantage, they can be selected for over many generations; however, their removal may not cause a major change in cell function. Similarly, a naïve person would find it very difficult to understand the advantage of a cup-holder unless they tested the driving performance of people drinking hot coffee. The point is that the selective advantages needed for proteins or mutations of proteins to be incorporated in all cells over thousands of generations are only in the 1% range that would not be readily observed by most assays. Darwinian selection will result in highly sophisticated machines robust to many perturbations that have finely tuned features.

Time to Extinction: It is useful to consider how long the genes coding for less-competitive proteins will remain in the gene pool.

Steady state system: In a stable ecosystem, the population of any given species will assume some average level over cyclic fluctuations. For example, the mouse population on a hypothetical island is on average 100,000 and they have a generation time of 6 months. Assume that a mutation in a protein gives the mice a 1% survival advantage as a heterozygote and 2% as a homozygote over the original protein in one generation. This means for the sake of calculation that for every 100 homozygous mutant mice in the first generation 102 will reach the second

(continued)

generation at the expense of the original mice. After 1000 generations and several cycles of population growth and contraction, the population of animals with the original gene will be very small (less than 3% of the total) and the number of homozygous animals will be even less (it should be pointed out that inbreeding of homozygous animals will increase the rate of decline).

Feast–famine cycles: More common in history is the expansion of the population of a given species and then a dramatic decline in that population due to the expansion of a predator population, change in habitat or a disease. In those cases, the selection pressures can be much more severe and over 90% of a given population can be lost in a brief period. Those situations are difficult to model because they select for markedly different factors, e.g. the decimation of buffalo populations by early settlers and of native Hawaiians by European diseases. The high selection pressures in those situations can often select for mutant genes in the population that would otherwise not be favorable, e.g. malaria selection of G6PD deficiency and sickle hemoglobin. In such cases, there may be clonal selection of individuals who have particular genes.

It is difficult to do evolutionary studies on mammals because of their long lifetimes; however, in bacterial systems it is possible to apply selective pressures and see mutations. For example, many recent studies have looked at the mutations in rate-limiting proteins in metabolic pathways that are produced upon dramatic changes in the temperature. Often, reproducible mutations in the key enzymes are found because those changes confer better performance at the new temperature. These experiments take a month or so because the bacterial doubling time is about 30 minutes and nearly 1000 generations will occur in that period (reviewed in Lindsay, 1995). As noted in the text box, if a mutation causes only a small decrease in the generation time, then in 1000 generations that mutation will dominate in the cell population.

In the same way, we can look at a cell or an organism that has a reasonable selective advantage over another cell type in the population, and with a few assumptions calculate the number of generations that must pass before the disadvantaged species is lost. If generation times of small mammals are on the order of months, several hundred million generations could have occurred in the 60 million years since dinosaurs walked the earth. Over that number of generations, even a gene that confers a weak selective advantage could displace other related genes. Thus, the organisms have tested many mutations of proteins and have shared DNA with other species through infections and other gene transfer mechanisms. For this reason, we will often need to have very quantitative assays of function to understand where a given protein plays a role in that function.

1.3   Robustness is Strongly Favored in the Evolution of Cell Functions

Cellular functions that enable species to survive and carry forward the genetic materials must be preserved through many different adverse conditions. The property of robustness is therefore strongly favored in cells. Robustness is defined as *the ability to adapt to and tolerate a variety of conditions*, such as the changes in: (1) number of proteins per cell, (2) salinity and pH, (3) temperature, (4) nutrient level, and (5) other environmental factors such as disease or predation. Robustness has a cost and very specialized systems adapted to an unusual environment can at times outcompete a more generally adaptable organism. For example, in caves where there are limited food resources, salamanders have evolved to become blind and to have no pigment in their skin. Producing pigment and eyes requires valuable energy and the animals without those features have a slight selective advantage. However, if the cave was suddenly opened to the light, the specialized salamanders would not compete well because they would be seen more easily by predators and would not be able to see predators to evade them. Therefore, a robust organism might not dominate in every situation, but over time with many different challenges and subtle environmental changes, robustness is favored over efficiency to survive.

Robustness is often dependent upon changes in the protein composition of cells through changes in gene expression. The mechanisms of regulated gene expression enable cells to express new proteins as needed. Specialized proteins can tune-down sensitivity to environmental perturbations, aid in repair processes after injury or other types of insults, or transform a cell from one phenotype to another. Different cell types will exhibit different expression programs, e.g. fibroblast vs. epithelial cell, and many of these programs are known. However, depending on the phase of the cell cycle, the same cell can also exhibit different expression patterns (Figure 1.2). For example, a fibroblast in division at M phase shuts down many complex functions, including growth, and drastically increases its ability to change shape – as though it were a different cell. In the case of fibroblasts growing in wounded tissue where damage to the environment is often severe, they must express many new proteins to enable them to rapidly divide and initiate healing.

The ability of single cells to change phenotype to survive an environmental challenge is critical for the propagation of the DNA. An example of this is the ability of cells to express heat shock proteins at high temperatures. Heat shock proteins aid in the refolding of heat-denatured proteins and thereby reduce the heat damage to the cell. The various environmental challenges that a robust cell should be able to live through (under normal circumstances) are outlined in detail below (Figure 1.3).

Figure 1.2

Cell phases and
differentiation with different
sets of complex functions
on and off.

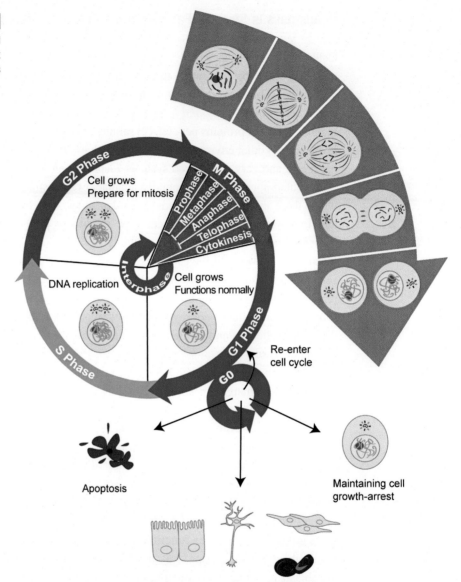

Differentiate based on different environmental cues

1.3.1   Number of Proteins per Cell

It seems that the number of molecules per cell of many proteins involved in a
specific function can vary over a wide range without altering that function
significantly. For example, during bacterial chemotaxis (the ability of bacteria to
move up a concentration gradient to a food source), the concentration of many

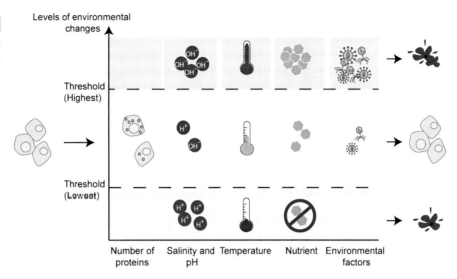

Figure 1.3

Robust cell adaptation in different environmental fluctuations.

proteins in the chemotactic pathway can vary over a thousandfold without altering the ability of the cells to undergo chemotaxis. There is a compromise in the rate of the response, however, when certain proteins are depleted. During development where cells are changing to a new type, the change in the cell protein composition can be incomplete and yet it will be able to function as the new cell. Further, there can be a major loss of many proteins during starvation without a major compromise of cell function. Although there are exceptions where the level of functional activity is directly proportional to the number of proteins per cell, in most cases the cell is robust to large changes in protein number. A rationale for this is that the production of protein is slow and often stochastic plus many proteins are concentrated at the site of function so that the concentration in cytoplasm is not particularly relevant. To turn off functions, cells often rely upon targeted proteases that can rapidly degrade critical proteins in a functional complex. Thus, the concentration of a given protein that is involved in an important function is not a good indication of the level of functional activity.

1.3.2   Salinity and pH

For bacteria and many fishes, the ability to withstand osmotic shock, as well as changes in ion content or pH of the surrounding medium, is critical to their survival. A sudden flood can decrease the salinity in a saltmarsh by over twofold and the organisms must therefore adapt rapidly, otherwise they will undergo lysis (rupture of the plasma membrane) when water moves into the cells. Along with osmotic pressure, additional factors such as mechanical tears and detergents can also cause cells to lyse. Mammalian cells possess a large reservoir of internal lipid that can be used to seal over a leak in the plasma membrane that results during lysis and this prevents the critical contents of the cell from leaking out. Surprisingly, a large fraction of lysed cells will reseal their membranes and survive that trauma. A consequence of this observation is that many cell functions are performed by proteins that are anchored to the membrane or to cytoskeletal structures and these will not diffuse out of the lysed cell. We will discuss cell volume control and osmotic pressure as a regulator of ion channels in later chapters.

Calculations of Numbers of Molecules/Cell and Concentration

Eukaryotic cells range in size from about 4 μm to millimeters (for larger syncytial cells). An average size for the common cells used in cell culture (mouse 3T3 cells) is 2700 μm^3 for a suspended cell (volume $4/3\pi r^3$ = 2700 μm^3 or 2.7×10^{-9} cm^3). In contrast, bacteria (prokaryotic cells) such as *Eschericia coli* are cylinders of about 2 μm in length and 0.8 μm in diameter (volume $l\pi r^2$ ~1 μm^3 or 1×10^{-12} cm^3). The concentration of cytoplasmic proteins is about 180 mg/ml. If we assume that the average molecular weight of proteins is 50 kDa, then the overall concentration is 3.6 mM or 2×10^{18} molecules per ml (alternatively, 8×10^9 molecules per eukaryotic cell or 2×10^6 molecules per prokaryotic cell). If we carry this approximate calculation further and introduce the number of different protein molecules in a cell (10,000 for the eukaryotic and 2000 for the prokaryotic), then the number of molecules of any given protein will be on the order of 10^5 and 10^3, respectively. When we divide these numbers by Avagadro's number and the cell volume in liters, the concentrations are about 10^{-7} and 10^{-6} M, respectively. These numbers have important implications for the functional organization of cells that will be discussed later.

1.3.3   Temperature

Freeze–thaw cycles are common for organisms in polar climates and cells must be able to survive freezing or organisms will be lost. At low temperature some filament systems such as microtubules will disassemble which will block mitosis (cell division to form two daughter cells). In tropical climates, excess heat poses a different set of problems. Denaturation, or unfolding, of proteins may result from excessive heat, and this can lead to an excess of non-functional protein and insoluble aggregates that the cell must clear. Expression of a family of proteins known as heat shock proteins (HSPs) is stimulated at high temperatures and under conditions of stress to the organism. This highlights the need for cells to have intrinsic mechanisms that protect wider cellular functions against temperature fluctuations. Antifreeze proteins help to reduce ice crystal damage upon freezing and HSPs help to refold proteins that are denatured at high temperatures.

1.3.4 Nutrient Level

To preserve viability, fluctuations in nutrient level must be accompanied by changes in cell metabolic levels. One way cells address this need is to store nutrients such as fat and glycogen away for hard times; however, they will also alter metabolic pathways as determined by the nutrients available to them. For example, homeostatic mechanisms are in place to enable transitions from carbohydrate to fat metabolism. Shifting from metabolizing nutrients in the gut to stored material and recycling components from its own cells is important during starvation. Switching from one type of metabolism to another involves a major change in not only the enzymes but also potentially in the shape of the mitochondrion, the organelle responsible for the oxidative metabolism that produces more energy.

1.3.5 Other Environmental Factors

This is a catch-all category that lumps together other factors that are critical for organism survival. Parasites (viruses, bacteria, or larger organisms) and other infectious agents all contribute to the function of complex organisms, often synergistically. For example, fungi and bacteria cover most of the human body surfaces (internal and external). Although they are required for normal tissue and organ function, they are also associated with diseases. Hosts have developed tolerance for parasites within a range of their densities and parasitic activities. Beyond the normal range, the symbiotic relationship breaks down. Most major parasites produce chronic infections that partially weaken the host over years. In the meantime, the parasites will spread their DNA to other hosts. Viruses and bacteria are beneficial, however, in that they can carry genetic information from one organism to another and provide a way to propagate new and better proteins throughout a population. When a robust function is selected over others, the ability of bacteria to share DNA and the ability of viruses to carry DNA between organisms assures that any selective advantage will be shared among different organisms or cell types over time.

Although it is obvious that different species will have different ways of doing things, there are complex functions that are shared. These provide the foundation for most biological systems; e.g. the flow of information from DNA to RNA to protein is the paradigm utilized in virtually all species. From an engineering perspective, there are a number of complex functions of cellular machines (e.g. DNA replication, mRNA translation into protein) that are essentially the same throughout the biological kingdom. These complex functions often depend upon mechanisms that are observed across species, even though some of the protein components involved in these mechanisms might be different. The notion of shared mechanisms means that while the steps or methods of carrying out a function are retained across species, improvements in the components that carry out these steps can still occur. These modifications will provide some species a selective advantage. For example, the ribosome is a basic machine that has been optimized in bacteria to function under a wide variety of conditions. Humans and other mammals have similar ribosomes, but some components are adapted to a mammalian environment and respond to different control mechanisms. However, improvements to the basic machine of the ribosome can potentially be transferred to human ribosomes, such as heat tolerance. This is again analogous to the car industry. Different manufacturers have produced cars with different designs, but the basic mechanisms that allow cars to accelerate, brake and turn are essentially the same, albeit with different parts. For example, a Honda disc brake will be made of different parts than a Ford disc brake, yet they both work in the same way to slow down the car. Hence the function is robust and has been retained in many car designs. As functions or mechanisms are shared, we can look at the steps involved in a given function in a single cell type to understand how molecular components would interact to produce similar quantifiable functions in other cells. These shared functions are described, in this text, as "functional modules" which we define as *a minimal set of molecular components working together to yield a measurable activity or a function.* Understanding functional modules in one system will enable the understanding of similar functions in other systems.

1.4 The Cell State and its Environment have an Effect on Cellular Functions

The ability to change state is an important feature of a robust system. Just as humans respond differently to the same stimulus when we are in different contexts, so too will cells. How cells respond will be based both on their history and their current context. Cell state (which can be defined by the set of functional modules that are active in producing the observed phenotype) is changed by the activation of signaling pathways, movement through the cell cycle, and a complex set of environmental cues. For example, cells on fibronectin, which is a common component of the extracellular matrix, will express a different group of proteins than when the

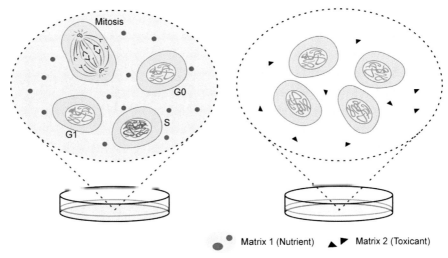

Figure 1.4

Mixture of cells
response to stimuli.

same cells are grown on a different matrix component such as laminin. This means that the same cell type can exhibit multiple cell states. Although two cells often appear to be of different types if they exhibit different cellular responses to a given stimulus, they may in fact be the same cell type in different cell states. Furthermore, these cell states can change without variation in protein content. An important implication of this is that biologists can get contradictory results from the same cells if the cells are in different states (Figure 1.4). This is particularly relevant in experiments that involve a large number of cells with a mixture of states, a biological noise. In this case, an average result will be observed instead of the more deterministic insights that could be attained if all cells were in a single state.

The definition of cell state requires the definition of the active functions, but only a subset of the active functions really define the cell state. Regardless of cell type or cell state, there are basic functions that should be active to keep the cell alive and healthy. For example, ATP is needed to maintain ion balances across the plasma membrane and to produce replacement proteins that are lost with time. In a similar way, we need to breathe and pump blood to stay alive whether we are sleeping or running. Thus, basic functions do not define the cell state and can be considered as state-independent. The activated functions that are state-specific usually depend upon stimuli or control mechanisms. When circulating hormones reach different tissues, they often activate different functions that are state-specific. The number of different functions that can define cell states is limited and probably approaches the number of different cell types or several hundred. It is therefore possible to define and understand those complex functions in a reasonable period of time so that we can imagine defining many of the normal cell states (perhaps there will not be a good description in our lifetime, but probably in our children's lifetimes).

1.5   Eukaryotic Cells are Organized into Compartments

There are several teleological explanations for creating compartments in eukaryotic cells such as the membranous organelles: (1) nucleus, (2) endoplasmic reticulum, (3) Golgi apparatus, (4) mitochondria, (5) endosomes, (6) lysosomes, and (7) peroxisomes. Many functions can occur within a single compartment. Further, the compartments communicate with the surrounding cytoplasm through the generation of signaling molecules and/or the recruitment of cytoplasmic proteins. In many cases, fission or fusion is facilitated (membrane dynamics) or the compartment is activated for transport along the cytoskeleton. Examples of the former include vesicle budding and fusion, tubulovesicular membrane extension and fusion, and lipid exchange through the cytoplasm. The dynamics of membranes makes it difficult to rigidly define each compartment because many components are shared at times and organelles are able to self-assemble when mixed with other compartments. In addition, compartments can be local concentrations of components attached to the cytoskeleton that are open to the cytoplasm. Communication by diffusion of biochemical signals, or motor protein-driven transport along cytoskeletal filaments is relevant particularly between cellular compartments that are not membrane bounded, such as poly-ribosomes, endoplasm, ectoplasm and cell extensions such as lamellipodia, filopodia, and flagella.

One rationale for compartments is that it is useful to localize molecular components, and therefore their functions, within compartments as this will limit the functional impact of damaging environmental perturbations to the compartment itself. From the perspective of Darwinian selection, this will confer robustness by allowing functional modules to adapt, and reducing their sensitivity to environmental perturbations. Exclusion of components involved in other functions simplifies the engineering of processes and control systems, and minimizes the crosstalk between functions. For example, biochemical signaling pathways can include similar kinase phosphorylation sites in different membrane-bounded compartments without a crosstalk between distinct signaling pathways, because the kinases (enzymes that transfer phosphate from ATP to specific amino acids in proteins) themselves will not normally cross membrane barriers. In many cases redundancy exists, with a small set of components shared between functions but in different compartments. This allows the cells to have backup sets of components in case one fails. Compartmentalization is the critical difference between prokaryotic and eukaryotic cells and clearly enables more complex functions to be performed by a single cell as part of a tissue. Multicellular organisms need this capability to support tissues, and this explains why only a limited number of functions can be performed by multicellular prokaryotic biofilms. In this case, although bacterial biofilms or other structured colonies can act as multicellular organisms, they still lack compartmentalization in the individual cells, and therefore are unable to facilitate the multiplicity of functions of complex eukaryotes.

Communication between membrane-bound compartments is often complex. For example, signals and materials must somehow be transported through the hydrophobic barrier created by lipid bilayers. An exception to this is found in the nucleus, which is enclosed by a double lipid bilayer that contains stable pores, which permit the passage of various molecules. Furthermore, internal compartments pose major problems in ion balance and volume control. This is particularly relevant because the vast majority of the cell's bilayer membrane (over 95% in most cases) is present as internal compartments separated from the plasma membrane. To move, fuse and divide membranes rapidly and specifically during the cell cycle is a major engineering feat; however, it is the efficient handling of these internal compartments that makes many specialized functions possible in a multicellular organism.

1.6 Coordination of Functions Can Yield Emergent Properties

Many complex functions are the result of multiple tasks that need to be completed in sequence or in neighboring regions of a cell. These functions culminate in emergent properties, i.e. when *multiple functional modules are coordinated to perform a multistep task*. Without the proper coordination of the functional modules, the complex function cannot be completed. Taking the complex function of DNA replication as an example, the formation of two identical double strands from one takes many steps. At a basic level, the existing double-stranded DNA must first be separated into the two strands that have opposite polarities by helicases. One DNA polymerase can move with the helicases to assemble a complementary strand of one of the original strands, called the forward strand. However, another DNA polymerase must move in the opposite direction on the other original strand (lagging strand) because of chemical constraints. In a roughly periodic fashion, a new DNA polymerase will assemble on the emerging lagging strand and assemble the complementary strand until it encounters the end of the previous complementary strand. Then the polymerase will disassemble and the two ends of the new complementary strand will be joined in a separate step. In this brief description of the process, many details were left out that were involved in proofreading and repairing errors, joining these segments with others, etc. All of these steps are highly orchestrated and must all occur before the cell can proceed in the cell cycle. It is, indeed, similar to an automobile assembly line with different tasks being performed by separate workers (functional modules) in a coordinated fashion and with inspections/repairs by other workers before the final product can be accepted. At this point of our understanding, it is not necessarily clear how the coordination of functional modules actually occurs in many contexts (whether through force, position or timing), but there is a lot of engineering needed to create the robust emergent properties of the system.

1.7 SUMMARY

Cells have evolved over billions of years through constant mutation and selection to become very robust machines for the propagation of their DNA. As robust machines, they compartmentalize many important functions into organelles such as the nucleus, endoplasmic reticulum, Golgi apparatus, mitochondria, endosomes, lysosomes, and centrioles. Many compartments are bounded by lipid bilayer membranes with hydrophobic cores that restrict movements of ions and charged molecules. Robustness is conferred by highly engineered functional modules that perform the needed task under many conditions and by carrying components needed to survive many challenges in reserve. Cells respond to environmental perturbations via adaptation such as receptor recycling or compartmentalization, repair themselves upon insults via redundancy in functional modules, and can transform a cell from one type to another by differentiation or from one cell state to another in different environments. Cells have developed functions to vigorously confront a number of environmental hazards such as feast or famine, flood or drought, cold or heat, and bacterial or viral infections. Proteins are concentrated in cells so that functional modules can be activated readily. Although theoretically there are many possible ways to perform functions, a surprisingly uniform set of basic mechanisms are shared by most eukaryotes. This feature makes the dissection and understanding of a function in one system applicable to many systems. A functional understanding of how a cell works as a robust machine involves breaking down complex functions into coordinated functional modules that produce the emergent properties. Considering the cell as a robust machine with a standard set of functional modules can be useful for many biomedical applications.

1.8 PROBLEMS

1. Calculate the number of generations to extinction of a gene for a native protein in mice that does not confer protection to a lethal disease in the presence of a mutant that will protect against the disease. Assume that 90% of the homozygous native protein animals die, whereas 60% of the heterozygous animals and 10% of the homozygous mutant animals die if an epidemic of the disease strikes. When the epidemic first strikes, the population is 10,000 with 20% of the mutant gene. Because the natural resources are not limiting, the animals that survive the disease

will double in number each generation. The disease strikes once in each generation on average. (Remember that you can't have fractions of animals and if the calculations say that 7.9 animals will survive, then in reality only 7 do survive. Less than 1 means 0 or extinction.) To give a hint, there are 400 homozygous mutant, 3200 heterozygous, and 6400 homozygous native animals before the first epidemic; and after the first epidemic, the number drops to 360 homozygous mutant, 1280 heterozygous, and 640 homozygous native animals.

Answer:

Native		Hetero												
1423		1553						Mutant						
1423		1553					424							
142		621					382							
284		1242					764 (1810/2290 = 0.395)							
70		876					1506 (1016/2452 = .207)							
20		644					2774 (684/3438 = .0995)							
6		492					5018 (504/5516 = .0457)							
0		192					9495 (192/9687 = 0.020)							
0		153					17091							
122		30763												
98		55373												
78	62	49	39	31	24	19	15	12	9	7	5	4	3	2	1	0

Answer: 26 generations

2. The HIV virion is a sphere of about 100 nm in outside diameter and we need to know how many proteins plus RNA can be packed inside. Calculate the size of a 50 kDa protein assuming a density of 1.3 g/cc, assuming that it is a sphere. Calculate the size of two copies of the HIV RNA (9000 bases each and assume that the average molecular weight of a base is 400 daltons and density is 1.4 g/cc). Remember that the lipid bilayer is 5 nm thick and membrane proteins normally add about 5 nm to its thickness. What is the approximate number of 50 kDa proteins that can fit into a single virus particle after you subtract the volume of the membrane and the two copies of the RNA?

Answer:

(a) Calculate the total volume of the viron. Remember that there are 10 nm taken off of each side of the sphere, so your actual diameter is 80 nm, and your radius is 40 nm. The volume of the interior space is approximately 268,028 nm^3.

(b) Calculate the volume of the RNA.

$1\,\text{Da} = 1.65 \times 10^{-24}\,\text{g}.$

Volume = 4242.85 nm^3 (or 4.24×10^{-18} cm^3). Remember there are two copies of the RNA, so that the total volume of RNA is 8486 nm^3.

(c) Total volume for left in the viron= 268,028 − 8486 = 259,596 nm^3.

(d) Calculate the volume of the protein = 63.5 nm^3 per protein.

(e) 259,596 total volume/63.5 per protein = **4088 copies** of the protein can fit in the viron.

3. If we mutagenize a million cells of an immortal cell line *in vitro* and then culture it at a high temperature for a year through repeated passages, what fraction of the cells will contain a mutation that decreases the doubling time from 24 hours to 23 hours on average? Assume that the mutation occurred in a single cell initially and was randomly distributed in the population of cells over the year.

Answer: 1 in 10^6 cells has a doubling time of 23 hours and will have gone through 381 doublings in a year, whereas the control will have 365 doublings. The 16 extra doublings will increase the mutant population 65,536-fold over the control cells. Thus, the mutant went from 0.0001% to 6.5% of the population. In another 6 months, the mutant line would be over 90% of the population.

4. Parasite infection will apply evolutionary pressure on humans and over time mutations will arise to inhibit the parasite even if those mutations are detrimental to the human.

   For example, malaria caused by *Plasmodium* infection is widespread in Africa. Sickle cell anemia, a single base pair mutation inherited disease, confers a survival advantage against malaria but in the homozygous case it also causes death in about 40% of individuals below reproductive age (this number is now dropping). On the other hand, none of the homozygous individuals died of malaria, whereas over the past 10,000 years on average about 20% of the wild-type and 10% of the heterozygous individuals died from malaria infections before a reproductive age. What percentage of the hemoglobin gene in the population will there be at steady state? (Ana Ferreira et al., 2011.)

Answer: In the case of the fraction of sickle cell genes in the population at steady state, we can postulate that when 4% of the sickle cell population will die in each generation, the fraction of the sickle cell genes in the population will stabilize. If we look at the fraction of sickle cell-containing individuals who will die in each generation, it will be 2% of the heterozygotes and 16% of the homozygotes. When the overall fraction of the sickle cell population (heterozygous X and homozygous, can be approximated by X^2) that is lost per year equals the 4% of the normal population $(1 - X - X^2)$ that is lost per year, the fraction of sickle cell genes will stabilize. Thus, $0.04\,(1 - X - X^2) = 0.02\,X + 0.16\,X^2$ and we can solve for X. To make this calculation, the fraction of the population with a sickle cell gene will contain both heterozygotes and homozygotes needs to be defined as:

$1 = 0.5 \, X + 4 \, X^2$ or about 0.3 or 30% of the population.

5. In a large population of bacteria ($> 10^7$), it is believed that most proteins in the bacterial genome are expressed in some cells. Assuming that there are 5000 genes in the bacterial genome, how would you experimentally test this hypothesis?

Answer: Assuming that the expression patterns are stable, then the cloning of 50,000 individual cells should have at least one clone with the protein of interest. An alternative approach is to isolate the mRNA from the large population of cells and amplify the mRNA for the gene of interest to determine if it is present even at a low copy number.

6. Consider one chromosome has 6 million base pairs. At mitosis, this chromosome is condensed into 2 μm in length and 100 nm in diameter microfiber. How many fold is the DNA condensed during mitosis? At G1 phase, the chromatin fiber has a diameter of 30 nm. For chromatin with a length of 25 nm, there are about 15 nucleosomes, and each nucleosome contains about 167 base pairs. How many fold is the DNA condensed in the G1 phase? (One base pair is about 3.4 Å.)

Answer: The DNA with 6 million base pairs would have a length of 2.04 mm and thus there is a 1000-fold condensation of the DNA. In the second part, there is a 34-fold condensation in G1 phase as in the 25 nm length of the 30 nm solenoid there are 851 nm of DNA (2505 base pairs).

7. Here are two strategies for propagating a virus and we would like to know how many (animals/people) are infected after 15 weeks in each case as well as the number of viruses. Show your work for partial credit.

(A) Mild pathogen strategy: the pathogen compromises performance only mildly in healthy individuals such that they continue to have a normal number of daily contacts with other individuals. In this case, we assume that there is a population of 8 million individuals and one individual is infected with the disease initially. After infection, there is an incubation period of 3 weeks before symptoms are evident; the infection lasts for 2 weeks and the contagious period is in the first 5 days. Assume that the infected individuals encounter 50 other individuals per day and the average probability of infection per encounter is 4%. Because infected individuals can continue to work, they encounter the same number of other individuals per day. The average number of pathogens per infected individual is 10^{12}.

Answer: After 3 weeks 10 people, 6 weeks 100 people and so on, 15 weeks 100,000.

8. Same conditions as above.

(B) Carrier strategy: this pathogen is carried in a host animal and then can pass to humans. In this case, we assume that there is one infected carrier animal (a population of 8 million and one infected carrier animal). There are two scenarios here, one for the carrier and another for the human population. The carrier will infect five other carriers in the first week and then die. Each infected carrier will only become infectious after two weeks and then will infect five other carriers in the next week before dying. For the human population, each carrier will infect four humans while they are infectious. Each infected human will incubate for 3 weeks and then will be infectious for 2 days before being confined to bed, where they have a 25% chance of dying. During each of the two infectious days the infected people will encounter 50 people and will infect 2. With the more severe infection, there is a higher average number of pathogen per infection or 10^{13}.

Answer: Carrier: after 3 weeks 25 carriers, 6 weeks 125 carriers, and so on, 15 weeks 15,625 carriers.

Humans: after 3 weeks 16 from humans and 20 from carriers, after 6 weeks 144 from humans and 100 from carriers, after 9 weeks 500 from carriers and 976 from humans and so on, until after 15 weeks 33,616 from humans and 12,500 from carriers (total 46,116).

9. From the McDonald et al. (2002) article, it is evident that the virus DNA can be transported to the microtubule organizing center over a period of a couple hours. In figures 1B and 5A (non-injected cell), what fraction of the green particles over the whole cell got to the perinuclear region (within 5 μm of the nucleus)? What fraction of green particles is within 5 μm of the nucleus in the injected cell (dynein inhibited)? What fraction of the green particles are also red in figure 5? In figure 6, what fraction of the green particles contained DNA? Using these numbers and assuming that 20% of the particles with DNA near the nucleus can actually integrate that DNA into the genome, how many viral particles are needed to get an infected normal cell versus a dynein-inhibited cell (on average)?

Answer: ***Answers to this problem will probably vary because you actually have to count the number of the dots in the pictures. Your number will probably vary due to things such as the quality of your printout, how sharp your eyes are, how you determine exactly what is perinuclear, etc. What is important is that you take the percentages that you get and you carry those numbers through the rest of the problem.***

1(a) In normal cells, I counted ~36 perinuclear dots, and 14 dots that were elsewhere, giving me 36/50 total = 72%

1(b) In the dynein-inhibited cells, I counted ~13 perinuclear, and 55 non-perinuclear, giving me 13/68 total, or about 20%

1(c) I counted ~3/5 green dots also being red dots (the yellow in these pictures means that the red and green are overlapping, and that both are there).

1(d) 4/11 by my count, or 36% have DNA in them.

1(e) Using the answers from above:

   If 20% of (d) can infect a cell, how many virons are needed to infect a cell?

   (d) is 36% of (a).

   (a) is 72% the total number of virons needed.

   *We know that one copy of the viron getting into the nucleus and infecting the genome is all that is necessary for a cell to be permanently infected with the HIV. Therefore, we can back-calculate for the total number of virons needed to infect one cell.

   X * 72% * 36% * 20% = 1, this gives us 19.2 virons, but as we can't have 0.2 virons, we assume we need 20 virons to infect a normal cell.

   Repeat this calculation for the dynein-inhibited cells:

   X * 20% * 36% * 20% = 1, which will give us 70 virons necessary to infect a dynein-inhibited cell.

   p.s. Several people have complained how this is fairly subjective and even ridiculous to have to count the dots in order to get our data, but this is how it actually is in the real world.

10. If we assume that viral particles are cleared by macrophages in the spleen at the rate of 10% per hour and that they bind to CD4 lymphocytes at a rate of 5% per hour initially, then how many viral particles are needed to infect one cell on average based on the calculations in problem #3.

Answer: Macrophages clear 10% out of blood.

   ~5% of virons in bloodstream bind to target cells.

   The mathematical logic is similar to question 1 above.

   Y = number of virons in the blood stream necessary to infect a cell, X from above is the number of virons needed a cell.

   Y * 90% * 5% = X.

   When we plug in the X's from above, we get 444 virons to infect a normal cell, and 1556 virons for the dynein-inhibited cells.

1.9 REFERENCES

Ferreira, A., Marguti, I., Bechmann, I., et al. (2011). Sickle hemoglobin confers tolerance to Plasmodium infection. *Cell* 145(3): 398–409. doi:10.1016/j.cell.2011.03.049

Gibson, D.G., Benders, G.A., Andrews-Pfannkoch, C., et al. 2008. Complete chemical synthesis, assembly, and cloning of a *Mycoplasma genitalium* genome. *Science* 319: 1215–1220.

Huxley, J. 1958. Introduction to 150th Anniversary Edition of *Origin of the Species* by Charles Darwin. New York, NY: Signet Classics.

Lindsay, J.A. 1995. Is thermophily a transferrable property in bacteria? *Crit Rev Microbiol* 21(3): 165–174.

McDonald, D., Vodicka, M. A., Lucero, G., et al. 2002. Visualization of the intracellular behavior of HIV in living cells. *J Cell Biol* 159: 441–452. www.ncbi.nlm.nih.gov/entrez/query.fcgi?cmd=Retrieve&db=PubMed&dopt=Citation&list_uids=12417576

1.10 FURTHER READING

Cellular organization: www.mechanobio.info/topics/cellular-organization/

Differentiation guided by environmental factors: www.mechanobio.info/physiological-relevance-of-mechanobiology/

DNA replication: www.mechanobio.info/topics/genome-regulation/dna-replication/

The extracellular matrix and basal lamina: www.mechanobio.info/topics/mechanosignaling/cell–matrix-adhesion/

2 Complex Functions of Robust Machines with Emergent Properties

Cellular functions have evolved over time to better enable organisms to survive major environmental challenges (see Chapter 1). Improvements to basic functions or new functions that more successfully supported survival were preserved and shared between organisms, either by infections or by interbreeding. Because the environment is constantly changing, the complex functions have evolved to be robust so that they will work under many different conditions and the design principles of robust devices (see text box below) can be used to understand them. *A complex function involves multiple functional modules coordinated to perform a complex task.* In this chapter, we will describe a general approach to understanding the features of complex functions and how the functional modules that underlie complex functions may be linked to yield the desired emergent properties across different scales. An emergent property is the outcome of many steps and many modules in a complex function. We suggest that models of complex functions violating the principles of robust devices are likely to be wrong. Furthermore, we will describe why it is important to quantitatively measure the performance of complex biological functions (outputs) under different circumstances (inputs) to better test models of different complex functions.

2.1 Principles of Robust Machines with Standard Functions

We propose that cells are highly engineered, robust machines. They are also very small machines where all actions are dominated by diffusion. The stochastic nature of diffusion processes can account for some of the observed biological variability. However, biological systems have devised mechanisms to use diffusion to drive muscle movements and to direct the shaping of the organism. Evolutionary refinements of the basic mechanisms of biological systems enable them to reproducibly create organisms of diverse shapes using the information encoded in the DNA. Thus, fruit flies with the same DNA will look the same as well as identical twins. In the case of the twins the development took many years and yet they often do look identical. Thus, the robustness of the biological systems

Figure 2.1

Principles of robust
machines.
(a) Compartmentalization
of complex functions;
(b) functional coupling
should be simple.

a.

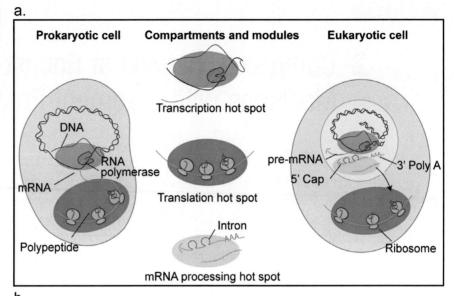

b.

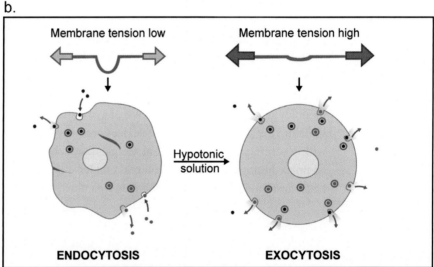

extends to making reliable decisions using inherently noisy stochastic processes. Because it seems that the cells that form the basis of biological systems are robust machines, the complex functions that they utilize should follow the design principles of robust devices (Figure 2.1, from Thomas et al., 2004). This will be a theme throughout much of the book that will help to direct our speculations about biological mechanisms because models that violate the principles of robust machines will likely be wrong.

Consider the basic process of protein synthesis on the ribosome using the mRNA template and loaded transfer RNAs (tRNAs). Because this process involves the sequential addition of amino acids to create a complete protein, it can be likened to assembly-line processes that add specific components in a defined series to produce a complete automobile or computer. What makes the assembly of proteins at tens of amino acids per second relatively more difficult than the assembly line is that there are 20 different tRNAs randomly colliding with the ribosome and the correct one needs to be chosen from those in less than 0.1 seconds. It is as if the assembly-line worker would need to choose the correct part from a stream of randomly arrayed parts and then assemble it before the next part could be added. What makes the biological systems remarkable is that they have found ways to rapidly determine if the randomly sampled part is the correct one before proceeding. This can best be done if the incremental steps in the process are each checked in a proofreading system. Thus, biological systems have found ways to suppress errors in important functions, which is similar to the problems facing computers as the features become smaller and the error rate for each computation or memory event increases.

An important aspect of biological systems that aids in robustness is the general use a set of standard functions. The array of different functional modules in a given cell constitutes a tool set that will operate in standard ways. Cells can draw upon those standard tools for desired tasks. New or modified tools can evolve to improve the performance or better couple the tool to the specific cell's needs. Implicit in this concept is the hypothesis that the same complex function may be performed in the same or a closely analogous way in multiple cell systems. Further, when a complex function is understood in one cell system, then it is easier to understand it in another cell system. This has been the underlying premise of many books on biochemistry that describe complex functions of transcription or translation in bacteria as an aid to researchers, who wish to understand transcription and translation in eukaryotic systems. Here we likewise suggest that it is important to understand each complex function in at least one cell type because that will make it easier to understand how similar complex functions work in other cell types. In this regard, model cell systems are useful because they will perform a specialized complex function more extensively in a specialized cell than in other cells, but the complex function can then be understood in less specialized cells based upon the understanding from the specialized cell. A subcellular *in vivo* understanding of each complex function is important because many complex functions rely critically upon relative positioning and timing that are not seen at a multicellular level and are difficult to reproduce *in vitro* at a molecular level. The development of a standard set of functions is an important component in the robustness of biological systems.

2.2 Compartmentalization of Complex Functions (Localized and Modular)

If we apply the first axiom of the design of robust devices to cells, then the majority of subcellular complex functions can be treated as localized, modular entities that communicate with other complex functions by signaling pathways (Figure 2.1). This is a logical concept for membrane-bounded compartments such as mitochondria, lysosomes, etc., but can be extended to the nucleoli where ribosomes are produced, transcription factories, polyribosomes and motile functions such as filopodia or periodically contracting lamellipodia. In each of those cases, a complex function can operate in a confined region of cytoplasm to fulfill a set of tasks. Such independent complex functions can respond to perturbations without direct interference from other functional activities. The transcription of DNA into RNA involves extensive motions of those long polymers that would naturally interfere with the actin filaments and microtubules involved in cell motility. This implies two aspects of the cellular compartments: (1) localization, and (2) modularity. The small size of most proteins (2–20 nm) means that the functions can be performed within limited spatial and temporal scales (localized). If functions were to involve much larger scales, then there can be significant interference between complex functions (e.g. the loss of membrane attachment for an enzyme can dramatically increase the number of unwanted sites that it alters). In addition, functions are modular and the organelle that performs a complex function in one cell type can be easily adapted to perform the same complex function in a different cell background by the propagation of that organelle. Thus, the operational aspects of the functions should be isolated so that they can do the desired task; but control and communication signals between compartments are needed to integrate functions in the cell context.

Knowing the complex functions performed in each of the major cell compartments provides a perspective on how various tasks are apportioned in cells (for more details please see cell biology textbooks, such as Alberts et al. 2014).

Principles of Robust Complex Functions

1. Compartmentalization of complex functions (localized and modular).
2. Simple controls of complex function are best (physical or biochemical).
3. Complex functions are either on or off (rates are typically not continually adjusted).
4. Term limits (complex functions turn off after relatively short periods and need to be turned on).
5. Cyclic in nature.
6. Backups or variations of important complex functions are common.

2.2.1 Nucleus

The nucleus contains a large amount of DNA (2 m of DNA is packed into a human cell nucleus of 6–8 μm in diameter) that would be difficult to manage in cytoplasm when other cytoplasmic filaments and membranes are present. The nucleus is one of the most prominent compartments and is the site where three critical complex functions are performed: (1) replication of the DNA to provide two complete copies for the daughter cells; (2) transcription of DNA into RNA for the production of mRNA and other RNA species; and (3) assembly of ribosomes that must then be transported to the cytoplasm to enable translation of mRNA into protein (Figure 2.2). The large volume of material transported between the nucleus and cytoplasm means that a transport mechanism is needed to link the two compartments. An unusual feature of the nucleus is the way in which it is separated from the cytoplasm. At the end of mitosis, the endoplasmic reticulum (ER) membrane is wrapped around the nuclear material as a double membrane, which keeps the nucleus topologically in the cytoplasmic space. Pores in the double membrane enable the transport of large RNA–protein complexes such as ribosomes and mRNA particles that are several million daltons in mass. The pores do not block salts from moving into the nucleus. Recent studies show that the transport of mRNA through the pores is by an active process. In mitosis, the nuclear membrane is resorbed into the ER and condensed chromosomes are transported by cytoplasmic microtubules into the daughter cells. Thus, the nucleus confines DNA

Figure 2.2

Functions in the nucleus: DNA replication, transcription of DNA into RNA, ribosome assembly in nucleoli, transport of mRNA out and many proteins in, organization of DNA into chromatin.

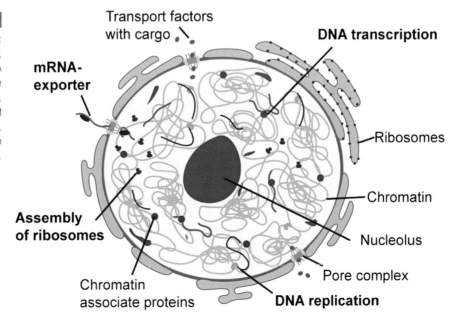

and RNA transcripts, along with their associated proteins, in a separate cytoplasmic space and this prevents these long polymers from becoming entangled in cytoskeletal filaments. This compartmentalization also allows the DNA and RNA to be processed with very specialized modular machinery for replication, transcription, RNA processing, and ribosome assembly.

2.2.2   Endoplasmic Reticulum

In most mammalian cells, the endoplasmic reticulum (ER) is the largest membrane compartment and contains 50% or more of the cell membrane lipid. Membrane lipids, soluble and secreted proteins as well as lysosomal proteins are all synthesized in the ER. Because the lumen of the ER is outside the cytoplasm, the ER can have a distinct ion composition, and often serves as a reservoir for intracellular calcium. When organelle membranes eventually fuse with the plasma membrane in exocytosis, the processed material from the ER lumen is secreted or incorporated into the extracellular leaflet of the plasma membrane bilayer. To accommodate the large volume of membrane within the cell, ER is typically organized in tubules of about 0.2 μm in diameter that are normally spread throughout the cell by microtubule motors that follow the microtubule array. The isolated but disseminated calcium store in the ER is critical for regulating complex functions in the cytoplasm. In addition, the extensive distribution of the ER and its associated complex functions in large areas of the cytoplasm provides an internal communication network throughout the cell (see Lippincott-Schwartz & Phair, 2010).

Another complex function performed in the ER is protein translation (Figure 2.3). Here, polypeptide chains with a signal sequence translocate into the ER lumen as they are formed on docked ribosomes in a relatively energy intensive process. Once in the ER lumen, there are proteins, called chaperone proteins, that help the polypeptides fold properly and initial carbohydrates that inhibit non-specific binding are added to many proteins and lipids. Lipid molecules are synthesized on the ER cytoplasmic surface and some are destined for the external surface of the plasma membrane. They need to be transported into the lumen as others are transported back to the cytoplasm to maintain a balance. Due to its central role in many synthetic and signaling activities, ER is a very dynamic compartment that can accept newly synthesized components, sort them, and pass them on to their proper destination.

2.2.3   Golgi Apparatus

Although smaller than the ER (about 15% of the total membrane), the Golgi is the primary destination for most of the components synthesized in the ER

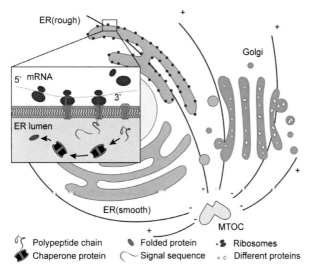

Figure 2.3

Protein translation and translocation in ER and Golgi apparatus.

(Figure 2.3). Further processing of the proteins and lipids adds appropriate carbohydrates. A critical complex function is the sorting of components to be transported to the plasma membrane from those to be transported to lysosomes, secretory vesicles or back to the ER. Unlike the ER, the Golgi is organized into relatively flat stacks of membrane sheets in a region near the nucleus. Processing is progressive from one stack to another and there is a debate about whether movements between membranes depend upon the flow of membrane or by formation of vesicles at one stack that then fuse to the next stack. Microtubules and microtubule motors help to organize a central Golgi around the microtubule organizing center(s) that anchors the microtubules. When microtubules are depolymerized by drugs, multiple small Golgi stacks form near the ER at sites where ER components are concentrated for transport to the Golgi. The small Golgi are capable of processing proteins, but at suboptimal rates. The Golgi apparatus is the primary site for carbohydrate modification of cell surface glycoproteins, soluble cytokines, and extracellular matrices. Secreted, large carbohydrate domains are porous polymers that help to lubricate cell–cell contact regions and by their shear mass keep the membranes apart while allowing small molecules to reach the cell surface.

2.2.4   Mitochondria

Mitochondria presumably originated as invading bacteria that developed a symbiotic relationship within eukaryotic cells. They retain very little DNA and cannot live independently. Mitochondria produce ATP for the cell by the more

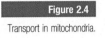

Figure 2.4

Transport in mitochondria.

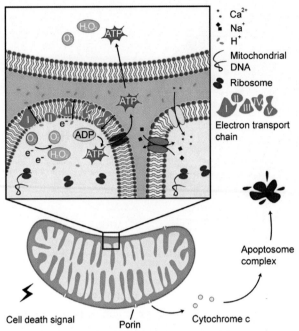

efficient pathway of oxidative phosphorylation (more efficient than anaerobic sugar metabolism). Mitochondria have a double membrane that tightly regulates metabolite, proton, ion, and ATP transport (Figure 2.4). The inner membrane has a gradient of protons and electric potential to help transduce energy from the oxidation of carbohydrates into ATP synthesis. Changes in metabolism produce dramatic changes in mitochondrial morphology that belie complex mechanisms for mitochondrial fragmentation and transport. In addition, mitochondria provide a trigger for cell suicide by catalyzing apoptotic cell death pathways when their outer membrane is opened and cytochrome c leaks into the cytoplasm. Cell lines have been generated without mitochondria; however, they are relatively weak because of the greater efficiency afforded by mitochondria. Mitochondria have become a target of drug delivery research to specifically induce cancer cell death.

2.2.5   Endosomes

Cells take up nutrients, hormone signals, and exchange plasma membrane proteins by pulling in portions of the plasma membrane as endocytic vesicles (Figure 2.5). Using fluid-phase or plasma membrane markers, researchers have found that an area of membrane equal to the plasma membrane is taken up every

Figure 2.5

Endosomes are formed
by several different
pathways: clathrin-
dependent, caveolin-
dependent, and
independent pathways, but
most endocytosed
membrane (> 95%) is
recycled back to the plasma
membrane.

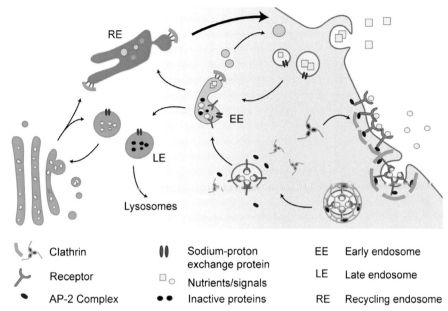

Figure 2.5

Endosomes are formed by several different pathways: clathrin-dependent, caveolin-dependent, and independent pathways, but most endocytosed membrane (> 95%) is recycled back to the plasma membrane.

⟍	Clathrin	‖	Sodium-proton exchange protein	EE	Early endosome
⟍	Receptor	☐ ○	Nutrients/signals	LE	Late endosome
●	AP-2 Complex	● ●	Inactive proteins	RE	Recycling endosome

30–60 minutes in actively growing cells. Early endosomes form by several different mechanisms including by fission of recently endocytosed vesicles and the folding back of membrane veils onto the cell surface to form macropinosomes. Proteins are sorted in early endosomes to go either back to the plasma membrane or to move on in the endocytic pathway eventually to lysosomes. Small vesicles and membrane tubules carry membrane back to the plasma membrane and the residue destined for late endosomes moves by microtubule motors on microtubules to an area of the Golgi called the trans-Golgi network. Late endosomes further sort inactive proteins or nutrients for movement on to the lysosomes, whereas many components including most lipids move back to the plasma membrane. Sodium-proton exchange proteins acidify the lumen of the endosomes to aid in the aggregation of many proteins that become neutral at acidic pH. A complex set of factors including membrane curvature, pH, protein concentration, and bilayer asymmetry go into the sorting decisions. It is clear that the sorting process is critical for refreshing plasma membrane proteins, hormone receptors, and for the uptake of nutrients. It is also the path of entry for many viruses. Endosomes are also an important means for cells to adapt to environments via receptor recycling to dampen cellular responses to extracellular signals, e.g. in pain management.

Over the past decade, the endosome has been exploited for delivery of drugs and genes to cells. Further, technologies to block endosome uptake of virus and other pathogens have also been developed.

Figure 2.6

Lysosomes and peroxisomes are important for the breakdown of proteins or lipids and the oxidation of fatty acids, respectively.

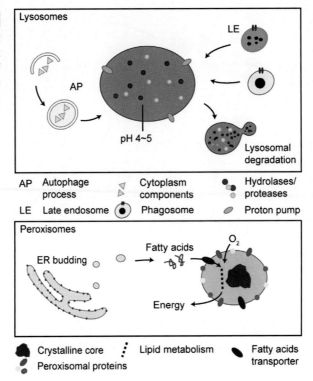

2.2.6 Lysosomes

Although the targeted degradation of many proteins by the ubiquitin pathway occurs in the cytoplasm, the lysosome also has the components needed to be a major recycling center in the cell (Figure 2.6). The acidic lysosome (pH 4–5) contains many hydrolases and proteases that are only active at low pH. If they leak into the cytoplasm, the neutral pH will inhibit their activity and minimize the damage they would otherwise cause. The factors that target material to the lysosome are poorly understood; however, in addition to material from late endosomes, aggregates of improperly folded proteins will be covered with membrane and fused to lysosomes in the process of autophagy. As in the macroscopic world, there is less interest in the garbage disposal process in cells than in other processes.

2.2.7 Peroxisomes

These small spherical organelles are necessary for cell survival because mutations of proteins involved in their formation and function will kill cells. Approximately

100 of these vesicles are normally found in cells and they have a critical role in the metabolism of lipids (Figure 2.6). This is a very specialized compartment that was formed for a particularly complex function and there appears to be no redundancy. Metabolic organs like the liver contain highly specialized hepatocytes that are full of peroxisomes.

2.2.8 Non-membranous Compartments

The cytoplasm is segmented into additional compartments by the organization of actin filaments, microtubules, and intermediate filaments and their movements (Figure 2.7). Membranes are not part of the formation of these compartments and small components can freely diffuse between them. However, large cytoplasmic vesicles are sequestered in endoplasm, the core of cytoplasm around the nucleus that contains all of the microtubules responsible for the transport of vesicular organelles. In contrast, the peripheral ectoplasm is normally a thin layer that is organized by actin filaments that assemble at the plasma membrane normally and move inward where they depolymerize. Vesicular organelles in the endoplasm are physically unable to reach the plasma membrane because of the much smaller gaps in the actin network of the ectoplasm. This keeps apart components that could otherwise interact. Hormone signals can rapidly alter the volume of the ectoplasm and thereby rapidly increase or decrease secretory vesicle interactions with the plasma membrane.

There is considerable structure to the cytoplasm and many of the components of cells are concentrated in specific cytoplasmic regions. Molecular components that are involved in a particular complex function are often co-localized as a robust functional module. Those modules may not formally be compartments, but they can often act as a compartment. For example, many metabolic enzymes are bound

Figure 2.7

Non-membranous compartments and movements.

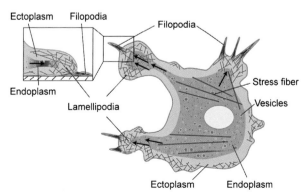

to the surface of the ER and it is postulated that they are organized there in complexes where the next enzyme in a pathway is positioned to rapidly accept the product from the previous enzyme.

2.2.8.1 Implications of Compartments for Modeling of Complex Functions

With this in mind, it is obvious that eukaryotic cells have developed ways to integrate many stochastic events to produce a reliable or deterministic outcome. *The localized and modular nature of the compartmentalized complex functions that is stochastic but yields deterministic emergent properties make the complex functions resilient to local disturbances and are thus robust.* For example, transcription and translation and protein modifications are different complex functions occurring in different compartments. Within each compartment, the stochastic nature of the complex function is suppressed by error-checking and by integrative processes that make the output less sensitive to fluctuations in individual events. When integrative signaling components exceed a certain threshold, a decision is made that constitutes a deterministic output. Segregating different complex functions into different physical compartments increases the decision points that will filter out the noise or compensate for the failure of an individual event. Even when a complex function in one compartment is less efficient, it need not drastically affect the complex functions in other compartments (Figure 2.8).

Figure 2.8

Deterministic cellular behaviors from stochastic complex functions – emergent properties: input signals will trigger modular complex functions in different compartments that are physically separated.

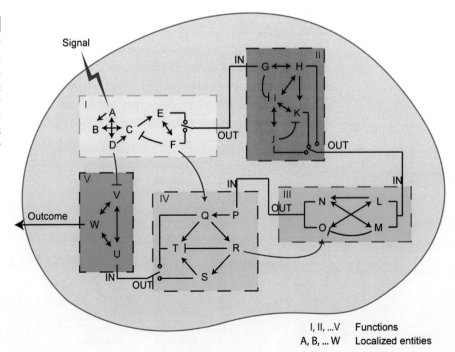

2.3 Compartments Communicate to Coordinate Functions for Emergent Properties

The functional modules that give rise to a particular complex function are themselves driven by modular components that are linked spatially and temporally in a compartment. However, there is often coordination between compartments in a particular context that gives rise to emergent (observable) properties of the system through forces, positioning, and chemical signals to enable many multifaceted tasks to be completed. For example, the development of adhesions in the cell periphery communicates with the nucleus and causes alterations in the pattern of genes that are made into mRNA that will ultimately change the protein components. For a given cell, we should be able to predict its behavior if we know which complex cellular functions are active and how those functions communicate with other functions of the cell to produce emergent properties. Because a finite number of complex functions are active in any given cell (probably in the range of 100s), the task of being able to predict the cellular behavior is feasible. This needs to be qualified by the realization that each complex function is stochastic and will produce a range of outputs. In the example of the peripheral adhesions affecting the nuclear transcription activity, there are chemical and physical links that contribute to the control of transcription (Figure 2.9). Proteins in the adhesions are modified and move to the nucleus where they will alter transcriptional patterns. Further, adhesion proteins can activate kinases that will stimulate entry into the cell cycle, which will also produce signals to the nucleus that affect transcription as well as altering cell metabolism. As we understand the functional compartments and the ways that they communicate with each other, we can develop diagrams to model how the pattern of inputs can affect the specific set of functions that are activated and the pattern of outputs that will cause further changes.

Figure 2.9

Linked DNA transcription. Mechanical links from the actin cytoskeleton apply forces to the nuclear membrane that then produce chemical signals inside the nucleus that alter DNA transcription.

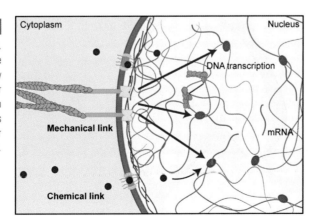

2.4   Complex Functions Should Not Be Overly Complex

Another design axiom is that the *function should be as simple as possible*. Why use a multistep process when a single-step process can do the same thing? For example, consider the problem of DNA supercoiling (twisted from the normal helical state). The supercoiling of DNA is relaxed by the enzyme topoisomerase 1 (Topo1) that breaks a phosphate–sugar bond in one strand allowing the DNA to relax the torsion by rotation around the other strand and then it restores the phosphate–sugar bond. Hypothetically, a specialized protein could sense twist in the DNA, bind to DNA and then recruit Topo1 to cause the bond breakage and reformation. The simpler scheme that appears to be true is that Topo1 has both sensory and catalytic activities and can act alone. There are other cases where the mechanosensing and the enzymatic activities reside in different proteins, but in this case the one protein can perform both functions.

Further, signaling and communication between different systems also should be by the simplest mechanism. In the case of the endocytosis and exocytosis systems, there must be communication to regulate the surface area of the cell. This is a difficult task because there are multiple secretory pathways adding membrane and multiple endocytic pathways taking it inside the cell. Hypothetically, the hundreds of proteins involved in exocytosis and endocytosis could communicate through a master regulator that would balance the outward and inward flows. Alternatively, the physical tension in the membrane could directly control the rate of endocytosis, decreasing endocytosis with high tensions and increasing it with low. Indeed, several studies have shown that tension in the membrane controls the rate of endocytosis and hypotonic shock (or a rise in tension) can trigger rapid exocytosis. Thus, the surface area will automatically adjust to maintain a proper balance. From this and other cases, the simplest signaling systems are often physical or mechanical in nature. It is common to start with the simplest hypothesis to explain the results and only move to a more complex hypothesis when the simple one is proven wrong.

2.5   Complex Functions are Largely Digital (On/Off) and Automatically Turn Off (Term Limits)

In most biological systems the major complex functions are digital (either on or off) and do not adjust the rate. Whether it is cell contraction, the production of mRNA or synthesis of proteins, the function is on for only a limited period and

Figure 2.10

Complex functions are largely ON or OFF at a given time and controlled by (a) threshold and (b) frequency of activation.

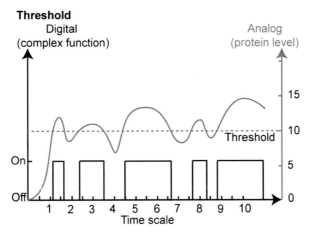

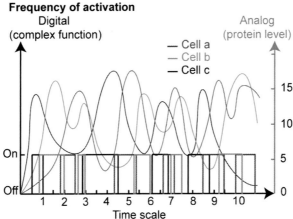

Figure 2.10

Complex functions are largely ON or OFF at a given time and controlled by (a) threshold and (b) frequency of activation.

then turns off as if it had term limits. If more is needed, then the function is turned on again until the desired output level is reached. The activation of the function is then tied to a sensory system that only needs to measure one level, i.e. it is a digital switch that will activate the process if the conditions are correct. In the On state, the complex function operates in a standard way that is altered by the availability of substrates or other factors, but there does not appear to be continual adjustment in the rate through complex feedback mechanisms. This is much simpler than to continually monitor the level. For example, mRNA production to produce a critical protein should be activated when the level of that protein falls below a certain threshold (Figure 2.10). An alternative method would be to keep basal mRNA production On all of the time and adjust the amount of mRNA that is being made to keep in tune with the level of the protein. The latter method is much more complex and could fail in many more ways than the first method. In the first method, there will naturally be larger fluctuations in the level of the protein, but cells are not sensitive to the exact levels of most proteins. Further, the length of the On state affects how often the function needs to be turned on, but if mutations alter the On time, the system will easily compensate. Thus, the cell can perform a complex function multiple times to produce the same level of output as if the complex function is always on but the rate is adjusted. The limitation comes if there is a major demand that exceeds the output rate of the functional modules, which should then cause an increase in the capacity through an increase in the number of modules.

Recent analyses of a number of complex functions including motile, metabolic and growth functions find that they are digital when studied at a single cell level. When analyzed at a population level, there are intermediate levels (analog), but that is the result of the fact that the complex function is only active in a fraction of the cells in the population. The advantage of regulating complex functions and functional modules in a digital fashion is that cells do not have to develop an analog regulator of rate. Only the On signal is needed and that can be tuned to the level of the output parameter that is needed. This necessarily results in a range of functional outputs when observed at a single cell level in a population of cells in different states. If we look at the levels of a given protein in single cells, there are often wide variations (particularly in bacteria) that make it seem like something is wrong. As robust systems, the levels of a single protein are not often important (there are exceptions) and the cells in a population can be in many different states (different phases of the cell cycle, senescent, or in a different morphology). Thus, a digital control of protein production and many other parameters enables cells to perform properly and the digital control systems are relatively easier to design as well as being more robust.

2.6 Term Limits (Complex Functions Automatically Turn Off)

In the previous section, we suggested that the cell only needed to turn on the complex functions. This is true if a complex function is designed to automatically turn off (Figure 2.11). In the political realm, term limits regulate the length of time that any given politician can be in office. Similarly, cells find it useful to automatically turn off a complex function after a period of time. The complex function then will not automatically run and the cell must activate it again for further output. As most complex functions utilize energy or resources, the organism conserves energy or resources by having complex functions that will automatically turn off. Turning on the complex function in a robust way to keep

Figure 2.11

Term limits of a complex function: a vesicle activated to move only a short distance stops automatically until another signal activates it again to continue moving along microtubule.

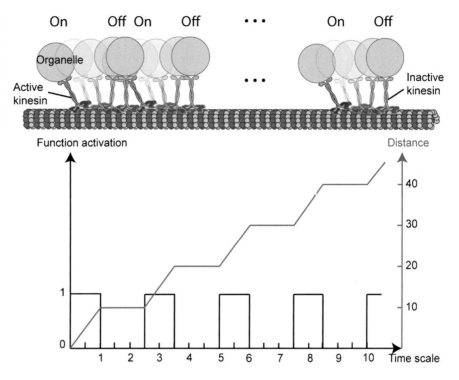

Figure 2.11

Term limits of a complex function: a vesicle activated to move only a short distance stops automatically until another signal activates it again to continue moving along microtubule.

the level of a complex function's product at its proper level is important. For example, motile functions are normally characterized by intermittent activity. Organelle movements on microtubules are saltatory in that the vesicular organelles only move a short distance before they stop and then they start moving again. Actin-based motility is similar in that edge protrusions are limited and stochastic. Cells migrate by a series of limited extension events and often stop moving, round up and restart migration. Similarly in transcription, there is evidence of production of a bolus of mRNA and then a stop to wait for another activation signal. In all of these cases, the cells use an automatic turn-off mechanism rather than developing an analog system to regulate the output. Automatic turn-off has a number of obvious consequences for people who are studying complex cellular functions *in vitro*. Because the default state is off and the system is only on for a limited period, it is difficult to study *in vitro* as the *in vivo* activation signals might not be preserved in test tubes. Experimenters have developed ways to override the off switches in several cases, but that gives an unrealistic view of the normal cell functions. My own studies of vesicle motility *in vitro* have always been hampered by our inability to develop *in vitro* motility assays that would be relevant to cellular vesicle transport. If we activated movement, we were overriding the normal off controls and the signals that altered motility *in vivo* didn't work properly. In many cases, the off switch is an important integral part of the function, whether it be the hydrolysis of GTP by a small G protein or dephosphorylation of critical phosphates by a soluble phosphatase. This design feature of biological functions is perhaps underappreciated and runs counter to our thinking. However, the recent hybrid vehicles testify to the elegance of such a design by rapidly turning off the engine when the ON signal (foot on the gas pedal) is removed and activating the generator that recovers energy from braking the car. Basically, robust systems should not persist in an On state unless they absolutely need to be on for the survival of the organism (e.g. heart and lungs).

2.7 Complex Functions are Cyclic in Nature

Many complex functions within the cell need to coordinate events temporally and spatially in a way that would be very difficult to accomplish should the functions occur continuously. Also, if the complex functions are to automatically turn off, then it is logical to build the overall level of activity around cycles (Figure 2.12). At a teleological level, it is easier to reach the goal for major functions through a series of small steps. By going through a cycle of activity and then evaluating the outcome, the level of product can be evaluated and the function can be activated again or not. In stochastic systems, the redundancy of multiple cycles to reach a larger goal helps to reduce errors. In cell motility, there are repeated extension and retraction events (steps) that are linked but are often separated in time and the important modular components are separated in space. Those cycles can incrementally displace extracellular matrix or another cell. After the first cycle, the cell needs to decide to make the next step and will not make further steps when they are not needed. Through cycles, the cell can incrementally move toward a goal and can stop midway or change direction rapidly. This will decrease the overall error rate, particularly when the stochastic nature of cells means that a single cycle of activation of a function has a high probability of error, i.e. the probability of repeatedly activating a function incorrectly is exponentially less than the probability of doing it once.

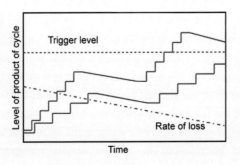

Figure 2.12

Cycles of complex functions are integrated to activate downstream process when the level of product reaches above a threshold.

Cycles as Integration Devices

One rationale for the use of cycles such as kinase/phosphatase, depolarization/repolarization, nuclear import/export, etc. is that they can be tied to an integration signal. In other words, the inherent noise of a diffusive system needs to be integrated over time for a cell to make a high-fidelity decision and integration of the stochastic functions to make major decisions creates a deterministic system. If activation of an enzyme is needed to signal a change in cell phenotype, then one activation event that has a significant probability of happening in a period, X, should not trigger the process; however, if multiple activation events occur in the same period, then the change in cell phenotype should be triggered. The period is often defined by the level of an inactivating enzyme (phosphatase for a kinase, repolarization for depolarization and GTPase activating proteins for small G proteins). For example, in neurons, depolarization is accompanied by an influx of calcium that will be relatively short-lived and limited. Calcium binds to calmodulin that then binds to the calmodulin-dependent (Cam) kinase. Because the Cam kinase requires 12 calmodulins to be bound before it is active and it has a slow On rate, several depolarizations must occur before it can be saturated. Between depolarizations, the calcium level is very low and bound calcium–calmodulins will unbind and lose calcium at a slow rate. Thus, the Cam kinase activity will be related to the frequency of depolarization (how often the neuron is fired) and will aid in the memory of firing (typically reinforcement of synapses occurs through biochemical changes that result from Cam kinase activation). Integration in these systems comes from the requirement that the activation signal must occur multiple times in a limited time period. Adjusting the strength of the activation events or their frequency, will adjust the probability of activation as will alterations in the inactivation rate. This is like the noise filter in an analog to digital converter to remove noise (see Coultrap & Bayer, 2012).

Cycles can be used for integration that will also reduce the error rate of stochastic processes. A good way to get reliability with a stochastic, error-prone process is to repeat it multiple times and use the integrated activity to make a reliable decision. As explained in the text box, both the strength of the activation signal and the frequency of activation are important elements in setting the threshold for a signal that is reliably above the level that would be activated by stochastic variation. This type of integration is important for major changes in cell state such as growth, differentiation, or apoptosis.

We have been considering how cellular functions communicate in a robust way because each complex function produces signals that can cause the activation of other functions. Whether it is the cell pulling on its neighbor or a neuron depolarizing, those functions generate signals as well as the obvious issues of creating force in a tissue and transmitting a neuronal message. In the case of the cell pulling, the signal will activate growth if the tension is too high from tissue stretch, and in the neuron, the repeated depolarization will cause the synapse to be strengthened. Missing here are the details for given examples. Because an investigator wants to see an effect, they will adjust the level of the stimulus to the point where that effect is seen. A more complete understanding of the control process requires quantitative data and modeling, e.g. how the thresholds of frequency and strength of stimulus are related. Knowing the cycles is only the first step in understanding how they actually are involved in higher-order functions.

2.8   Backups or Variations of Important Complex Functions

With a complex robust device it is common to have multiple systems that can perform essential functions. In the automobile, braking is essential and there are typically two sets of brakes and if they fail, the engine can be turned off in a manual transmission or the automatic transmission moved to park (not recommended except in major emergencies). In the cellular context, it is essential to send out new membrane proteins and normally the proteins to be secreted are transported from the Golgi to the plasma membrane on microtubules by the microtubule motor, kinesin (Figure 2.13). If drugs are added to depolymerize microtubules, then microtubule transport ceases but the proteins are still secreted through an alternative pathway, albeit after a delay to set up the alternative pathway. Similarly, there are many instances where the removal of a protein involved in an essential function results in only the slight compromise of that function because a related protein or another pathway takes over. From the analyses of mouse knockouts, many of the proteins that were thought to be critical (myosins, small g-proteins, and kinesins) did not cause embryo death or even severe abnormalities. In many cases, proteins with overlapping capabilities assumed the roles of the deleted proteins, but in some cases, alternative complex functions were activated to fill in the gap. This serves to emphasize that it is important to know which specific function is actually active (i.e. whether an alternative function has taken over) and to be able to quantitatively measure the functional parameters (rates, forces, and dynamics).

Although not strictly an alternative pathway, there are situations where adverse conditions cause cells to activate repair mechanisms. For example, high temperature tends to denature proteins, which not only reduces protein activity

Figure 2.13

A variety of motors with
overlapping capabilities.

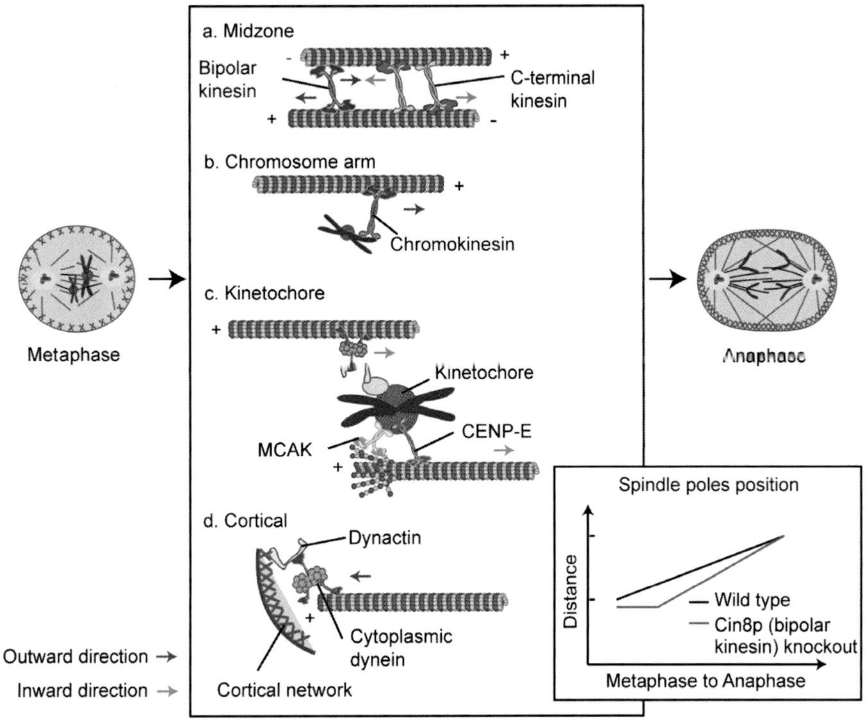

but also stimulates further protein synthesis. In addition, an increase in the level
of damaged or denatured protein will activate a heat-shock response to increase
the level of heat-shock proteins, chaperonins, which help to refold denatured
proteins. A variety of stresses such as chemicals or overstretching of cells will
also evoke a heat-shock response because those also tend to produce denatured
proteins. When normal synthesis is not sufficient to provide the needed active
proteins and there are many denatured proteins, the cellular machine can
activate alternative (helper) complex functions to aid in producing more active
proteins. This is economical and enables the cell to use the activation of sensory
machinery to regulate the level of helper functions that are needed. Heat-shock
proteins (HSPs) are not normally expressed in high amounts because they
require energy and are not needed in normal circumstances. In the example
above, the level of protein denaturation is the critical parameter that determines
the level of transcription and translation of HSPs. Thus, the loss of many active
proteins will constitute a signal to stimulate an alternative pathway to produce
active proteins.

2.9 SUMMARY   In considering the design of complex cellular functions, the principles of robust machines appear to apply. Complex functions are driven by coordinated functional modules to yield observable activities or emergent properties. Compartmentalization of complex functions occurs through the formation of membrane-bound organelles as well as through the local concentration of components in cytoplasm. When choosing the most robust way to perform a function, the simplest is usually the best. Complex functions are digital in nature and will automatically turn off unless activated by control signals. This means that the default state is off and that conserves cell resources. Cycles are critical in complex functions because they can be integrated to give a reliable activation signal. How complex functions communicate with each other through signaling pathways is a critical issue that was not discussed extensively here but will be considered in the next two chapters that describe coordination in more complex functions and in different cell states or phases.

2.10 PROBLEMS

1. For a cell to grow, it must keep the ionic environment inside the cell high in potassium plus magnesium but low in both calcium and sodium. It needs to expel CO_2 from metabolism and import O_2. How would you design a system to robustly control the cellular ionic environment? What are the inputs and the outputs? Do you need a separate regulatory system? Please take a look at the review of how to engineer a robust system by Thomas et al. (2004).

   Answer: The article suggests that the basic design of the biological systems is similar to that of robust devices, called Axiomatic Design. Two major principles apply to both: namely, have independence of functional requirements, and minimize the content of the design (make it simple). For independence, there is a separate system that transports O_2 and CO_2 from the one that maintains the proper ion balance. Hemoglobin and red blood cells serve to transport O_2 and CO_2 through the circulatory system. Hemoglobin binds O_2 in lungs where CO_2 is released with the aid of carbonic anhydrase in the red cell. Further, CO_2 binds to hemoglobin and knocks off O_2 in the periphery and O_2 binds to hemoglobin and knocks of CO_2 in the lungs – a very simple system. On the other hand, the cellular microenvironment is primarily maintained by the Na/K ATPase that transports sodium out and potassium into cells with energy derived from ATP hydrolysis. The Na/K ion gradient is then an energy source for transport of other ions, but in special circumstances, they have dedicated ATP-transport systems.

2. How can there be integration over time of the level of an activity? We take as an example the integration of synapse firing to produce a signal to reinforce a synapse.

If a cell depolarization causes the entry of 1000 calcium ions in a synapse of 10 μm^3, then the concentration will be X. There are 10,000 calmodulin molecules and 500 Cam kinase molecules that can bind 12 calcium–calmodulins. There are other binding sites for calcium–calmodulin and thus, after each depolarization, on average one calcium–calmodulin binds to the 500 Cam kinases. The half-time for separation of calcium–calmodulin from a complex with the Cam kinase is a minute normally but increases to an hour if the 12:1 complex is reached. If there are seven depolarizations in the first minute and none until the fifth minute, then what will be the minimum number of depolarizations in the fifth minute to give about 50% of the 12:1 complex of calcium–calmodulin:Cam kinase? Assume that the distribution of calcium–calmodulins per Cam kinase follows a Poisson distribution.

Answer: The Poisson distribution describes the number of molecules bound per Cam kinase based on the average and from the standard curves about 10 calcium–calmodulins bound per kinase will give 20% of the 12:1 complex and virtually all of those need to be bound in the fifth minute because the others will be over 90% dissociated.

3. Assume that a kinase is activated by a transient mechanical force. Using an analog of the system described above, explain how you could determine the average force over time with an abundant substrate for the kinase and a phosphatase.

Answer: Assume that the phosphatase activity was a constant probability of dephosphorylation per unit time. Then the level of the phosphorylated protein would increase with increasing mechanical force.

4. If transcription is an On/Off process with an average On time of 5 minutes and an Off time of 60 minutes, then what will be the average concentration of a protein that lives for two days before it is precipitously degraded? Assume that the rate of mRNA synthesis is 50 per minute and that the number of proteins made per mRNA is 200. The volume of the cell is about 2000 μm^3.

Answer: Over two days there will be 44.3 cycles on average or 221 minutes of transcription giving 11,050 mRNAs or 2.21×10^6 proteins (divide by Avagadro's number, 6.023×10^{23} molecules/mole to get the number of moles) or 3.67×10^{-18} moles in 2×10^{-12} liters or 1.84 micromolar.

5. In the past, protein knockouts and localization studies have been used to understand which proteins are involved with which functions. Can you define the mechanism of even a relatively simple function such as clathrin-dependent endocytosis if you have a list of the proteins involved and their location in the cell at the light microscope level (± 300 nm)? In this case there are about 70 proteins involved and many of them have been crystalized so that their atomic structures are known. If this is not enough, what additional information would you need to describe the function at a molecular level?

Answer: You cannot determine function because there is no order to the protein activities. By knowing when the protein arrives at the site and when it leaves, you can develop a crude model that then can be refined by experimental tests.

6. Why is the definition of a short temporal step (in a small subcellular meso-scale space) and measuring emergent properties with the aid of inhibitors of that step more informative in understanding the causative relationship between the underlying modular components and the cellular phenotype than a gene knockout?

Answer: Implicit in this question is the assumption that the cellular phenotype is based upon a cyclic process. In that case the gene knockout can alter the phenotype by breaking the cycle at any point. Inhibiting the process acutely in a given location at a given point in the cycle will inform where that activity is occurring and how it fits within the larger cycle.

7. Because most robust functions are localized in membranous organelles or within domains of the cytoplasm, how do they communicate with other functions to produce coordinated effects? Describe a system that cells actually use to communicate between functions (a) when the functions are contained within different cytoplasmic organelles, or (b) when the functions are in different cytoplasmic domains.

Answer:

(a) When the functions are contained within different cytoplasmic organelles, signaling molecules such as calcium or cyclic AMP can be introduced into the cytoplasm by one organelle and they need then to be taken up by the other organelle or alter its function from the outside.

(b) When the functions are in different cytoplasmic domains, even proteins activated at one site can freely diffuse to the other domain unless there are size restrictions.

8. Why is the automatic turn-off of functional units important for the functions of filopodia extension, transport of individual vesicles, and cortical contractions in epithelial monolayers?

Answer: In the case of filopodial extension, it is important that the filopodium extends for the proper length and does not continue to extend in the absence of a positive signal to extend further (it will only work as a sensory probe over a limited distance). In the case of vesicle transport, the vesicle will diffuse to many regions of the cytoplasm and only when it is in the correct regions should it move. By automatically turning off motility, the vesicle will need to be in a proper situation to be activated. In the case of cortical contractions of epithelial cell monolayers, the purpose of the contraction is to typically alter the morphology of the monolayer. Because it is difficult to measure the morphology rapidly, the cessation of contraction is needed to enable sensory mechanisms to work and either stop or reactivate the contraction. In all of these cases, the default pathway involves less activity and is conservative of cell resources.

2.11 REFERENCES

Alberts, B., Johnson, A., Lewis, J., et al. 2014. *Molecular Biology of the Cell*, 6th edition. New York, NY: Garland Science.

Coultrap, S.J. & Bayer, K.U. 2012. CaMKII regulation in information processing and storage. *Trends Neurosci* 35(10): 607–618.

Lippincott-Schwartz, J. & Phair, R.D. 2010. Lipids and cholesterol as regulators of traffic in the endomembrane system. *Annu Rev Biophys* 39: 559–578.

Thomas, J.D., Lee, T., & Suh, N.P. 2004. A function-based framework for understanding biological systems. *Annu Rev Biophys Biomol Struct* 33: 75–93.

2.12 FURTHER READING

Cellular organization: www.mechanobio.info/topics/cellular-organization/

DNA replication: www.mechanobio.info/topics/genome-regulation/dna-replication/

Genome regulation: www.mechanobio.info/topics/genome-regulation/

Membrane trafficking and endocytosis: www.mechanobio.info/topics/cellular-organization/membrane/membrane-trafficking/

Nucleus: www.mechanobio.info/topics/cellular-organization/nucleus/

3 Integrated Complex Functions with Dynamic Feedback

In previous chapters, we described the cell as a robust machine that carries out complex functions through the activity of distinct functional modules. At the next level, it is important to understand how the cell accomplishes these complex functions by coordinating the activity of multiple functional modules. Clathrin-dependent endocytosis and plasma membrane protein synthesis are just two complex functions that result from an integrated sequence of events. For a good understanding of functions, we suggest that they should be dissected at the nanometer (protein–protein complex) level. The production of almost anything in the macroscopic world needs a plan that outlines the number of steps in its production, as well as a decision tree to provide for modifications and/or quality control. This will ensure that the final product meets all required standards. In the production of cars or computers, components are added in a series of steps that often occur in a prescribed order and depend upon the satisfactory completion of each step before subsequent steps can commence. Similarly, complex functions in cells, including clathrin-dependent endocytosis and plasma membrane protein synthesis, all occur in a series of often repetitive steps that depend upon the sequential completion of prior steps. This results in a classical 'if/then' decision tree that will ensure that the complex function is completed properly. In some cases, there are multiple complex functions that the cell can choose depending upon the conditions (e.g. if a cell encounters another cell, it will change its behavior). For us to understand any given integrated complex function, we need to define the steps (which are typically driven by distinct functional modules) and the critical components needed to complete the task.

To understand the multiple levels of complex functions, it is useful here to consider an easily applicable example, again using cars. To raise or lower a car window, several machine elements come into play. Only a basic understanding of the mechanism is required to raise and lower the window. However, to be able to repair a car window, the mechanism must be understood in much greater detail. Here, a complete list of the parts, right down to the screws that hold them in place, may be required. The task of repairing the mechanism is made more complicated when parts differ between car brands, even though they have similar roles in the overall function. An even greater understanding is reached by looking inside the motors that drive the mechanism, and dissecting each element of each part. Such details may be required if that part needs repair or you wish to understand how this will affect the overall complex function.

Similarly in cells, complex functions can be broken down to increasingly greater levels of detail, starting with the function itself, and continuing down to the amino acids that make up the protein components (atomic level). To understand these events, we suggest that they are best analyzed at the level of the proteins rather than the atomic level. In this text, we will focus primarily on whole proteins and protein aggregates (tens of nanometers) because this is the level at which there is relevant information about the physical and biochemical aspects of the function.

Enough is known about several cellular complex functions to describe them in physical and biochemical terms. In these cases, we are then able to build rudimentary process engineering diagrams and this is useful to understand how the process is actually accomplished by the cell. For this chapter, we will explore complex functions that are largely linear in their progression, such as clathrin-mediated endocytosis and membrane glycoprotein secretion. In both of those cases, there is a clear end-point and we can postulate a series of steps that must be completed in order to reach that end-point. When steps are not completed properly branching will occur. For example, depolymerization of microtubules will cause the Golgi to disperse from the normal position at the centriole and form multiple small Golgi in many parts of the cell. Similarly, targeted inhibition of actin polymerization will block inward movement of a clathrin-coated vesicle and encourage other endocytic mechanisms. To understand the importance of each step in a given mechanism, and to fully appreciate the consequences of an uncompleted step, it is useful to look at the details of complex functions such as clathrin-dependent endocytosis (Weinberg & Drubin, 2010) and membrane glycoprotein production. In the former case, a detailed model has been developed based upon the more than 60 different proteins that have been localized to the site of endocytosis, their biochemical functions, and the time course of their coming and going (Liu et al., 2009).

3.1   How to Approach a Complex Function

Understanding how a complex function is carried out generally starts with knowing the goal of the function, and the context in which it will be performed, which can be described by the 'dependent variables' required. In the case of clathrin-dependent endocytosis, the goal is for cells to bring membrane and external material into the cell. In terms of the context, this function must be carried out where energy is available to bend the membrane and to form an internal vesicle. The energy source (ATP) is a dependent variable that is considered as a constant in most cases. Other dependent variables include the lipid, PIP2, the availability of clathrin, actin and the other proteins involved in the process. Because the function is readily repeated in cells, and the endocytosis can be monitored dynamically, we do not need to redesign it from scratch in order to understand endocytosis. Instead, we can deconstruct it through reverse engineering. A similar approach is taken in various industries today, in particular, in computer hardware design. To understand how a particular device is made, a company may start with the end product of a rival company, take it apart to understand how each piece fits, and attempt to enhance the design. The classical way to reverse engineer cell functions is to remove or alter proteins involved in the function from the cell.

Dependent Variables Link Complex Functions

What is a Dependent Variable? The simplest example is ATP production, because a high ATP concentration is needed for many complex functions. Further, the rate of many functions depends upon ATP production maintaining normal levels of ATP. For the complex function of protein translation, the dependent variables include mRNA production and transport, tRNA production and loading with the appropriate amino acid, GTP and ATP production, ribosome production and transport, as well as a number of other proteins involved in initiation of translation (Figure 3.1).

  How should the Dependent Variables be treated in describing a complex function? The simplest way to describe a complex function is to assume that the dependent variables are supplied (mRNA, tRNA-aa, ATP, etc.) and that one only needs then to describe how the complex function works under optimal conditions. This is obviously a first approximation of the complex function because a deeper understanding of the complex function should involve understanding how it is robust to changes in mRNA, ATP, ribosome and other component levels. Thus, moving from the first description under optimal conditions to physiologically relevant perturbations is a logical way to proceed.

Figure 3.1

Linked complex functions – protein translation assumes dependent variables that are the products of other linked complex functions.

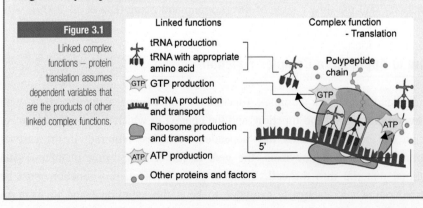

As the blueprint for cellular machinery is present in the genetic information of the cell, we can also gain insight into complex functions by understanding proteins that are encoded by it. When possible, the analysis of the mRNAs present in the cell will tell which proteins are actually being made. Still, it is difficult to know which proteins might be part of the function. In classical genetic screens, the DNA is mutated by chemical treatment or ionizing radiation and the resulting mutant organisms are screened for the loss of the function of interest. Once mutants are found, the DNA is sequenced and the mutated protein is identified. For example, when exploring clathrin-dependent endocytosis, researchers performed genetic screens looking for the loss of endocytosis and established a set of mutated genes that altered endocytosis. Those proteins were then further analyzed by comparing their sequences with other known proteins. Antibodies to the proteins were used to label cells and find the location of the proteins. From these data, one can formulate a basic hypothesis of how the proteins might participate in the function. Much greater refinement of that hypothesis can come from the understanding of the location of all of the proteins during the function.

These are just some tools and approaches that can be applied to dissecting complex functions. As the tools become more sophisticated, and greater insight is attained, however, hypotheses will change. This has occurred, for example, in the case of clathrin-dependent endocytosis. In the beginning of the twenty-first century, it was hypothesized that clathrin-dependent endocytosis occurred through three basic steps, namely assembly of the coat, vesicle formation, and coat disassembly. However, with new methods to measure dynamic fluorescence, the relative timing of many proteins at the endocytic site was determined. Thus, the complexity of the process and the number of postulated steps increased. It follows then that super-resolution microscopy and other new technologies will take the complexity and our understanding to an even higher level. The important issue in this process is that we should choose the level of understanding that fits our needs. If we want to alter the course of a genetic disease where a critical protein is mutated, then we will need to understand the process in much greater detail than if we only want to control the level of endocytosis of certain ligands through that pathway.

When we take apart a device to better understand how it works, we often find parts that have no obvious role in the function. The simplest way to gain insight into their role is to reassemble the device without the part and see how the device functions differently. A similar approach can be taken to understanding complex functions in the cell. Using methods of gene knockout, or protein knockdown, proteins can be removed from the cell to see how the function is altered. Because most devices are much simpler than cells, the loss of any given part will typically cause an effect. However, in cells the loss of a protein will not necessarily alter the function that it is linked to in a clear way. This may either be because the protein is only engaged to help the function in specialized circumstances, or there is redundancy for this proteins activity. As functions are better understood and a given protein is linked to a specific step in the function, the process of understanding the role of the protein is easier. This means that assays for intermediate steps in the function are needed and sometimes those are difficult to devise. The functions that we have chosen to discuss are ones where there are tools to approach intermediate steps. For example, in dynamic functions like endocytosis, proteins often only associate with the functional complex for a short period and then leave. Knowing when a protein is present makes it easier to postulate the role that the protein may play in the process and to devise an assay to test that hypothesis.

3.2   Steps in a Complex Function

Most complex cellular functions result from the sequential completion of different steps, which usually are linked to specific functional modules. In Chapter 1 we defined a functional module as *a minimal set of molecular components working together to yield a measurable activity or a function.* In many contexts these modules are distinct protein complexes that accomplish a single step in a complex function. In some cases the steps will be carried out by a large number of proteins working together. This is the case with clathrin-dependent endocytosis that requires a sequential assembly of proteins at the site of endocytosis that is initiated by adaptor protein aggregation (see Figure 3.2). Once initiated, the process of endocytosis proceeds at a rate determined by the availability of the proteins and tension in the membrane that resists the bending to form the vesicle. Here, the binding and release of different components to a core complex is critical for the completion of several steps.

The issue of defining steps in processes like endocytosis is difficult because many copies of different proteins must bind, perform a function, and leave. Further, different regions of the same vesicle can be at different stages of endocytosis. In Figure 3.2 clathrin-dependent endocytosis is broken down into seven steps. These steps are somewhat arbitrary and do not refer directly to distinct protein-binding events or modifications. A complete analysis of the steps should include a description of the bases for the binding of each protein, the consequences of protein binding for the overall function and the

Figure 3.2

Clathrin-mediated endocytosis is a complex function with sequential steps or linked functional modules.

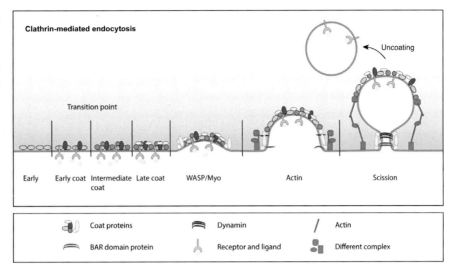

mechanism of release of each protein. This is a major task, but once completed would provide a template for understanding clathrin-dependent endocytosis in other cells. That level of understanding would also enable one to engineer a repair of endocytosis or enhancements that could help when the process is inhibited in disease states.

In processes where many proteins collectively carry out a task, there is an expectation that protein binding should be coupled with the modification (conformational or enzymatic) of at least one of the proteins in the complex. These modifications would then enable the function to proceed and subsequent steps to begin. The assembly of a focal adhesion is another example of a functional module involving the coordinated aggregation of a large number of proteins. Focal adhesions allow cells to adhere to the extracellular matrix, pull on the underlying surface, and then release when they need to move. From the analysis of the many protein–protein binding interactions that occur during focal adhesion assembly and function, there is evidence that one type of interaction response dominates. In particular, where protein A binds protein B and then recruits protein C, a modification of either A or B will occur (Zaidel-Bar et al., 2007). Protein modifications upon binding provide for the input of energy to drive the function toward completion. Dissociation will often involve another kinase, dissociation of a product or the binding of next protein in the process. As complex functions are probed at this level of detail, there are often unexpected steps that emerge, when cells missing associated proteins are analyzed.

3.3   Decision-making in Complex Functions

In a relatively linear complex function, the decision to proceed to the next step typically depends upon the proper completion of the previous step, as this will enable the binding of components involved in the next step. For non-linear functions, however, such as the processing of endocytosed vesicles for either recycling or movement to lysosomes, subsequent steps involve much more than single protein binding and complex modification, but rather movement of proteins or a signal from one compartment to another, or other major rearrangements of cellular components.

If we look at endocytosis by measuring the relative time course of the binding and release of individual proteins to the site of clathrin-dependent endocytosis (Figure 3.2), then it becomes clear that a number of distinct functional modules are assembled at the site. The early components create a cluster that recruits cargo, membrane-bound proteins as well as early coat proteins. Other components can then bind to form an intermediate coat. This is followed by the recruitment of clathrin, and subsequently the assembly of actin, which drives the coat inward. Finally, BAR-domain-containing proteins can assemble a ring around the neck and mechanically assist the scission of the newly formed vesicle.

Checkpoints in Complex Functions

The idea that certain steps need to be completed before the next step starts is common and this is an effective checkpoint. For quality control and coordination of subsequent actions, the early components need to assemble and often be processed by phosphorylation or another type of reaction. The assembly of the early components will lead to bending of the membrane plus other modifications and that will activate binding of the next set of components. Once the actin pushes the clathrin coat far enough into the cytoplasm, a bare region of membrane can bind the BAR proteins. Mutations that block clathrin endocytosis can provide important clues because they may cause the endocytosis to stop at a certain step. Knockdown of proteins to decrease their cytoplasmic concentration can slow or stall endocytosis at a given stage. Thus, there can be a series of checks at different points to know each time that the right components are in place before proceeding. Physical chemical (e.g. membrane curvature) as well as chemical parameters are commonly used as checkpoints.

As mentioned earlier, functions like endocytosis can be viewed at extremely high detail. With sensitive quantitative assays, analyses of clathrin-dependent endocytosis would likely reveal additional proteins that are part of the complex function, such as protein kinases or other enzymes that could interact with the proteins in the clathrin coat only briefly. Many of these will probably be explainable as requirements that were not imagined such as the need for energy input to dissociate a particular complex. Others will likely enable the function to continue in the face of environmental challenges such as changes in pH or ion levels, i.e. making endocytosis more robust.

An alternative approach to understanding complex functions is to consider how the distinct functional modules are linked together. 'Linked functions' are defined here as *those where the performance of one function is linked to the activation of the next function.* In other words, we consider how modules are coordinated to perform steps that culminate in the overall process. The functional modules may be coordinated in time or space and often involve physical parameters such as force. Again, we can look at the automobile engine to understand this concept, as it has many modules that perform distinct steps in order to achieve a common goal – the conversion of chemical energy from gasoline to thermal energy. Here, each module is physically linked to enable the cyclic uptake of a gas–air mixture, compression, ignition, expansion, and release in a coordinated fashion. The emergent property is the conversion of thermal energy to mechanical energy through the rotation of the engine drive shaft. No one step in the overall function can be isolated and much care is needed to ensure that the steps are carried out sequentially. Many biological functions occur in a similar fashion, where timing (often in the form of reaction rate), geometry (spatial distance between and membrane binding of components), and physical force in the protein complexes all help to coordinate their activity and to accomplish the final function.

In the examples discussed thus far, the mechanics of the system also play important roles in controlling the direct links between functional modules. For example, mechanical deformation of the membrane is important during endocytosis, and this is mediated by clathrin coat assembly followed by actin polymerization. To physically bend the membrane requires these proteins to do work and the amount of work is related to the bending stiffness of the membrane (a constant) and the tension in the membrane that tries to flatten it. Membrane tension levels can vary over 10-fold during the cell cycle of fibroblasts. At high membrane tension, the clathrin coat will not bend the membrane sufficiently for the actin assembly module to bind to the site and do its job. Thus, the endocytosis stalls at high membrane tension. This is one example of the linkage of physical parameters to a function. When the function involves a major physical change, mechanical properties are linked to biochemical properties, as they obviously influence each other. Thus, any model of such a complex function needs to include the mechanical factors as a load or accelerator that will modify the rate of certain biochemical steps.

Importantly, complex functions do not exist in isolation of each other. This raises the question of how different complex functions that produce similar or competing emergent properties are coordinated or regulated together. Cells will often use a system of biochemical regulation to coordinate multiple processes. What is required, then, is a common factor related to each of the functions that, when modified, will serve as a trigger to initiate the biochemical pathways. These common factors are often mechanical. For example, clathrin-dependent endocytosis must be coordinated together with functionally similar endocytic processes such as caveolae, macropinocytosis, and clathrin-independent endocytosis. In this case, the activities of each complex function are linked to the physical properties of the membrane, and indeed, membrane tension has been shown to serve as a physical break on all of the processes. Membrane tension is sensitive to the area of the plasma membrane, the cytoskeleton binding to the plasma membrane and cell volume (Sheetz, 2001). This means that while each complex function is regulated biochemically, it is the mechanical property of membrane tension that provides the very simple mechanism to coordinate the different processes of endocytosis and maintain a constant membrane area for the whole cell.

Mechanical or physical properties of the system therefore provide direct ways to coordinate cellular activities without complex biochemistry. By linking the biochemistry of other processes to membrane tension, then it is possible to use the physical parameter to regulate not only the rate of endocytosis, but also other factors such as cell polarization and volume regulation.

It should be noted that this does not clarify the mechanisms controlling the activation of the complex function of endocytosis. Control of complex functions is dependent upon many aspects of the cell, such as the phase in the cell cycle or other cellular states. Endocytosis depends upon the level of exocytosis as well as the binding of ligands to receptors that are taken up by clathrin-mediated endocytosis. Hence, there is a dynamic feedback between important cell factors and these complex functions.

If the function is blocked or dramatically slowed for one reason or another, then alternative pathways may be activated. However, the time required to activate these pathways may be relatively long. In the case of secretion of glycoproteins by transport from the Golgi to the plasma membrane, the normal microtubule-dependent pathway of secretion will be blocked by the depolymerization of the microtubules. The rate of secretion of glycoproteins will drop immediately after microtubule depolymerisation, but will recover in about an hour when the Golgi has dispersed to multiple sites in the periphery. Thus, the block of a function can activate an alternative pathway, but the new pathway may rely upon major changes in cell organization. If so, then there can be a major lag before the cells will activate other pathways that can accomplish the same function.

3.4 Integrated Complex Functions

Integrated functions, which involve the movement of a protein between several cytoplasmic compartments or intracellular organelles, and may require a variety of other complex functions as steps in the process, are considered far more complex to describe than those explored thus far. Conceptually, the different complex functions are modular tools that should be considered as semi-independent entities with important links to the rest of the process.

Glycoprotein synthesis is an example of an integrated complex function because it integrates other complex functions such as coat-dependent vesicle formation, carbohydrate addition and protein sorting. Several steps in this complex function are well established; for example, we have a good understanding of how a protein enters the ER during protein synthesis and how the final glycoprotein is exocytosed in the end. However, the individual steps in between, and their coordination, are poorly understood. The primary method to understand the process has been to knock out or otherwise remove components in the biochemical pathways involved. When one path to production is blocked, the robust nature of the process often means that another pathway can take over. This issue is encountered when attempting to dissect any robust complex function, and makes analysis difficult because the second path will necessarily be different than the first. However, with the new generation of fluorescent probes and super-resolution microscopy, the kinetic parameters can be measured directly. This should allow a clearer view of the actual process and how it will change with physiologically relevant perturbations.

Glycoprotein production is an important complex function in cells and is involved in the processing of many different molecules that range in size from glycolipids (5–10 nm) to some of the glycans (1000 nm when fully glycosylated). The initial stages in glycoprotein production are relatively clear in that they involve the docking of a signal recognition peptide to the ER entry complex, and the subsequent insertion of the relevant portions of a protein into the ER (Figure 3.3). However, the subsequent steps can be quite different, as the mechanism will differ depending on the specific proteins involved. For the majority of proteins destined for the plasma membrane, the next steps are the removal of the signal peptide and preliminary glycosylation (Geva & Schuldiner, 2014). When those modifications are completed, the protein can be concentrated into vesicles coated with CopII, which is a protein that helps to pinch off vesicles and facilitates

Initial stage of the glycoprotein production.

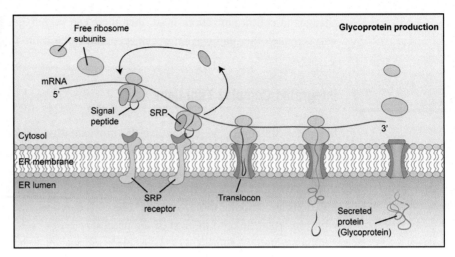

Subsequent stages in glycoprotein production.

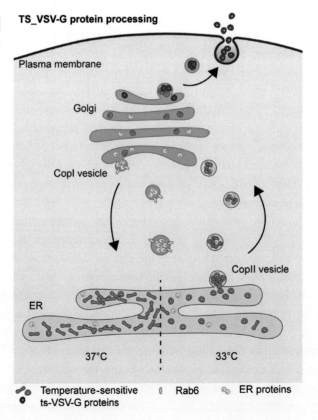

the movement of the vesicle from the ER to the Golgi. CopII-dependent vesicle formation is itself a complex function that is analogous to clathrin-dependent vesicle formation in that the energy for the concentration of the membrane proteins and for membrane bending comes from coat assembly. Many of the details are different, however, with unique mechanisms that control changes to the size of the vesicle, in order to accommodate larger proteins or protein complexes. Although much effort has been expended to define the parameters that determine which protein will be concentrated into a CopII vesicle, no clear biochemical signal has been found. There is a general belief that the length of hydrophobic alpha helices, as well as perhaps charged amino acids, will influence the decision. Another possibility is that a coat protein will bind to the glycoproteins to concentrate them into a CopII vesicle. The advantage of a multimeric coat protein complex is that the free energies of many weak interactions can sum to produce a strong interaction. A classic example is the case of the temperature-sensitive vesicular stomatitis viral G (ts-VSV-G) protein that has been used in many studies of membrane trafficking. At 37°C, the ts-VSV-G protein is retained in the ER and a large concentration of it can be produced in the ER at that temperature. At the permissive temperature of 33°C, however, the ts-VSV-G protein can move out of the ER to the Golgi and on to the plasma membrane. The most likely explanation is that the protein changes conformation at the non-permissive temperature and thereby loses an important interaction required to package the ts-VSV-G protein into the CopII vesicles, because recent results show that the ts-VSV-G protein does not aggregate in the ER at the non-permissive temperature. Thus, the fourth major step (following protein insertion into the ER, protein folding, and initial carbohydrate processing) in the integrated complex function of membrane glycoprotein synthesis involves a complex function of coated vesicle formation.

Because the Golgi and ER can mix and then can separate themselves, a robust pathway must exist to carry material in the reverse direction, i.e. from the Golgi to the ER. This has been attributed to another coat protein, CopI, which is a major component in retrograde vesicular transport. However, an additional process, which occurs independently of CopI and involves the membrane-bound GTPase, Rab6, and perhaps tubular membranes, has also been attributed to this complex function (Geva & Schuldiner, 2014). In this case the two parallel pathways provide redundancy to the complex function, with one pathway able to take over whenever a mutation or conditions compromise the other pathway. While such situations are commonly observed in robust complex functions, they serve to highlight how difficult it is to understand and define the particular steps involved. When investigating the CopI pathway, an obvious approach is to use a knockout of the Rab6 pathway as the background cell line. However, for this approach to work, it is assumed that a third or fourth pathway does not exist.

Small GTP Binding Proteins

The general class of small GTP binding proteins can act as signaling proteins and they are activated by GTP exchange factors (GEFs) and are inactivated by GTPase activating proteins (GAPs). They play major roles in actin-based motility (Rho, Rac, and Cdc42) as well as in membrane trafficking (Arfs and Rabs). Presumably, the binding of a specific Rab to a vesicle will target it for binding to the next membrane compartment (e.g. ER to Golgi).

An important aspect of many membrane traffic pathways is that they are governed by mass action, which means that an inefficient exit process will become more efficient as the concentration of the protein builds up in the compartment. For example, the ts-VSV-G protein will become very concentrated in the ER at the non-permissive temperature and will tend to leak through low-affinity exit pathways as a result. That will increase the level of ts-VSV-G leaking into the next compartment. Thus, building the system so that it will be sensitive to changes in the concentration of the precursor will make the production of the product more robust to conditions that compromise the rate of the process. In more complex processes, the recovery or reverse pathways such as CopI and Rab6 movement from Golgi to ER can be equally important for the overall function and should be analyzed carefully when looking at the overall process.

3.4.1   Non-linear Complex Functions in Robust Integrated Systems

Because the integrated complex function of protein secretion is vital to the organism's survival, it needs to be robust to a variety of perturbations. In terms of compartmentalization of functions, the ER and the Golgi make sequential modifications to the membrane glycoproteins. However, they can reform if mixed and there is a lot of bidirectional communication between them. Proper glycosylation can occur when the compartments are mixed, because the specific pattern of carbohydrate addition is defined by the specificity of the enzyme, irrespective of position. This makes the system quite flexible and there is even traffic from the cell surface back to the Golgi for some proteins to be reglycosylated. The rate of adding the proper carbohydrates will be increased if the membrane glycoprotein moves from enzymes that add the early carbohydrates (in the ER) to those that add the late carbohydrates (Golgi). There are further subcompartments to the Golgi and the portion nearest the exit site is the portion where the enzymes that add the final carbohydrates reside. Thus, compartmentalization in this system is not rigid and appears to self-sort according to the flow from the ER to the plasma membrane. As pointed out in the first chapter, the increase in efficiency provided by the sorting of the carbohydrate-modifying enzymes need not be very great before it will dominate in the population.

In terms of alternative pathways, we have discussed the matter of the two recovery systems for transport from Golgi to ER and the two pathways from the Golgi to the plasma membrane. The emphasis is on the multiple pathways of forming and fusing vesicles. Relevant to the alternative pathways issue is that membrane movements between compartments can be considered like water flows in a water fountain with several sequential pools. If the outflow of one pool is slowed, then it increases in volume and the volume of the downstream pool decreases. In a similar way, slowing membrane flow from the ER to the Golgi by expressing the temperature-sensitive mutant of the virus, tsVSV-g at the restricted temperature, will cause an increase in the size of the ER. When such high volumes in the ER occur, the leakage of membrane to the Golgi will increase through an increase in the tsVSV-g proteins that are non-specifically incorporated into the vesicles (a form of mass action) or through ER–Golgi fusion events that may be more likely. Thus, the communication between compartments can occur by several mechanisms that constitute alternative pathways.

Within the secretory system there is a need for integration of many different complex functions and it is a daunting task to understand how the system will function normally, much less under different physiological perturbations (Figure 3.4). Mass balance (synthesized membrane glycoproteins that enter must come out) needs to be maintained and components recycled back to compartments when they have not been properly processed. Because the ER, Golgi and secretory vesicle compartments can shrink or swell dramatically, there is considerable flexibility. In thinking about the integrative aspects of such complex systems, it is important to keep in mind the relative timescales of the functional complexes and the integrated complex functions. Whereas functional complexes work on a second timescale or the speed of many enzymes, the integrated complex functions require many minutes or even hours to produce the final product. In addition, there will be many different aspects of the cell that will influence the integrated complex functions. Those aspects will often appear as modifications of enzymes in the process that may introduce additional steps or modify existing steps. Keep in mind that there will be a limited range of operation for non-linear complex functions and outside of that range the cell will not normally survive. In such cases, apoptosis or

Figure 3.4

Robust integrated system through coordination and dynamic feedback of linked complex functions.

Microtubule-dependent transport

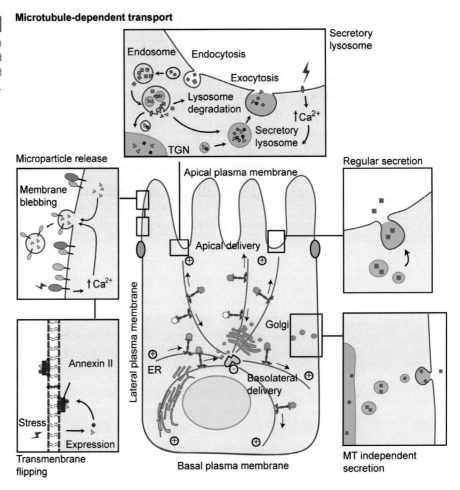

other forms of premature cell death may be initiated (for example, when hormone receptors are not processed properly and fail to reach the plasma membrane), and this will provide an indirect effect on the function of interest. Thus, it is important to have detailed assays for intermediate steps in a complex process to know where perturbations are acting.

From the viewpoint of the cell, all of these steps and decisions for alternative steps occur seamlessly at the level of single molecules and adapt rapidly to changes in metabolism or the cell cycle. Although this appears dizzyingly complex at first, the number of possibilities is finite and no more complex than systems designed by human engineers. What is needed is a means of describing these processes that encompasses the complexity and the robustness in an understandable way. Until a universal convention has been adopted, we will individually describe the different steps as best we can. Linking the steps and the alternative pathways to a given process and other processes along with feedback controls goes beyond the level of any current program, although we could make considerable progress if we focused on a few well-defined cell states.

3.5 SUMMARY

For us to understand complex or integrated complex functions, the processes need to be broken into steps that then are linked in the overall processes. In the example of clathrin-dependent endocytosis, over 60 different proteins have been mapped to the function and their relative timing at the site is known. The overall process has been divided into seven steps, but it is likely that further understanding of the process will reveal more steps or at least additional parallel activities. In working out the steps involved in such a function, it is useful to start by considering all of the tasks that must be performed to reach the overall goal and how the cell could possibly accomplish them. With such detailed plans and knowledge of where and when different proteins are involved, it is useful to build models of the process and to then test the predictions of those models to increase our understanding. With integrated complex functions such as glycoprotein secretion, the process involves many complex functions in multiple cell compartments that can fuse and reform readily. Treating the individual complex functions as isolated tools enables one to model the whole process as a series of steps performed by separate tools that then receive inputs from important cellular parameters and from the final output of the system. Starting from a rudimentary understanding of each of the complex functions provides the basis for modeling the overall process. Robust features of the system include flexibility in the boundaries of compartments, number of proteins needed for the cell to function, parallel alternative pathways, and the intermittent activation of each function dependent upon need. The If/Then nature of the cellular decision tree does not fit with our notions that A should lead linearly to B, but does fit with the way that modern manufacturing functions through the use of feedback controls from the environment or in-line quality checks to modify production. Once individual complex functions are understood at a detailed level and their relationship to other complex functions is known in integrated complex functions under normal conditions, the processes should be further studied in different cell states and with different physiological perturbations to better understand the mechanisms actually involved.

3.6 PROBLEMS

1. Based upon the Weinberg and Drubin (2012) paper, please describe the steps that you imagine will be needed to perform clathrin-mediated endocytosis of the

transferrin receptor in mammalian fibroblasts. Assume that there will be at least 60 proteins involved. Base your answer on the necessary changes that must occur during endocytosis of a specific subset of membrane proteins.

Answer: There is no single answer to this question. In the paper, they identify seven major stages in the process and highlight 23 different proteins in the process of endocytosis. In each of the stages there are many molecular steps that must occur for the various proteins to function in the stage. For example, there should be a step that initiates the recruitment of the Ede1p and Syd1P. The second step would be the recruitment of cargo. Once they are assembled, the third step of recruitment of the Ap2 Complex starts. Within the early coat stage, there are likely several enzymatic steps involved in determining that the right cargo is present and that the additional proteins, Chc1p/Ck1p, Yap1801/Yap1802p and Pal1p are recruited properly. The point of considering the process as a series of steps is that the phosphorylation/dephosphorylation and other reversible post-translation modifications are likely critical elements of component assembly in an ordered fashion. Irreversible modifications such as proteolysis or ATP cleavage can move the process forward in a directional fashion. A complete answer is not now available but there are likely as many or more steps than the number of proteins involved. As we understand the process in more detail those steps will become evident initially as logical questions that arise from thinking about how the system could work as a robust and repeatable cycle.

2. In a very extensive temperature-sensitive screen of mutations in yeast, the investigators found 100 different proteins that, when mutated, caused a block to clathrin-mediated endocytosis at the non-permissive temperature. Does this mean that Drubin's lab missed 40 proteins directly involved in clathrin-mediated endocytosis? (Remember that the cells can be propagated at the permissive temperature and then the temperature is raised for the assay of endocytosis.)

Answer: No, the fact that endocytosis doesn't continue at the non-permissive temperature simply means that a dependent parameter (ATP, normal membrane tension, osmotic balance, etc.) is altered by the alteration of the activity of the mutant protein. Usually, the first step in the characterization of the 100 altered proteins is to determine the localization of the protein.

3. If you add a drug to selectively block CopII vesicle formation and not CopI, then how will the size of the Golgi and ER change over a short time period? The CopII vesicles are involved in forming vesicles that move from the ER to the Golgi, whereas CopI vesicles move material from the Golgi to the ER.

Answer: In membrane traffic, the inhibition of movement from one compartment to another decreases the size of the acceptor compartment and increases the size of the donor compartment. In this case, the drug will block movement of membrane from the ER; therefore, the ER will increase in size and the Golgi will decrease because the exit pathway from the Golgi is not inhibited.

4. Imagine that you have cells lacking a protein that is thought to be important for the fusion of exocytic vesicles containing plasma membrane proteins with the plasma membrane. How would you test your hypothesis with the ts-VSV-G protein tagged with green fluorescent protein (assume that the ts-VSV-G protein is secreted by the normal membrane glycoprotein pathway)? For background, see Hirschberg et al. (1998).

Answer: Because the classical experiments in the Hirschberg paper determined the rate constants for ts-VSV-G protein moving between compartments including movement to the plasma membrane, the rate of movement from the Golgi to the plasma membrane could be measured. The mutant cell transfected with ts-VSV-G should show a decreased rate of movement from Golgi to plasma membrane but a normal rate of movement from ER to Golgi.

5. In the Hirschberg et al. (1998) paper, they suggest that the rate of movement of the VSV-G–GFP protein from the ER to the Golgi is 2.8% of the total ER protein per minute. What does this tell you about the process of movement from the ER to the Golgi ((1) random, (2) dependent upon an exclusive transport system that would transport no other membrane proteins, or (3) a conveyor belt that moves proteins from synthesis through processing to the exit vesicle)?

Answer: Because the fraction of the protein that moves from the ER to the Golgi is constant at 2.8% per minute, the process of movement appears to draw randomly from the proteins in the ER. With either the exclusive transport or conveyor belt mechanisms, there should be a very high rate of movement initially as the bolus was released that would decrease with time.

6. Take the function of translation and describe the steps in the function. Also, do you see definite decision points where there are If/Then decisions made that can produce different outcomes?

Answer: In eukaryotic cells, (1) the mRNA is transported to the right location in the cytoplasm, (2) the initiation complex is formed that in itself requires several substeps, (3) elongation occurs in a cyclic process with some quality control both in proofreading the amino acids incorporated, looking for signal sequences and the proper folding of the protein, and (4) the termination of the process and release of protein plus ribosome. If a signal sequence is produced, then the protein–RNA–ribosome complex targets to the ER and the protein is injected into the ER. If the protein is not folding properly or some other problem occurs causing premature truncation of the protein, then the mRNA will be degraded.

7. Take the integrated complex function of endocytosis and processing to lysosomes and describe it in an analogous way to membrane protein secretion.

Answer: In both cases, the proteins are being moved from a source to a sink compartment through several intermediate steps. Although the processing is different in that the endocytosis processing is selecting for inactive or abnormal proteins whereas secretion is selecting for proteins that have been correctly modified enzymatically, there are still a series of selection steps in the transport processes.

8. Consider the complex function of DNA replication and provide a list of the steps in the process, starting with the activation of the replication complex. Which steps are primarily dependent upon physical parameters and which are biochemical?

Answer: This list is partial in the sense that each of these steps can be broken into smaller steps. Further, the biochemical versus physical parameters sensed in each step are open to interpretation in that many aspects of the physical parameters such as the length of the RNA primers are manifest in biochemical interactions.

1. Activation of replication.
2. Binding of initiator complex to origin site (only two old strands).
3. Helicase binding and opening of loop.
4. Binding of two polymerase complexes for leading strands.
5. Primase binding and synthesis of RNA primer.
6. Okazaki fragment polymerase binding.
7. Degradation of RNA primer (length of primer is likely defined mechanically).
8. Completion of fragment synthesis and annealing.

3.7 REFERENCES

Geva, Y. & Schuldiner, M. 2014. The back and forth of cargo exit from the endoplasmic reticulum. *Curr Biol* 24(3): R130–136.

Hirschberg, K., Miller, C.M., Ellenberg, J. et al. 1998. Kinetic analysis of secretory protein traffic and characterization of Golgi to plasma membrane transport intermediates in living cells. *J Cell Biol* 143: 1485–1503.

Liu, J., Sun, Y., Drubin, D.G., & Oster, G.F. 2009. The mechanochemistry of endocytosis. *PLoS Biol* 7(9): e1000204.

Sheetz, M.P. (2001). Cell control by membrane–cytoskeleton adhesion. *Nat Rev Mol Cell Biol* 2(5): 392–396.

Weinberg, J. & Drubin, D.G. 2012. Clathrin-mediated endocytosis in budding yeast. *Trends Cell Biol* 22(1): 1–13.

Zaidel-Bar, R., Itzkovitz, S., Ma'ayan, A., Iyengar, R., & Geiger, B. 2007. Functional atlas of the integrin adhesome. *Nat Cell Biol* 9(8): 858–867.

4 Cells Exhibit Multiple States, Each with Different Functions

We are well aware that organisms exhibit different states of awareness, from sleep to heightened tension. Cells can likewise be in different states (phases) where different response networks and different functions are active. The different phases of the cell cycle are obvious examples of how the same cell can exhibit different behaviors when stimulated in a defined way without changing its differentiation state. Furthermore, a given function may have different activators and different levels of activity in different phases. Thus, it is important to control the cell phase when studying any given function. With a large population of cells in a variety of cell phases, the assay of a function will potentially give a mixture of activity levels. There are many different possible phases for cells, including changes in type of motility or metabolic activity in addition to the well-characterized phases of the cell cycle, senescence, or apoptosis. It is important to consider how the cell decides to transition from one phase to another. As was discussed in the previous chapter, the inability of a cell to complete a desired function can be a stimulus to change to an alternative pathway that can involve a change in phase. Further, phase changes can also be induced by the internal cell clock, or external stimuli. Another issue is that widely different cell types can exhibit very similar phases (e.g. mitosis is similar for different types of cells). Thus, many of the characteristic functions associated with different phases are present in many cell types but need specialized conditions to become activated. The same function may have varied outputs in different cell types, but the basic elements would be common to all cells. Often the transitions between phases occur in a very short time and constitute a concerted change. The probability that cells in a population will be in the same phase can be controlled in the lab using patterned substrates and defined conditions. However, single-cell studies can provide a better understanding of the cell phase when a given subcellular function is analyzed. Using defined molecular markers for specific phases, it is possible to follow phase changes at a single-cell level in multicellular tissues.

Defining Cell Phase (State)

Each type of cell can often display multiple phases or states in which integrated sets of functions are active and others are specifically inactivated. A transition from one phase to another typically involves a concerted change in many cell functions. Changes in phase often do not require protein synthesis, but they are often accompanied by changes in protein expression patterns. Many inputs are integrated over time before activation of a major phase change. Where studied, phase changes are effected by proteolysis of selected proteins, activation of receptor tyrosine kinases or G-proteins, high levels of cAMP or other major signals.

Analysis of Functions (Multiple Levels of Phase Dependence)

1. An integrated cellular function should be described in the context of cell phase and the requirements for that function, e.g. protein synthesis is dependent upon cell phase and requires mRNA, ribosomes, energy, aminoacyl tRNAs, protein folding complexes, and post-translational processing (see below also).
2. At the next level, each requirement is in turn dependent upon a set of parameters; e.g. mRNA levels depend upon synthesis and processing, transport to the cytoplasm, and degradation rate (plus in neurons the mRNAs are packaged for transport down axons), which vary with cell phase.
3. Phase-dependent changes reach to the molecular level, which requires several iterations of this process; e.g. transport to the cytoplasm requires Ran-GDP to bind to the mature mRNA and facilitate movement through the nuclear pore complex. Further, the nuclear pore complex involves the assembly of over 100 proteins to form the fully functional complex.

4.1 The Role of Cell State or Phase in Cell Functions

Because different cellular functions will directly impact each other, the activation of a single function cannot be viewed in isolation. Generally, cells exist with distinct subsets of their functions active, and can transition between phases with different active functions. This is how cells control their behavior, as we will discuss below. The communication between the different functions can produce the desired response to stimuli. This acknowledges the integrated nature of cellular functions and indicates that different responses will occur as a result of differences in phase. Cell phase can be viewed from various perspectives. For example, cells are often called activated or quiescent. In an activated phase, a large but finite number of functional pathways and protein modules are active in any cell. In a quiescent phase, many functions are inactive but some low levels of metabolism and protein production are needed for maintenance. In general, a given type of cell phase can occur in a variety of phenotypic cell backgrounds (e.g. an endothelial cell or a fibroblast). In cellular terms, the signals that the cell receives from the environment will define the functions that are activated. These signals may be generated as the cell tests its environment, or they can originate from external sources like circulating hormones. There are also defined phases of the cell cycle (growth 1, S phase for DNA synthesis and others) as well as apoptosis and aging (senescence). It is also highly likely that for specific cell types, many additional phases exist that are yet to be defined.

Because cells behave in predictable and reproducible ways, there is an order to the activation of many protein functions and this is critical to their coordination. However, to systematically describe the relationships between various functional modules and pathways, we need to define or standardize the cell phase (or state) in which the pathway is analyzed. Our understanding of membrane trafficking indicates that epithelial cells will secrete membrane glycoproteins via different pathways depending upon whether they are bound to neighboring cells or only to the extracellular matrix. For the biochemical analysis of a given function, it is critical that all of the cells being analyzed are in the same phase, as this will ensure the same functions are operative under the same controls.

There are often dynamic changes in cell phase, and these are induced either through cycles of activity, or through hormone activation. For example, cells commonly cycle between a set of well-defined phases during their growth. Similarly, during the mitosis phase of the cell cycle, many cell functions are repressed in order for the cell to focus on the critical task of separating the daughter chromosomes and moving them into the daughter cells. Naturally, cells must cycle through several different phases of the cell cycle to enable them to replenish, rebuild, and refurbish for the next mitotic phase. These cyclic patterns of cell phase changes naturally lead to cycles that can be periodic, and this gives rise to what can be described as a cellular clock. In many other cell functions, there are bursts of activity that are usually limited in time and extent, and the cell often needs to go into another phase for recovery. In the case of hormone activation, the organism often has a crisis or event that stimulates the release of a hormone to activate many cells and subsequently change their phase in synchrony. Hormone activation is transient because the endocytosis of hormone receptors often downregulates the cellular response to even the continued presence of the hormone. Fibroblasts, for example, will migrate up a gradient of hormone and will transition to a different phase in response to the hormone. In addition, the cellular response to hormones can be altered by changing the composition or structure of the extracellular matrix. In addition, 3D matrices support a different type of motility compared to 2D matrices. Thus, many factors can affect cell phase even cyclically and it is important to define the hallmarks of the different phases in these processes.

Nature vs. Nurture

There is a role for both the cellular composition and the environment of the cell in cell function. This is similar to the nature versus nurture question in human development and probably the correct answer lies in the middle as well. Environmental factors, including matrices, neighboring cells, and hormones, can influence the cell composition, but cell type will limit the range of alterations. The environment and the composition will both interplay to cause the cell to move into a given state that will involve functional cycles. Those cycles may not be synchronous and will cause time-dependent changes in cell behavior. In turn, cell behavior will influence the cell environment and composition over time.

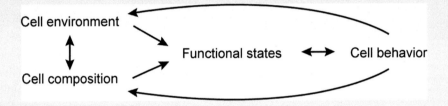

Example

If fibroblasts are moved from a poly-lysine-coated surface to a collagen-coated surface, they will develop a different composition as new proteins are made over periods of many minutes to hours. If the collagen is thick, cell contraction can alter the shape of the collagen, which will in turn alter the mechanical structure of the cell environment and the cell composition.

Previously, the integrated behavior of cells has been described in the context of organism development. In such studies, protein expression patterns were analyzed using modern sequencing approaches that underlie the fields of bioinformatics and proteomics. In tissues such as the kidney, cells in different positions along the kidney tubules display important differences in function as evidenced through the expression of different transmembrane ion channels and pumps. These proteins span the membrane and allow transport of solutes and other material filtered by the kidneys. Upon stimulation by a diuretic, many different types of cells are activated to increase their transport activities, which by definition involves a change of cell phase. Although each cell along a kidney tubule is of the same type, their composition will differ because of the signals received from their local environment in the tissue. Despite these differences, the cells can respond to the same hormone signal and change from one phase to another collectively; e.g. from an inactive to an active phase. Well-defined protein kinase/phosphatase signaling pathways are involved in such phase changes and some of these phase changes have been modeled. These models are established based upon known rate constants of the kinases and phosphatases in different cell phases. Cell composition differences revealed upon the bioinformatics analyses give the nature of the system, stimulation by chemical or environmental signals that evoke short-term changes in cell phase can be considered as nurture. With long-term environmental changes, the composition of the cells (nature) can change and short-term stimuli may have different outcomes as a result.

4.2   Cell Phase versus Linked Complex Functions

It is attractive to think of the complex functions of a cell as being linked to one another in a similar fashion to an assembly line, where the product of one function is needed for the next function. However, this understanding fails to consider the complex tasks of cell phases, which involve higher-order organization patterns that are needed to perform complex cellular or tissue processes, e.g. cell migration or tissue contractions. In those processes, there is a need for cell polarization, force-sensing, and directional force generation. Necessarily, in cell migration there is coordination of several different complex functions to enable the cell to move in a directional fashion; however, the important motile functions of cell contractility and adhesion assembly are spatially separated in the cytoplasm and are regulated

Figure 4.1

Cell phases involve different sets of activated complex functions.

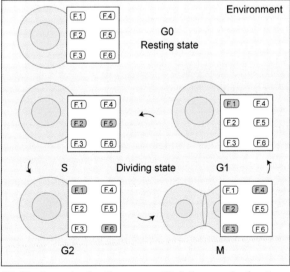

differently in other complex cell functions. For example, cells can migrate in a fibroblastic mode, aggregate mode, or amoeboid mode and all three modes of motility involve actin dynamics and myosin contraction. The details are very different between the modes, but amoeboid migration of macrophages and slime mold cells share many common features that distinguish amoeboid motility from fibroblastic motility. Thus, the coordination of functions often needs to change between different phases.

4.3   Cell Phases in Cultured Cells

When investigating a specific cell phase, and the complex functions active in that phase, researchers will often use an immortalized cell line such as 3T3 fibroblasts or HeLa cells. Cultured cells refers to cells grown *in vitro* or in a controlled lab environment, and they can either be wild-type or mutant lines. Wild-type cells reflect the 'normal' properties of the cell type from which they were derived, whereas mutant cells have had specific genes or proteins added, removed, or modified to change an observable and measurable property. In the majority of studies researchers will use immortalized or established cell lines. While these cells were originally derived from a live tissue or blood sample, they have been immortalized either through 'transformation' or repeated culturing to select for cells that will continually grow. When cells are transformed, DNA from viruses or bacteria may be introduced into the cell, or oncogenes such as Ras expressed. This produces a cell line that displays the properties of a cancer cell, and may therefore exhibit significant differences in the cell functions that are active. Cells that are immortalized through multiple passages in a non-confluent phase to select for spontaneous immortalization (e.g. the 3T3 procedure for immortalizing mouse fibroblast lines) are more closely aligned to their original cell type in terms of active functions; however, even these cells do not exhibit a complete array of wild-type cell functions.

A benefit of using cell lines is that they can be selected for several functions that have properties similar to wild-type cells. Still, cell lines drift over time and faster-growing altered cell lines can take over a culture. For this reason researchers must be wary of contamination of their culture. If even one HeLa cell got into a culture, over time the faster growth rate of HeLa cells would enable them to become the major cell in the culture. Indeed, many investigators have been surprised to find that their rat or mouse cell lines were actually a HeLa line. As might be expected, this is a major embarrassment to the lab if they publish results on the wrong cell line. The process of immortalization can introduce significant changes in cell phenotype and will naturally select for growth potential over differentiated phenotype. Thus, immortalized lines are useful for studying cells with a reproducible protein composition where a limited number of cell phases and phase changes can be described. However, the conditions and timing of the culture are critical for reproducibly studying cells in the same phase.

One of the main benefits to studying stable and immortalized cell lines is that the phase of the cells can be defined, to a certain extent, using environmental controls introduced through the culture conditions. Here the phase of the cell cycle will also play a role. Various culture conditions can be adjusted to help control cell phase, including the density of the cells, the composition of the substrate on which the cells are grown, and the composition of the culture media. In terms of cell density, the phase of cells will change dramatically once the cells become confluent (confluency is a visual measure of cell density, percent confluency is the percentage of the surface covered by cells). In terms of substrate composition, cells will produce matrix molecules and capture them from the medium to produce an altered surface matrix with new components such as collagen, fibronectin, or laminin. The matrix composition and organization on the substrate will dramatically affect cell phase. Finally, culture media may contain a variety of growth hormones or factors (either from fetal bovine serum or artificially added) that will cyclically activate the cells. Importantly, these factors can interact with the matrix to determine whether a given phase is activated. For example, cells are attracted to a chemotactic peptide on one substrate and repulsed by the same peptide on another substrate. Hormone and matrix responsiveness will also depend upon the phase of the cell cycle; and in an unsynchronized culture, approximately 50% of the cells are in G1, 33% in S, 15% in G2 and 1–2% in M phase (these are very approximate numbers and G2 is often variable). Although there appears to be a dizzying number of cell phases, it is possible in some cases to use single-cell assays to determine if a desired function is active in those cells. Motility functions are easier to identify in this way. For example, the cell culture conditions can be standardized using stamped patterns of matrix proteins in defined shapes. This will constrain cells to a single morphology, and help to limit the number of possible phases the cells can enter.

With so many cell phases, and so many methods of altering these phases, it is easy to see why there are many seemingly contradictory observations in the literature. Now, we have the technologies needed to characterize many functions reproducibly at a single-cell level and those studies are giving the most reproducible and reliable results in other labs. If those functional assays can be measured in cells where the cell phase is defined, then it will be possible to develop an understanding of which functions are active in that phase. With time it may be clear that certain groups of functions are always active together and then knowing the activity of a marker function can indicate whether the group is active, simplifying the system.

4.4   Protein Composition and Phase Behaviors

The protein composition of cells is thought to be a good predictor of cell behavior, but other factors are often more important. In other words, it is believed that cells of the same type (sharing the same protein composition) should behave the same while cells of different types should behave differently. Yet many cells derived from the same precursor (i.e. cells with the same protein composition) can exhibit dramatically different behaviors, even when they are growing in similar environments. For example, cells plated on fibronectin will spread rapidly without contraction on the surface while others will contract almost continually and there is no known difference in the cells. On the other hand, different cell types can exhibit the same behavior or cell phase. For example, the rapid spreading on fibronectin is very similar in fibroblasts and endothelial cells. Further, there are many similarities between the motile behaviors of neuronal growth cones, migratory fibroblasts and endothelial cells. One way to resolve this apparent paradox is to suggest that there are a limited number of complex functions controlling motility that can be used by many, if not most, cells under certain circumstances. For example, the complex functions that drive actomyosin-dependent motility may be considered 'tools' that are found in many

Figure 4.2

Cell phases are defined by limited sets of complex functions.

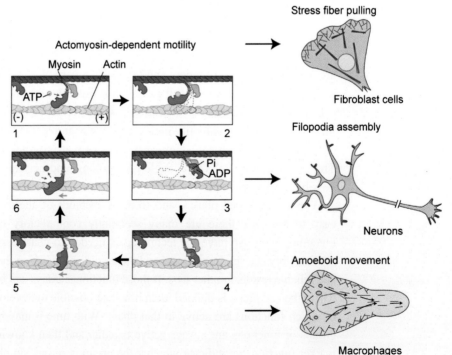

cell types and there is a limited tool set that is available. Cells utilizing actomyosin-dependent motility to drive a given function will still receive and produce biochemical signals based upon the local microenvironment, and these signals can differ in different regions of the surface to cause different motility in an otherwise homogeneous cell population. Depending on the local signals received, the activation of the function will differ, and so will the cell's behavior. Thus, we suggest that many cells share a common set of tools and that differentiation results from the expression of proteins for a few specialized tools, as well as different control and signaling pathways that activate those tools for specialized complex functions. To go back to the car analogy, there are many common functions between a sedan and a four-wheel drive vehicle and the differences are really in a few specialized functions. Further, they will behave similarly on paved roads, but off-road they will behave totally differently. Similarly, cells of different cell types have many common functions, but when faced with specialized environments can behave remarkably differently because of specialized phase behaviors.

4.5   Analysis of Functions and Cell Phases

Genomics and proteomics studies have helped to define whether proteins are involved in a given function and which components they bind to, respectively. The list of proteins involved in specific functions is continually growing, and yet for most functions these lists are still incomplete. Even if each protein component is known, their roles in the function are often poorly understood. This is especially the case when they play a signaling or communication role that influences other functions. This was noted previously for clathrin-dependent endocytosis.

In genomics studies the protein in question is removed from the cells by mutating or deleting the gene that encodes them, and then investigating how the function is changed. Often changes occur even though the protein is not part of the complex responsible for the function. Because the protein is not directly associated with the complex, these changes are considered indirect. Some indirect effects can be through altering a protein several steps removed from the actual functional module involved (e.g. causing modification in the level of a required component). Similarly, proteomic analyses, which normally measure binding of domains of one protein with other proteins, can identify protein–protein interactions unrelated to a functional module that contains one of the proteins. Thus, it is important to use multiple criteria to define the proteins in functional modules involved in a given complex function.

Linkage of a protein to a function through genetics or proteomics needs to be followed up with a quantitative analysis of the function under conditions where the protein is acutely inactivated or activated, as in a phase change. To control the assay of function, it is very useful to have an activation signal (e.g. temperature shift, ligand activation, inhibitory drug washout, small molecule activator, photo-switch or other rapid means of activation). Quantitative, rapid and local assays will aid in determining how closely a given protein is involved in the function. With quantitative information, it is often possible to determine whether the loss of a protein will alter the activation, the rate, or the inactivation of a function. The quantitative analysis of the functional parameters in a defined phase is particularly useful when devising a model of the function.

Experimental techniques to modify and investigate the phase of a cell are also readily applied in studies seeking to understand specific complex functions. Changes in cell phase can arise from major alterations in cell morphology or the cytoskeleton. For example, microtubules, which are a major component of the cytoskeleton, may undergo depolymerization, which can alter the cell phase. When microtubules are lost, a variety of functions are altered. Cell migration is often blocked in addition to changes in cell contractility, adhesion dynamics, Golgi distribution, and a host of other aspects of the cell. These changes are potentially directly linked to microtubule-dependent transport of material. Two major classes of motor proteins use ATP to transport materials along micro-tubule tracks in opposite directions. Kinesin moves toward the plus ends of microtubules away from the centrosome, and dynein moves toward the minus ends anchored typically to the centrosome. By labeling proteins with GFP (green fluorescent protein), researchers have observed movements of different proteins to distinct sites. To understand each function that is altered by micro-tubule depolymerization will require a different set of protein movements be followed with and without microtubules as the function is analyzed. Many of the effects on function may be several steps removed from microtubule-dependent transport.

In a different cell type, the normal epithelial monolayer, there is high polarity and tight packing of cuboidal cells, as well as the presence of tight junctions between neighboring cells that creates a cell barrier between a luminal and a basolateral space. Because different membrane proteins are on the luminal and basolateral surfaces, membrane glycoprotein secretion needs to be controlled such that one set of proteins goes to the luminal (apical) side and another set of membrane glycoproteins goes to the basolateral or basement membrane side. At the Golgi, the membrane glycoproteins for the apical surface are packaged in separate vesicles from those for the basolateral surface. Those different vesicles follow separate paths and fuse by different mechanisms with the two surfaces.

Figure 4.3

Analysis of functions.

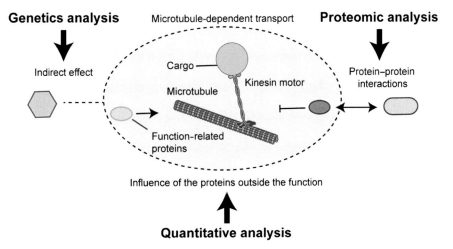

Each additional feature of the process of secretion presumably entails another set of proteins and/or feedback mechanisms that can modify the motility of secretory vesicles. Thus, the hypothesis that normal secretion involves microtubule-dependent transport may be correct, but other mechanisms of protein secretion could dominate in other phases, for example, during disease, after microtubule loss, or after hormone stimulation.

Because of these many caveats, there is no accepted paradigm for the analysis of phases or the functions that are active in those phases. If a set of parallel functions have readily quantifiable intermediates and endpoints, it is possible to define activities in different phases and to understand how transitions occur. However, most cellular functions do not lend themselves to such an analysis. Because of the integrated nature of different functions, it is insufficient to quantify changes in one function without a similar knowledge of other functions. For each phase, there will need to be a linkage diagram between functions that is similar to the metabolic charts that have been generated throughout the literature. Furthermore, an analysis of the function from the point of view of a robust system should be performed.

4.6   Definition of Cell Cycle Phases

When we talk of phases in the fibroblast, it is natural to think of the cell cycle and the distinct phases: interphase [G0 (quiescent), G1 (growth 1), S (DNA synthesis), G2 (growth 2)], and M (mitosis). Different functions are emphasized in each of these phases and it is useful to define those.

Phases of Mitosis

Prophase: Nuclear envelope disassembly, chromosome condensation, cessation of synthesis and membrane dynamics, cessation of motility (cell becomes round), centrosome duplication, ER separation.

Metaphase: Spindle formation, chromosome alignment.

Anaphase: Chromosome separation, chromosome movement, spindle elongation.

Telophase: Formation of the cleavage furrow, breakdown of the spindle fibers, reformation of the nuclear envelope, and the start of cell spreading on the surface.

G0, Quiescent. A non-growth phase characterized by limited synthesis of proteins, lipids, and carbohydrates to maintain life. Ion and volume homeostasis are maintained; motility and cell shape changes can occur in response to environmental signals. This phase can be elicited *in vitro* by removing serum from the culture medium.

G1, Growth phase 1. Characterized by a rapid synthesis of proteins, lipids, and carbohydrates, ion homeostasis, volume increase with growth, motility and shape changes that occur in response to environmental signals.

S, DNA synthesis. Some synthesis of proteins, lipids and carbohydrates, but the replication process will silence the DNA transiently for transcription and the usual patterns of activity will be altered.

G2, Growth phase 2. Similar to G1, but twice the DNA content (usually shorter than G1).

M, Mitosis. Involves the division of the cell DNA and contents into the two daughter cells. No protein or lipid synthesis occurs, ion and volume homeostasis. Can be broken into stages of prophase, metaphase, anaphase, telophase, and cytokinesis.

Of the major phases in the cell cycle, the G0 phase is perhaps the simplest and it is produced *in vitro* by withdrawing serum and serum hormones. Many aspects of cell metabolism are inactivated during this phase and protein synthesis is at a minimum. Serum growth hormones will activate cell growth *in vitro* via a class of proteins known as serum response factors (SRF). The transition from G0 to G1 is dramatic and many different mRNAs are transcribed during this event. This process has been studied extensively, and many functions are differentially activated in the two phases. Of particular interest is the activation of an SRF in the nucleus by the movement of a protein known as MRTF-A into the nucleus following an increase in polymerization of actin in response to growth hormone activation (Olson & Nordheim, 2012). The transition from G1 to S is less

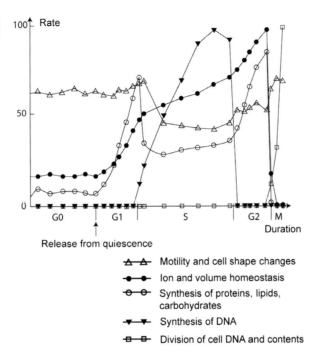

Figure 4.4 Functions in cell cycle.

dramatic, but there is necessarily a change in the pattern of transcription because replication, which occurs during the S phase, requires the silencing of transcription. In S phase, the pattern of transcription is different than in G1 or G2 in that the proteins responsible for DNA replication are produced early and the genes for the histone proteins are produced in the middle of the S phase.

4.6.1 Analysis of Mitosis

Analyzing the mitotic phase in detail within a cellular context, we can see that there are many coordinated functions that must occur concurrently. The orchestration of these events is very complicated and involves a number of linked processes to produce the desired emergent property. By looking at the disassembly of the nuclear envelope in prophase as an example, we can begin to understand some of the steps that make up this complex process. Here, chromosomes must be condensed as distinct units without any intertwining of the two DNA strands, which means that RNA synthesis must stop. With the synthetic processes halted during mitosis, the production of both new membrane and new protein stops. This signals a drop in secretion that is correlated with a rise in plasma membrane tension due to the lack of new lipids being integrated into the membrane while endocytosis

Figure 4.5

Disassembly of the nuclear envelope in prophase.

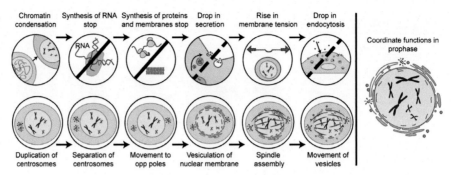

continues. The rise in tension in turn causes a decrease in the rate of endocytosis. Possibly related is a movement of membrane from the nuclear envelope into tubulovesicular membranes. Concomitant with the vesiculation of nuclear membrane, the lamins of the inner nuclear envelope are broken down. Prior to this, it was necessary for the centrosomes to duplicate and separate to the opposite poles of the cell as this allows the processes of Golgi and nuclear vesiculation to result in an equal distribution of membranous organelles into the daughter cells.

Exploring this phase at the more detailed molecular level, it becomes apparent that the list of steps and the required functions increase. For example, the motor proteins that form the tubulovesicular compartments from the nuclear membrane must bind to the nuclear membrane and be activated there. These motor proteins require an array of microtubules to move along, and these must first be organized from the two centrosomes that were duplicated in previous steps, and separated by motor activities to the opposite ends of the cell. Presumably the nuclear pore complexes and the links in the lamin networks must be broken apart in a way that allows them to be reassembled as well. Despite their complexity, it is extremely important to define all of the steps in a given process because they all play a role and involve different sets of proteins.

All of the functions that occur in metaphase take place in a matter of about 5 minutes. This adds further complexity because many steps must occur simultaneously and should not interfere with one another either physically or chemically. Protein activity is commonly regulated by the biochemical process of phosphorylation. This involves the addition of phosphate groups to the amino acids of a protein through the activity of broad class of proteins known as kinases. Indeed, phosphorylation is critical in controlling a variety of protein activities during mitosis and this provides a way of coordinating many different functions over time. For example, if one kinase can phosphorylate several different proteins that activate several functional pathways, then those functions can be correlated in time. If dephosphorylation is tied to the inactivation of several functions, then a phosphatase can be used to coordinately inactivate several functional complexes. In some instances a critical function, such as the alignment of the chromosomes at the metaphase plate, must be in place before the switch from the positioning to the separation of chromosomes (anaphase) can occur. In these cases, the completion of the step provides the required signal to inactivate each of the parallel functions that are working at the time and to activate the functions needed for the next phase. Until this signal is received, many other supporting functions may need to be kept active in the cell. This naturally raises the question of how a given cell makes the decision to transition from one phase to another, and it is not known in the case of metaphase to anaphase transition. In earlier phase transitions in the cell cycle, the targeted proteolysis of cyclin proteins occurs and that causes a very reliable switch to the next phase of the cell cycle. Whereas kinase/phosphatase systems can rapidly switch protein activities on or off, proteolysis provides a much longer loss of activity and can only be reversed when the synthesis of the protein is activated (typically a matter of hours). In the phase transitions of the cell cycle there are many different inputs that must contribute to the decision, including metabolite levels, cell density, hormone levels and numerous other factors, which makes it difficult to understand how the decision is actually made. We have chosen a set of more simple phase changes in cell spreading to consider the decision-making process.

4.7 How Do Cells Make Important Decisions to Change Phases?

The decision to transition from one cell phase to another is critical, yet in most cases the decision-making process has not been studied extensively. Here we will consider an example of phase transitions that are clearly dependent on the cell's processing of signals from the environment. In particular, we will analyze the process of fibroblasts growing on a fibronectin-coated surface and determine how the cell decides to grow versus apoptose. This is a process that will be considered in much more detail in Chapter 16, but for now we will use this system to illustrate many unexpected aspects of the different phases. Decisions to grow, die or differentiate are typically not single-step processes and the cell may need to go through several phases before it can make a critical decision. One way to describe this is as a series of 'If–Then' decisions that lead the cell to a particular phase. In the case of matrix-dependent fibroblast growth, these differentiated cells will only grow on certain matrices if those matrices are rigid. This probably reflects a function of those fibroblasts *in vivo*, which is to remodel matrix fibers that are often rigid. Furthermore, to grow fibroblasts for long periods involves the repeated suspension of cells using trypsin-EDTA and replating those cells with serum-containing medium that can coat the surface of the tissue culture dishes with matrix as well as stimulate the cells with growth hormones. Thus, the cells are selected over time for growth on rigid, matrix-coated substrates in the presence of growth hormones.

To understand the matrix-dependent growth of fibroblasts, it is very important to understand the steps that the cell goes through to arrive at the fibroblast growth phase because those steps are key in cancer, abnormal cell growth, and many disease processes. We can start with a cell in suspension where fibroblasts will not grow. They will die after a relatively short period of 1–2 days, although growth hormones can prolong their lifespan in suspension. This indicates that, *in vivo*, circulating growth hormones can activate or predispose a cell to growth, irrespective of the cell's immediate microenvironment. However, hormones cannot totally replace a matrix substrate. We will describe how a fibroblast responds to matrix properties and how this response leads to growth in an environment with insufficient hormone for growth without matrix stimulation.

4.8   What Does a Cell Need to Know About the Matrix for Growth?

Based upon experiments, there appear to be three major criteria that need to be met before a fibroblast can grow on a particular type of extracellular matrix, integrin clustering with matrix, a rigid matrix and sufficient distance between matrix sites (Iskratsch et al., 2014). First, matrix must be concentrated in a local region and attached to a stable substrate. Integrins need to be concentrated by binding to localized ligands so that actin will assemble on the integrin clusters and myosin contracts the actin to generate force on integrin-attached matrices. Second, the cell wants to know if the matrix is rigid before it develops a strong contact with the surface. Rigidity sensing is a complex cellular process that involves measuring the force required to displace the substrate to a constant distance (discussed in greater detail in Chapter 16). Third, the matrix sites must be separated by sufficient distance for force generation with sufficient density of molecules. For example, 50 μm^2 of fibronectin-coated surface is sufficient if it is distributed in 1 μm circles separated by 4 μm, but not if it is in an 8 μm diameter circle. Basically, cells need to contact rigid matrix attachment sites that are separated by about 20 μm in order to generate sufficient force on the matrix sites to stabilize the sites and grow. If the matrix proves suitable for growth, then additional steps are initiated that lead to the maturation of cell–matrix adhesions and the generation of signals that cause growth. In some cells additional signals may activate a separate set of functions that lead to polarization, or induce migration. It should be noted that the cell is

Figure 4.6

Cell spreading on matrix.

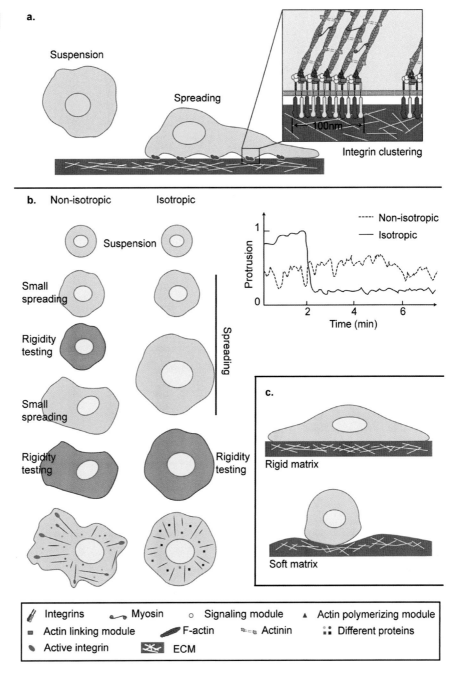

always maintaining some tension on the substrate. In an *in vitro* setting, this means that the release of matrix from the glass, or other major mechanical change, will result in a change in the cell. Several phase changes occur in this process and we will now explore them in greater detail.

4.9   Detailed Consideration of Suspension Phase and Transition to Spreading

A major change in cell phase is observed when a suspended cell adheres to a surface and begins to spread out. This is because a suspended cell has a totally different set of functions activated compared to a cell that has adhered to the matrix and underlying substrate. One of the main proteins that cells use to bind to the extracellular matrix is integrin (a dimeric membrane protein that binds to matrix outside the cell and then binds to the cytoskeleton on the inside). The affinity or avidity of integrin–matrix binding can be dramatically increased by the close spatial organization of matrix sites. For this reason, one of the major aspects of cell binding to extracellular matrix (cell–matrix adhesions) is the requirement for matrix sites to cluster together. Interestingly, matrix proteins can still activate suspended cells and this occurs when the cells encounter high local concentrations of matrix molecules. However, if the matrix molecules are soluble, or even small soluble aggregates, they will not support forces that will enable cell growth. Thus, the cell must test if the matrix molecules are attached to a substrate. Recent studies indicate that the cells polymerize actin from the integrin clusters, which acts as a handle to pull on the matrix clusters. If myosin pulls on the matrix via the attached actin, then a force transducer bound to the integrin cluster can signal to the cell that the matrix can generate force. There is a minimal number of immobilized matrix sites that must be bound by the cell to create a sufficient signal that activates cell spreading on the surface, i.e. a different phase.

Avidity

Avidity is the term used for an increase in the apparent affinity of matrix and ligands because of the spatial alignment of multiple binding sites. If two extracellular matrix ligands are spaced by about 60 nm and two integrins are bound to them, then the crosslinking of the integrins by a cytoplasmic protein of 60 nm will cause the binding affinity constant to be more than squared. Such increases in avidity require that the matrix spacing and the integrin spacing be matched in the bound configuration through the cytoplasmic protein complex. As they can become exponentially stronger if the number of crosslinked, liganded integrins increases, there is an obvious advantage for the cell to assemble large crosslinked adhesive structures. The rationale for the cell using this method of increasing the strength of contacts is that such adhesions can transmit high forces from the cytoskeleton to the matrix. Also, adhesions can be readily assembled by proper matrix spacing and readily disassembled by cytoplasmic proteins that reverse the binding of the cytoplasmic integrin crosslinker.

In determining the binding constant for protein A binding to protein B, there are two major factors: the enthalpy and the entropy changes upon binding. A major component of the entropy term involves the restriction of diffusion of the proteins in the bound complex. When the membrane proteins bind to matrix ligands, cytoplasmic proteins often crosslink the liganded membrane proteins and that will restrict their diffusion. As a result of crosslinking, the entropy penalty for the second membrane protein to release from ligand will be much lower; however, the enthalpy penalty for unbinding will be the same. In mathematical form, the strength of the bond formed by an organized array of ligands and receptors is more than the sum of the free energies of binding of the individual ligands and receptors and can be approximated by summing the binding enthalpies with only a fraction of the binding entropy penalties (hence, the small size of $T\Delta S2$). For example, with a K_D of 10^{-5} M for a single ligand–protein bond that two crosslinked proteins will have a combined K_D of less than 10^{-10} M, i.e. a much longer-lived interaction.

$$\Delta G = \Delta H - T\Delta S = \Delta H1 + \Delta H2 - T\Delta S1 - T\Delta S2$$

where G is free energy, H is enthalpy, T is temperature (°Kelvin), and S is entropy.

4.10 Two Different Modes of Spreading Constitute Two Different Phases

In the spreading phase, cells often display two different modes of spreading: non-isotropic and isotropic (Dubin-Thaler et al., 2004). In non-isotropic mode, they will spread in a limited area for 30–60 s and then stop to test the matrix before spreading further. In the isotropic mode, they will spread rapidly on the surface for a longer period without testing the matrix. The end of spreading then coincides with the depletion of the membrane area as the cell goes from the wrinkled spherical form in suspension to a flattened disc in the spread form. These two modes can be considered as two different phases because different motility functions are active and the cells respond to matrix signals in the non-isotropic but not in the isotropic. Clearly, not all motility functions are different in the two phases, as a similar form of actin polymerization is found in both; however, even that is different in the sense that actin polymerization will stop rapidly in the non-isotropic case but continues until it is stopped as the tension in the membrane rises in the isotropic. In different fibroblast cell lines, different fractions of cells spread non-isotropically versus isotropically. Factors such as serum deprivation can change the percentage of cells spreading in each mode. Thus, the signal to activate spreading can cause cells to go into one of two different phases of spreading.

There are several reasons why the two different modes of spreading should be considered as two different cell phases. Perhaps most important is the fact that experimental analyses of the two different modes will show that different proteins are active in different biochemical pathways. Thus, biochemical analyses of a mixed population of cells will give a mixture of the two phases. If a given function is to be understood, then that function should be active in all of the cells that are analyzed. Practically, this means that the simplest way to analyze a given function is to choose the phase of the cell in which that function is particularly active. In addition, the region of the cell for the analysis should be the region where the function is active. The second reason for considering these as two different phases is that integrins are active in the non-isotropic mode (evidenced by the formation of clusters of integrins and matrix in that mode) and not in the isotropic. In the non-isotropic mode, the cell stops spreading and periodically tests the surface matrix rigidity, whereas in the isotropic mode the cell will spread equally rapidly on matrix and on non-adherent surfaces. Thus, signaling and motility functions are dramatically different in the two modes, making it useful to consider them as different phases. However, even though this is a simple activation event that involves one ligand, fibronectin, that is uniformly coating the surface and integrin binding to it, we do not know how one phase is activated versus the other. This serves to illustrate how difficult it can be to have a homogeneous population of cells in the same phase at the same time for biochemical analyses of homogenates of millions of cells.

4.11   Contractile Spreading Enables the Cell to Sense Substrate Rigidity

Once isotropic spreading cells reach about two-thirds of their final spread area, they will test the rigidity of the substrate as the non-isotropic cells do continuously. If the matrix is not rigid, then the cell will not spread further and will soon round up. Although the process of rigidity-sensing can be activated under a number of conditions, there is a specialized contractile unit (about 2 μm in length) with myosin and actin that pinches the matrix substrate to test its rigidity (a more complete description is given in Chapter 16). Basically, the contractile unit pinches to a constant distance and measures the force produced. If the maximum displacement is reached and the force is not sufficient, then the adhesions rapidly disassemble. However, if the pinch creates a high force, then the cell will spread further and will only occasionally test the rigidity of the surface unless activated by a hormone such as epidermal growth factor (EGF). Thus, the signal from matrix rigidity-testing will determine if the cell can transition to a growth phase on rigid or to an apoptotic phase on soft (most normal fibroblasts are rigidity-dependent for growth). Isotropic spreading cells will start at a larger area when they test rigidity, but will reach the same spread area as non-isotropic cells after an hour, i.e. spread on rigid and rounded on soft surfaces. Thus, the cells using different phases can arrive at the same final shape.

4.11.1 Force and Adhesion Maturation

When the cell determines that a matrix is rigid, it will assemble more stable cell–matrix adhesions. This occurs through the continued generation of contractile force on cell adhesions in the next cell phase. Stable forces are needed to keep cells spread and adhesions presumably generate signals for cell growth and entry into the cell cycle if the required forces are present at the adhesion site. When the matrix contacts cover a small area, only small forces are generated; however, if patches of matrix molecules are separated by large distances (10–15 μm), the cell can produce large forces that stabilize the adhesions. It is believed that more separated adhesion sites correlate with larger forces because a greater number of bipolar myosin II filaments can assemble between the sites. Thus, continuous

Figure 4.7

Matrix area.

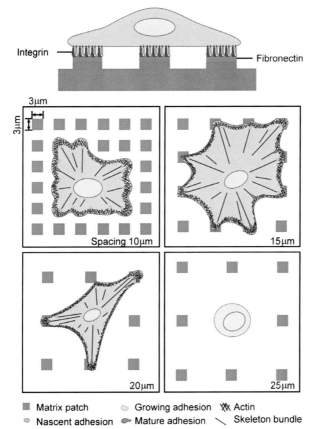

matrix is not as critical as having well separated (> 10 μm) sites of matrix attachment. The large forces generated on multiple displaced matrix sites enables the initial maturation of adhesions, and without the force generation, the adhesions will disassemble potentially causing the cell to apoptose. Force on adhesions is therefore necessary to keep the cell in a growth phase, and loss of force either from a short distance between adhesions or low density of matrix will cause fibroblasts to exit the cell cycle as they round up. Thus, simple mechanical factors can strongly influence the phase of the cell even though the cells have the same proteins and have gone through similar phases along the way.

4.12 Regional Specialization in Cells Can Involve Phase Changes

In considering the changes in phase that occur during cell spreading and migration, it is clear that many of the motility and matrix adhesion functions are activated locally. Thus, a cell can exhibit different behaviors in different cell regions, which raises the question of whether a cell can exhibit multiple phases simultaneously. We have no problem with this idea in epithelial cells where apical regions contain different membrane proteins and have different signaling functions than their basolateral regions. Similarly, we see in spreading cells that some regions of the edge will spread in an isotropic mode while adjacent regions are non-isotropic. Because subcellular phases can also be considered as complex functions that occupy a significant cell area, there is some ambiguity. The important point is that phase behaviors are common wherein a cell will activate a set of functions (often ordered) and inactivate others for an important purpose. These phase behaviors can only be understood through single-cell or subcellular analyses of relevant probes during the phase. For example, what are the elements needed for rigidity sensing and what are the signals that are generated for soft versus rigid matrices (how is it accomplished and what are the outputs)?

4.13 Many Integrated Complex Cell Functions Involve Multiple Cell Phases

We have chosen the specialized case of matrix adhesion formation as an example to understand the kind of steps and different cell phases that are needed to perform an integrated complex function in the cell. Because adhesion formation involves major changes in cell morphology, and potentially the activation of cell growth, the cells transition through several distinct phases where a subset of cell functions actively contribute to the final development of the adhesion. When we interfere with early steps in the process, there are often later consequences, as this interference can result in the cell being unable to form normal adhesions. For example, the growth of actin filaments from the early integrin clusters involves the formin protein, FHOD1. Depleting this protein will dramatically alter the behavior of lamellipodia and adhesion formation, forcing the cell down another path than normal fibroblast migration. At this point, the explanations for the purpose of each step in the overall function are speculations, but similar steps are found in a variety of environments with different cell types. What is missing in the overall view presented thus far is the diversity of pathways that can, over time, lead to a similar final cell phase. Normally, propagation of fibroblasts *in vitro* does not involve special treatment of the tissue culture plastic surface. Yet the cells still grow and have similar shapes to those plated on fibronectin-coated surfaces after a similar time. When cultured for hours, or even days, fibroblasts can express and secrete their own fibronectin and collagen, meaning the required microenvironment will still be assembled. The cells can also use soluble matrix proteins for this purpose. Thus, there are often alternative pathways that result in a similar final phase of growth. The practical consequence of this is that experiments to study functions in cells that occur over long periods often miss the intermediate steps in the functions. From the continuous analyses of complex functions that occur in cells in homogeneous phases, it is possible to understand how the cells reach their final phase. Further, following individual cells through alternative pathways makes for a more complete understanding of the overall process.

4.14 SUMMARY Different phases are needed for cells to block some functions while others are active. This is critical for a complex machine to multitask and control potentially interfering functions. Another reason for different phases is to provide resting phases where the complex signaling functions can be reset to a ground phase (similar to rebooting a computer). When in the resting phase, only those functions required for cell survival are kept active. At a practical level, it is important to study a cellular function in a cell where that function is clearly active in the majority of the cells. In some phases, given functions are primarily inactive and cannot be studied. If phases, and the transitions between them, can be controlled as in the case of the G0 to G1 transition, then it is possible to sort out the functions that are critical for one phase versus the other. This can also allow us to investigate how these phases or transitions might be modified in disease. Motility functions are relatively easy to study because they have clear hallmarks that enable localization of active functions to subcellular regions. Because there are a limited number of cell phases, it is possible to define them and to reproducibly activate them by defined stimuli. Studying cells in the same cell phase will make progress on understanding functional activities much easier.

4.15 PROBLEMS

1. In a study of hormone activation of a tyrosine kinase, the investigators used two different methods of analysis: (a) sodium dodecyl sulfate (SDS) gel electrophoresis to separate proteins from 10^6 cells followed by western blotting with an antibody to phosphotyrosine (pTyr), and (b) fixation of cells, permeabilization with detergent, and staining with a fluorescent antibody to pTyr. In the western blot, the level of phosphotyrosine increased by twofold relative to controls, but in the fluorescence microscope only 20% of the cells had anti-pTyr fluorescence intensities above controls. How would you interpret the experiment?

Answer: The level of pTyr in the sample reflects an average value over many cells and does not reflect subcellular mechanism. In the case of antibody staining, the high level of staining in a fraction of the cells indicates that the activation of phosphoryl-ation may be brief and cooperative such that cells will have different times for the peak of phosphorylation. Further, this result suggests that a cyclic process of activ-ation and inactivation may occur after hormone activation. Simple linear models of processes are commonly wrong and more complicated models are needed to explain actual behavior.

2. If you have two populations of cells, one with calf serum and one without, and you treat them with a fluorescent analog of a deoxynucleotide that will be incorporated into DNA, which group of cells will be sensitive to light that bleaches the fluorophore to cause DNA damage?

Answer: The cells in the presence of calf serum are most likely to be growing as a result of the growth hormones in the serum. With growth, they will synthesize DNA and the fluorescent analog will be incorporated into their DNA. With the non-serum cells, they will exit the growth pathways and will not synthesize DNA or be sensitive to light.

3. Recent studies show that cells on soft surfaces are not simulated by epidermal growth factor (EGF) to move whereas they are stimulated on rigid surfaces. Is this likely explained by:
   (a) the loss of EGF receptors on soft surfaces,
   (b) a disorganized cytoskeleton on soft surfaces,
   (c) an altered state on the soft surface, or
   (d) secretion of a protease on soft surfaces?

Answer: An altered state on the soft surface, (c).

4. When cells are treated with nocodazole to cause microtubule depolymerization, secretion of membrane glycoproteins ceases for about 30 minutes. During this period, the cells continue to endocytose membrane at the rate of 1% of the plasma membrane area per minute. Would the cells be expected to round up during nocodazole treatment?

Answer: Yes, the loss of surface area without the loss of cell volume will result in the rounding of the cells. An unrelated issue is that the nocodazole-induced depolymerization of microtubules will cause activation of myosin contraction through a separate enzymatic pathway that will also cause cell rounding.

5. During mitosis, cells cease exocytosis and endocytosis for about 10 minutes. If a population of cells is growing at a rate of one cell division per day and they are exposed to a toxin that must be endocytosed to cause death, then what fraction of the cells will die if a lethal level of toxin is in the medium for 5 minutes?

Answer: On average, the cells will be safe from the toxin for 10 out of 1440 minutes or about 0.6% of the time. However, the fact that the toxin will be present for 5 minutes means that only half of the time that it is non-endocytosing will be available or about 0.3% of the cells will survive.

6. If cells experience a period where there are low blood sugar levels, they will need to convert to more efficient oxidative metabolism. Which of the following changes are needed:
   (a) mitochondria should expand as a result of increased synthesis of mitochondrial membranes,
   (b) fat droplets should decrease due to increase degradation,
   (c) cells should become resistant to growth hormones, or
   (d) only 10% of the cells should change, because they can supply nutrients for the rest of the cells?

Answer:
(1) mitochondria should increase in size, and (2) fat droplets should decrease in size.

7. On different matrices, the same small molecule can cause a different response. In cells on one surface, the small molecule will be a chemoattractant, whereas on the other surface it will be a chemorepulsant. As the receptor protein is likely to be the same on both surfaces, how would you explain the different reactions to the same molecule?

Answer: One logical way to explain this is to suggest that the receptor activation by ligand is the same in both cases; however, the downstream signaling pathways are different. On the first surface, the ligand activates an actin polymerization pathway that causes migration toward the highest concentrations of the molecule. On the second, the cell has changed state such that receptor activation causes a retraction signal where it acts as a chemorepulsant. The switch can be as simple as the phosphorylation of an enzyme on one of the surfaces that then switches the downstream cell behavior.

8. In an otherwise homogeneous tissue, there are often adult stem cells that will be activated if growth conditions arise. What advantage is there to the organism to have only a portion of the cells responsive?

Answer: The simple answer is that the majority of the cells in the tissue are needed for normal tissue function and should not respond to a trauma signal that activates growth.

4.16   REFERENCES

Dubin-Thaler, B.J., Giannone, G., Döbereiner, H.-G. & Sheetz, M.P. 2004. Nanometer analysis of cell spreading on matrix-coated surfaces reveals two distinct cell states and STEPs. *Biophys J* 86(3): 1794–1806.

Iskratsch, T., Wolfenson, H. & Sheetz, M.P. 2014. Appreciating force and shape – the rise of mechanotransduction in cell biology. *Nat Rev Mol Cell Biol* 15(12): 825–833.

Olson, E.N. & Nordheim, A. 2010. Linking actin dynamics and gene transcription to drive cellular motile functions. *Nat Rev Mol Cell Biol* 11: 353–365.

4.17   FURTHER READING

Actin filaments are a force-sensing conduit for both internal and external forces: www.mechanobio.info/topics/cytoskeleton-dynamics/cytoskeleton/actin-filament/#conduit

Cell matrix adhesion: www.mechanobio.info/topics/mechanosignaling/cell–matrix-adhesion/

Cell polarization: www.mechanobio.info/topics/cellular-organization/establishment-of-cell-polarity/

Cell matrix receptors and integrin-mediated signalling: www.mechanobio.info/topics/mechanosignaling/cell–matrix-adhesion/integrin-mediated-signalling-pathway/

Clathrin-dependent endocytosis: www.mechanobio.info/topics/cellular-organization/membrane/membrane-trafficking/clathrin-mediated-endocytosis/

Extracellular matrix: www.mechanobio.info/topics/mechanosignaling/cell–matrix-adhesion/

Focal adhesion: www.mechanobio.info/topics/mechanosignaling/cell–matrix-adhesion/focal-adhesion/

Force generation and cell translocation: www.mechanobio.info/topics/cytoskeleton-dynamics/lamellipodium/lamellipodium-assembly/#force

Nucleus (nuclear envelope): www.mechanobio.info/topics/cellular-organization/nucleus/

5 Life at Low Reynolds Number and the Mesoscale Leads to Stochastic Phenomena

As noted earlier, cellular functions are always driven by the diffusion of molecular components at a basic level. Thus, it is important to understand why the low Reynolds number of cellular systems demands that basic movements of proteins occur by diffusion. We will go through the physical basis of the cellular systems and the implications of diffusion for cellular functions. In addition, cells of all sizes need to concentrate components in subcellular regions and organize their proteins and membranous organelles by transporting them from one place to another in the cytoplasm. Because diffusion basically limits the speed of transport of materials in the absence of energy-dependent active transport, it is important to remember that slow as well as very small life forms can rely upon diffusion for life processes. There is an active debate about when transport is best accomplished by diffusion or by a motor-dependent system. Both diffusive transport and active transport have been observed and we will consider the mechanism of both in this chapter. Physical constraints of diffusion in cell systems limit the size of life forms, but can be circumvented by reducing dimensionality as well as by active transport systems. As cellular dimensions reach beyond 100 μm, diffusive transport times become very long and animal cells typically utilize active transport systems supported by linear filaments to move materials rapidly. Membrane depolarization and pressure waves can move much faster than diffusion and are often used for signaling when actual material transport is not needed.

Diffusion Characteristics of Cells

1. Lack of persistence of movement; kinetic energy of moving object relative to thermal energy.
2. Diffusion coefficient in cytoplasm (50,000-dalton protein) $\sim 1 \times 10^{-7}$ cm^2/s.
3. Pore size in actin-rich regions of cytoplasm < 25 nm.
4. Pore size in microtubule-rich regions of cytoplasm > 250 nm.

(continued)

5. Time scale of protein domain motions 10^{-9-10} s.
6. Membrane diffusion coefficient (75,000-dalton protein) ~3×10^{-10} cm^2/s.
7. Transport by diffusion (proportional to concentration gradient and diffusion coefficient and inversely dependent upon distance squared, i.e. 2× distance results in 1/4× rate of transport).
8. Non-ideal diffusion is common in cells because the accessible volume is restricted (confined diffusion) or the medium is flowing (e.g. membrane flow and diffusion).

5.1   Thermal Energy versus Momentum at Low Reynolds Number

To understand cellular functions, it is necessary to understand the fundamental physics of molecules (proteins, lipids, and carbohydrates) at the nanometer level. As particles get smaller, their mass scales roughly as the cube of their size, and the velocities of particle movement also tend to decrease. Because the energy of a moving particle is one-half of the mass multiplied by the velocity squared, the energy in a moving small particle is much less than the energy of macroscopic particles. At some point, the particle size will be small enough such that the typical kinetic energy of particle movement will be less than its thermal energy or kT. At that point, the dominant factor in particle movements will be diffusive. In water, that point is larger than bacteria and smaller than fleas.

In mammalian cells, all subcellular functions rely upon stochastic fluctuations of water pressure to drive the conformational changes of proteins that then catalyze function. Cellular functions are different from similar functions in the macroscopic world because of the low Reynolds number (see definition below) at the subcellular length scale. Low Reynolds number refers to the relative scales of the object and the pressure fluctuations of water in this case. A practical consequence of life at low Reynolds number is that inertial movements are nonexistent. Instead, Brownian movements (see textbox) are diffusive and provide stochastic fluctuations that normally create disorder in cellular systems. However, the energy from Brownian movements can be harnessed by cells for productive activities. To harness thermal motion for productive work, proteins will sample a conformation space in a cyclic fashion where the cycle is biased by the binding of high-energy compounds such as ATP, their hydrolysis and then the release of lower-energy products like ADP.

Another way of thinking about Reynolds number is in terms of the relative magnitude of the energy in the moving particle and the magnitude of thermal fluctuations. The energy of a swimming bacterium (10 μm/s) is 1/2mv^2 (0.5×10^{-15} kg $\times 10^{-10}$ m^2/s^2 = 5×10^{-25} J), whereas the thermal energy of a molecule (4×10^{-21} J) would be 10,000-fold greater. Therefore, the momentum of the bacterium will be dissipated rapidly by the Brownian diffusion fluctuations of the water environment.

Definition of Reynolds Number and Brownian Motion

$$\text{Reynold's number} \qquad R = vL\rho/\eta \qquad (5.1)$$

See Berg (1993, pp. 75–77). For a fish with a density approximately that of water ($\rho = 1$ g/cm^3), a length of 10 cm (L), moving at a velocity of 100 cm/s (v) in water ($\eta = 0.01$ g/cm s), we calculate R to be about 10^5. In contrast, for a bacterium of the same density, length 1 µm (L = 10^{-4} cm), moving at a velocity of 10^{-3} cm/s through water, we calculate R to be 10^{-5}.

Brownian Motion

At low Reynolds number, the movements of objects are driven by random fluctuations of pressure from water molecules impacting on them that causes them to move randomly, designated as Brownian motion.

The original 'Life at Low Reynolds Number' by Edward Purcell was based upon considerations of the Navier–Stokes equation for flows dominated by viscosity, i.e. where the density multiplied by the velocity squared over the length term becomes much smaller than the viscosity (see Phillips et al. 2012, chapter 12.4.1). The ratio of these two terms is the Reynolds number (R). When R < 1, the movements are governed by viscosity and not by the momentum of the particle.

The implications of the low Reynolds number regime ($< 10^{-2}$) are wide-reaching and make irrelevant much of our intuition about the macroscopic world. It is so common to think of movements in the macroscopic world where momentum is important that it creeps into our consideration of subcellular phenomena. When thinking about muscle cells, for example, it is easy to imagine that once started, a contraction will continue because of momentum. However, although the macroscopic muscle has momentum, it is the diffusive movements of subcellular protein components that will determine whether the contraction will continue. Specifically, the diffusive movements of myosin under the load conditions determine the speed at which it will undergo a directional movement on actin filaments. This movement is driven by conformational changes in myosin following a cycle that is biased by ATP hydrolysis. Similarly, most enzymes undergo rapid changes in morphology that are constrained by or coupled to enzymatic reactions. In all cases, thermally driven diffusion causes the conformational changes of proteins on the timescales of subnano- to microseconds. Interactions with other molecules constrain diffusional movements and help to convert the conformational changes into productive activities on the micro- to millisecond timescale. Thus, much of the business of biology is done very rapidly and we typically only see the products of thousands to millions of individual reactions that underlie a given biological function. These diffusion-dependent changes and not momentum are the bases of biological motility because at low Reynolds number, movement stops when motors stop.

5.1.1 Viscous Drag on Particles

An important equation for understanding diffusion in general is the Einstein–Smoluchowski relation, which relates the friction coefficient of a particle moving through a medium to the diffusion coefficient of the particle in that medium. A case in point is a magnetic bead of 2 μm in diameter in a magnetic field (the force on the particle is F_x). The drift velocity of the particle (v_d) is related to the force by a constant called the frictional drag coefficient (ϕ_d):

$$vd\ \phi d = Fx \tag{5.2}$$

Because the drag is the same for diffusion as for externally applied forces, the diffusion coefficient can be derived from Equation (5.2) as

$$D = kT/\phi d \tag{5.3}$$

where kT is the energy from thermal vibrations. The generality of this relationship makes it possible to move directly from knowledge of the frictional drag coefficient to the diffusion coefficient.

As spherical particles move through water, they encounter viscous drag that limits diffusive steps. The Stokes' formula describes the relationship between viscous drag and spherical particle size.

$$f = 6\pi\eta r\,v \tag{5.4}$$

Note that this can be related to Equation (5.3) and $\phi_d = 6\pi\eta r$. Thus, the diffusion coefficient for a sphere is given by:

$$D_{sphere} = kT/6\pi\eta r \tag{5.5}$$

For a sphere of 1 μm diameter in water at room temperature $\phi_d = 9.5 \times 10^{-6}$ g/s and $D_{sphere} = 44 \times 10^{-8}$ cm^2/s = (6.6 μm)^2/s.

5.1.2 Diffusion of Small Particles

Brownian motion describes the basic motions of subcellular particles and all but the largest of cells in suspension. Brownian movements result from the local fluctuations in pressure on the particles. For a 1 μm particle, those fluctuations can produce movements of 3–8 μm in a second and the displacements can be nicely described by a Gaussian distribution around the point of origin. To get an idea about the important aspects of diffusion, we will analyze in detail the process of diffusion of drunken ants away from their hole after whisky has been spilled on it (in this example the ants can only move north or south from the hole).

One-dimensional diffusion (the objects in this case are drunken ants)

Assumptions:

1. Steps of r length occur at regular intervals (τ).
2. The direction of each step is equally likely to be + or – independent of previous steps.
3. Each object moves independent of other particles.

If 128 objects start at X = 0 and move r length, then after the first interval (τ) 64 ± 11 (11 is the statistical fluctuation given by $128^{1/2}$) will be at +r and 64 ± 11 at –r.

> After 2t, there will be 32 at –2r, 64 at 0, and 32 at +2r.
> After 3t, there will be 16 at –3r, 48 at –r, 48 at r, and 16 at +3r.
> After 4t, there will be 8 at –4r, 32 at –2r, 48 at 0, 32 at +2r, and 8 at +4r.
> After 5t, there will be 4 at –5r, 20 at –3r, 24 at –r, 24 at +r, 20 at +3r, and 4 at +5r.

Notice that with increasing time the width of the distribution increases (with the square root of time) and the height decreases (also with the square root of time). The distribution can be described as a Gaussian distribution (we will come back to this point).

5.1.2.1 Important Features of Diffusion

1. No net movement occurs (the average position is always zero).
2. Distribution is symmetrical (a corollary of 1).

5.1.3 Root-mean-square displacement $<\Delta X^2(n)>^{1/2}$

It is useful to describe the distribution of diffusing particles over time by the average of the square of the displacement, as it is not dependent upon direction (– direction and + direction).

Mathematically, the diffusion of particles in one dimension is given by

$$2D_1t = <\Delta X^2> \qquad (5.6)$$

Figure 5.1

Gaussian distribution.

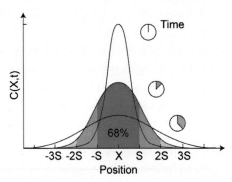

Gaussian distribution of diffusing particles

two dimensions is given by

$$4D_2t = <\Delta X^2 + \Delta Y^2> \qquad (5.7)$$

and three dimensions is given by

$$6D_3t = <\Delta X^2 + \Delta Y^2 + \Delta Z^2> \qquad (5.8)$$

where D is the diffusion coefficient, t is the elapsed time, and $<\Delta X^2>$, $<\Delta X^2 + \Delta Y^2>$, or $<\Delta X^2 + \Delta Y^2 + \Delta Z^2>$ is the average displacement squared. An important aspect of diffusion that is evident from this equation is that the average distance moved scales with the square of time.

5.1.3.1 Gaussian Distribution of Diffusing Particles

If all of the particles at time equals zero are at the origin, then the distribution of the particles after many elemental steps assumes a Gaussian distribution in one dimension.

$$P(x)dx = \left(1/(4\pi Dt)^{1/2}\right)e^{-x^2/4Dt}dx \qquad (5.9)$$

For a normal curve the fraction of the area that is within one standard deviation of the origin (s = $(2Dt)^{1/2}$) is approximately 68% of the total area. The probability that it has wandered more than two standard deviations in one direction is 4.5% and three is 0.26%. These numbers provide a useful way to estimate changes in the concentration of proteins upon diffusion.

5.1.3.2 Diffusive Transport

If we consider the case of one-dimensional diffusion, we can model the process of diffusive transport by considering what happens after one step to the concentrations of molecules that were originally in narrow slices at points X and X + δ, where δ is a single step. If there are more molecules at X than at X + δ, then more molecules will step to X + δ than vice versa. The concentration will increase at X + δ, and this means that there will be net movement of molecules, which is diffusive transport. Thus, diffusion can only produce a net transport down a concentration gradient and the rate of transport is directly proportional to the magnitude of the concentration gradient. The mathematical expression that describes the relationship between the concentration gradient, the diffusion coefficient and the transport rate is Fick's Law.

$$J_x = -D \, dC/dx \qquad (5.12)$$

where J_x is the flux in the x direction, D is the diffusion coefficient, and dC/dx is the concentration gradient in the x direction. Fick's Law can be applied simply when there is a tube connecting a large reservoir of concentration C_1 with a second reservoir of concentration C_2. The concentration gradient is then linear and the flux is constant through the tube.

5.1.4   Practical Implications of the Diffusion Equation

If we take the typical dimensions of a cell (volume of 4000 μm^3 or a cylinder 2 μm high with a diameter of 50 μm), then the distance for diffusion from one point to another in the cell will be on the order of 20 μm. With a typical protein diffusion coefficient of 10^{-7} cm^2/s, the time for diffusion would be on the order of 40 seconds. Thus, for rapidly diffusing components, diffusive transport can be used to bring proteins from sites of synthesis to sites of function.

To model the synthesis and assembly process, we will consider a simple case of synthesis (translation) at site A and assembly into a functional complex at site B in the cytoplasm. Proteins need to get from A to B for assembly. If they diffuse from A to B and are trapped at B, then the cell doesn't need to expend energy to move them. The cost that the cell pays for diffusive transport is a long time constant that relies on a concentration gradient. The cost for active transport is the need for ATP hydrolysis.

In the case of highly asymmetric cells such as neurons, the distance from the cell body to the tip of an axon is on the order of a meter in humans and many meters in larger animals. For our typical protein to diffuse over a meter requires approximately 10^{11} s. This translates to about 3000 years (10^{11} s/3.1 $\times$ 10^7 s/year = 3220 year). Thus, we can safely say that diffusion cannot supply material to the ends of axons.

The way that material is transported from the cell body to the ends of axons is by kinesin-driven transport along microtubules (see Chapter 9 for description of microtubules). Kinesin, like myosin, couples ATP hydrolysis to directional movement through a cycle of conformational changes in the motor–microtubule complex. The average time for a single ATPase cycle, which results in a single step along the microtubule, is several milliseconds. For that step to occur, the kinesin molecule must undergo a series of conformational changes that are local minima in the conformational energy of the molecule. Rapid intramolecular vibrations (pico- to nanosecond vibrations) enable the molecule to pass through the energy barriers between the different 3D conformations. ATP binding and hydrolysis lowers the energy barriers for conformational changes that will move the protein toward one end of the microtubule. Nevertheless, it is energy from thermal vibrations that drives the motor proteins over the energy barriers between the conformational states. For kinesin, one ATP hydrolysis cycle will cause a displacement of 8 nm

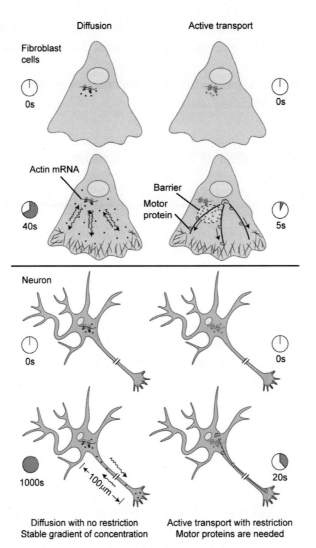

Figure 5.2

Diffusion and active transport.

against a force of about 5 pN. Thus, the mechanical energy output of a single motor cycle is about 40 picoNewton nanometers, whereas the energy of ATP hydrolysis is about 80 picoNewton nanometers (thermal energy, kT, is about 4 picoNewton nanometers). Diffusional movements in kinesin are therefore about 50% efficient in coupling ATP hydrolysis with the directional work.

5.2   Diffusion in Cells

The diffusion of components in cells has been analyzed (Luby-Phelps, 2000) and it is often similar to diffusion in water when components are smaller than the mesh size of

Figure 5.3

Diffusion in cells.

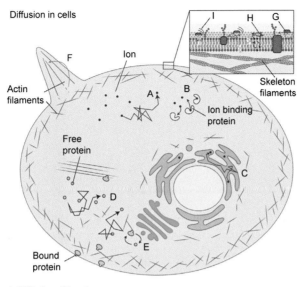

Diffusion in cells

A: Diffusion of free ions
B: Rapid exchange between free and immobilized ion
C: Diffusion of ion in internal cytoplasmic compartment
D: Diffusion of free proteins
E: Rapid exchange between free and bound state
F: Cytoskeleton diffusion
G: Diffusion of free proteins in the lipid layer
H: Lipids diffusion
I: Protein diffusion inhibited

the cytoskeleton. Major deviations from what is observed during diffusion through water do occur when components interact with cytoskeletal filaments or membranes in the cytoplasm. Still, diffusion in cells is confined by the plasma membrane and this will increase the effective concentration of the components. Components that move within cells vary drastically in size, from salts to large mRNAs. In all cases, the diffusional characteristics define how the components can be involved in various functions. We will consider several different cases as examples.

5.2.1 Ion and Metabolite Diffusion

The diffusion of ions and small metabolites in cells occurs largely unhindered; however, there are two major exceptions: cytoplasmic membranes and specific ion–protein interactions. Except for the nuclear membrane, which has open pores, intracellular membranes are not typically permeable to ions or metabolites. Mitochondria rely on an electrochemical gradient to produce ATP, endosome–lysosome compartments have an acid pH, and portions of the endoplasmic reticulum concentrate calcium. A practical consequence of internal membrane barriers is that the effective volume accessible to ions (as well as other cytoplasmic components) is significantly less than the cellular volume (20–40% less).

In the case of calcium, there are many proteins that strongly bind to it and this reduces the free concentration of calcium well below the overall cellular concentration. Calmodulin is a calcium binding protein with submicromolar affinity for calcium and it is present at concentrations in the range of 10 µM. Because many proteins possess calcium binding sites with high affinities, the diffusion of calcium in the cytoplasm is much slower than the diffusion of calcium in water. Of particular relevance are binding sites that are readily reversible such that the free calcium rapidly exchanges with the bound calcium. Because many of the binding proteins are associated with the cytoskeleton or internal membranes, there is effectively no diffusion in the bound state. In cases where there is rapid equilibrium between free and immobilized calcium, the diffusion coefficient is effectively the fraction of the time that the ion is free multiplied by the free diffusion coefficient. Diffusion of calcium in an internal cytoplasmic compartment such as the ER is constrained by the shape of the compartment, which in the case of the ER is tubulovesicular or essentially one-dimensional.

5.2.2 Protein Diffusion

Protein diffusion in the cytoplasm is normally only inhibited by 2–4-fold relative to isotonic saline. The viscosity of the cytoplasm is not very high for small molecules even though there is a high protein concentration of 13–20% (130 to 200 mg of protein per ml of cytoplasm). However, the cytoskeleton filaments can inhibit the movement of large proteins or protein complexes such as ribosomes when they are densely packed. In cases where proteins are binding strongly to the cytoskeleton or to membranes, there will most often be two populations of the protein evident in diffusion measurements; specifically, a rapidly diffusing component that diffuses as a free protein in the cytoplasm and a bound component. The bound component can exchange with the free component on the timescale of the lifetime in the bound state. If the exchange is very rapid, then there is only one diffusion coefficient that is again the fraction of the time in the free state multiplied by the diffusion coefficient in the free state.

5.2.3 Membrane Diffusion

The diffusion of proteins and lipids in membrane bilayers is considerably slower than for similar-sized components in the cytoplasm. This is due to the greater viscosity of the lipid bilayer, which is about 100 times greater than for water (i.e. about 1 poise). In a free lipid bilayer, lipids have diffusion coefficients of $D = 2\text{--}5 \ \mu m^2/s$ and free proteins have $D = 0.3\text{--}1.0 \ \mu m^2/s$. However, with the cytoskeleton apposed to the membrane in animal cells, the diffusion coefficient is reduced dramatically, by over 100-fold in some cases. The cytoskeleton–membrane interaction is often the basis for the inhibition of protein diffusion.

In erythrocytes, the actin filaments at the membrane help to organize membrane-binding proteins that restrict lateral diffusion. Although this is still a matter of debate, the erythrocyte membrane model may prove to be the most relevant when considering membrane diffusion. The majority of erythrocyte actin filaments are separated from the membrane surface by 10–30 nm. Thus, it may be other proteins that are crosslinked by actin at the membrane surface, like spectrin, that are really inhibiting lateral movements. The loss of actin would still appear to modulate lateral diffusion, as this disrupts the crosslinked network that these proteins normally form with actin filaments.

5.2.4 Diffusion of Cytoskeletal Proteins

Although the cytoskeleton is typically considered to be a structural component of cells, there is relatively rapid exchange of monomeric cytoskeletal proteins with those proteins in cytoskeletal filaments or larger complexes during normal cell dynamics. Because the timescale of free protein diffusion is much greater than the exchange rate in most cases, the cytoskeletal proteins will have an immobile filamentous form and a diffusing cytoplasmic form. Measurements of the diffusion coefficient in cytoplasm will also show the fraction of the protein that is free because the filament-bound fraction of a protein will appear immobile.

5.2.4.1 Non-ideal Diffusive Processes

The advent of computer-based imaging and the ability to apply single-photon detection methods to biological systems means that the detailed random walk of individual particles and even individual molecules can now be measured (using single molecule fluorescence methods). The analysis of the movement paths by Single Particle Tracking (SPT) is relatively straightforward. After determining the fraction of the particles that are immobile, the mobile fraction can be analyzed by SPT. It provides a detailed view of the diffusive movements of labeled cellular components. For ideal diffusion, a plot of the mean-squared displacement of a particle versus time should be a straight line with a slope of $2D_1$ for one-dimensional diffusion. However, the diffusion in cells is often not

Figure 5.4

Single particle tracking.

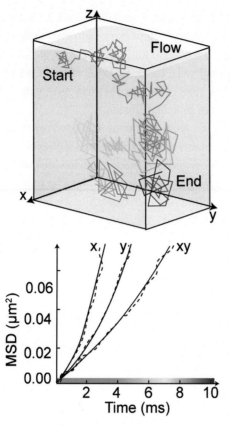

ideal in that the plots of the mean-squared displacement versus time (Equation 5.6) are non-linear. Two major types of non-ideal behaviors are observed in cells and these can provide insight into important aspects of cytoplasmic organization. The flow of the medium (cytoplasm or membrane) in which the particle is diffusing will give rise to an average velocity of movement, v. The mean-squared displacement of a particle in a flow will increase as $(vt)^2$, giving an overall mean-squared displacement versus time of

$$< \Delta X^2 >= 2D_1t + (vt)^2 \quad (5.10)$$

where v is the velocity of the flow. In these situations, one can easily document a flow that would otherwise be very difficult to determine because of the quadratic shape of the mean-squared displacement versus time plot.

Methods for Measuring Diffusion in Cells

There are a variety of ways to measure the diffusion coefficient of proteins or other components in cytoplasm or membranes. Most of the methods have major limitations and if the question is important, diffusion measurements should be made with multiple methods. Commonly used methods include: (1) fluorescence recovery after photobleaching (FRAP), (2) single particle tracking (SPT), (3) fluorescence correlation spectroscopy (FCS), (4) photo-activated fluorophore spreading rates, (5) dynamic light scattering (DLS), and (6) fluorescence depolarization. One of the oldest and still useful methods is FRAP. In that method, a fluorescently tagged molecule (typically a protein) in the cell is bleached in a local region and the rate of recovery of fluorescence by diffusion back of unbleached fluorophores gives the diffusion coefficient (Figure 5.5). It has the drawback that bleaching of fluorophores generates active oxygen species and free radicals that can alter protein function within about 10 nm (the technique of chromophore-assisted laser inactivation or CALI was developed to take advantage

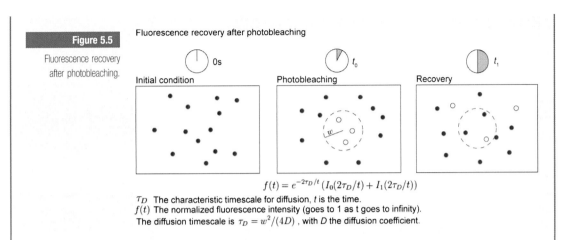

Figure 5.5

Fluorescence recovery after photobleaching.

Fluorescence recovery after photobleaching

| Initial condition | Photobleaching | Recovery |

$$f(t) = e^{-2\tau_D/t}\left(I_0(2\tau_D/t) + I_1(2\tau_D/t)\right)$$

τ_D The characteristic timescale for diffusion, t is the time.
$f(t)$ The normalized fluorescence intensity (goes to 1 as t goes to infinity).
The diffusion timescale is $\tau_D = w^2/(4D)$, with D the diffusion coefficient.

of that property). When it can be used, SPT is the richest way to measure diffusion because the details of the diffusion (binding events and restrictions to diffusion) are evident. The limitation of SPT is that the tracked label should not perturb the component of interest. For example, fluorescent proteins can be added to endogenous cellular proteins through encoding chimeric proteins, but the addition of the fluorescent protein domain to the protein of interest can alter that protein's behavior and function.

5.2.4.2 Diffusion within a Corral

The other common feature of diffusion in the cytoplasm or through membranes is that particles are restricted to particular regions and cannot enter other spaces because of physical barriers (membrane or cytoskeletal). There is no single equation to describe such diffusion but an approximation of the radius of the corral (r_c) comes from the equation

$$< \Delta X^2 > = \left(r_c^2\right)\left[1 - A_1 \exp\left(-4A_2Dt/\left(r_c^2\right)\right)\right] \qquad (5.11)$$

where A_1 and A_2 are constants determined by the corral geometry (Saxton & Jacobson, 1997; Fujiwara & Ritchie, 2002).

5.3 Diffusional Transport and Active Transport

Diffusive transport has a number of important implications for cellular processes because of its undirected nature. The most important is that the level of transport is proportional to the gradient of concentration. For any steady-state situation, a stable gradient will be found between the site of synthesis and utilization.

Figure 5.6

Diffusion within a corral.

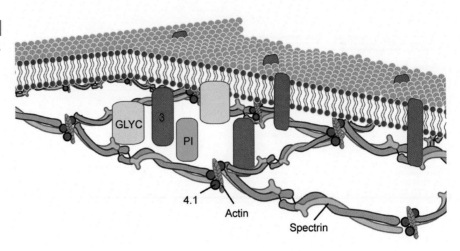

For example, there is a puzzle about why fibroblasts localize actin mRNA to the front of the cell and synthesize actin monomers there. Because actin filament assembly occurs preferentially at the leading edge, the concentration of G-actin is lower in this region and synthesis would help to replenish it. Thus, it makes sense for the actin to be synthesized there; however, the rate of actin filament assembly is orders of magnitude more rapid than the rate of actin synthesis. Further, diffusion toward the back of the cell would also occur with equal probability as it would toward the front. Because there is less actin filament assembly at the back of the cell, the concentration of G-actin would be the same at the back of the cell as at the site of synthesis. Thus, in terms of transport purposes, there is almost no advantage to the cell to place the site of synthesis near the site of utilization if diffusion to the rest of the cell is open. On the other hand, if diffusion to other parts of the cell is restricted by barriers, then efficient transport from source to sink (e.g. site of synthesis to site of filament assembly) would occur. The restriction of diffusion is very hard to accomplish for single proteins and has not been seen for actin, which is why this is a puzzle. However, restriction of diffusion is common for large aggregates in the cytoplasm. For example, observations of diffusion of dextrans in the cytoplasm show non-uniformity with larger particles (Luby-Phelps et al., 1987). The cortical actin region of the cytoplasm is restrictive of large particle diffusion, whereas the perinuclear region that is rich in microtubules and cytoplasmic membranes is not.

5.3.1 Active Transport

For rapid transport of material within cells, it is necessary to tie transport to an active transport mechanism. The critical elements of such a system are the motor, the transport complex, and the link between the two. Motor proteins in mammalian systems can move at rates up to about 5 μm/s. Plants have myosin motors that move 80 μm/s. Most movements are saltatory, which means that vesicles do not move continuously but rather move intermittently with diffusive periods in between. Consequently, the amount of displacement per unit of time is often less than the product of the observed vesicle velocity and elapsed time

$$d < vt \qquad\qquad (5.12)$$

5.3.2   Constraints of Diffusion

Because biological processes are occurring at low Reynolds number, diffusive movements are critical. For example, cells migrate by assembling the cytoskeleton at the front and disassembling it at the back of the cell and virtually all of the energy is utilized in filament assembly/disassembly rather than in friction with an *in vitro* environment. Diffusion processes are stochastic and that makes most biological processes stochastic, which would explain biological variability. Nevertheless, biological systems are highly reproducible because they average stochastic processes to make decisions reliably, such as in the shaping of the organism. Diffusion also constrains cellular functions and any proposed model to explain a given function needs to satisfy the criterion that diffusion is sufficiently rapid to explain the process; if not, there must be an active transport system.

5.4 SUMMARY   Diffusion drives all cellular processes. Further, momentum (inertia) is not relevant for subcellular processes because of the low Reynolds number at that scale. Active transport and motility (force generation) typically rely upon the coupling of the hydrolysis of high-energy compounds like ATP to cyclic protein motions that produce work. Because diffusion is a random process, diffusive movements of proteins will be random and diffusive transport requires a concentration gradient of the free protein, such as when a protein is disassembled from filaments at the source site and assembled into filaments at the sink site. For asymmetric cells like neurons that have processes longer than 100 μm, diffusive transport is too slow to support many mammalian processes and active transport of material by motor movements on filamentous tracks is needed. A variety of microscopic methods can measure the diffusion properties of material within cells that will define the barriers to movement and the frequency of protein binding to stationary structures plus any cytoplasmic flows. The important aspect of the stochastic nature of cellular functions is that they have a lot of noise and for the cells to reproducibly perform functions that shape the organism they need mechanisms to integrate forces and other factors so that the decisions are reproducible.

5.5 PROBLEMS

1. In an article by Dayel et al. (1999), they measure the diffusion of a green fluorescent protein by fluorescence recovery after photobleaching in the lumen of the endoplasmic reticulum (ER) and in cytoplasm and find a threefold higher diffusion coefficient in cytoplasm (1.5×10^{-7} cm^2/s) than in the ER. My lab has measured the diffusion of small (0.2 µm diameter) beads along microtubules (Wang & Sheetz, 2000) using single particle tracking and have found a one-dimensional diffusion constant of 2×10^{-10} cm^2/s. Two proteins (both 5 nm in diameter) are synthesized in cytoplasm at a rate of 1000 molecules per minute in a thin region of cytoplasm (many lamellipodia are only 0.2 µm thick) that is essentially two-dimensional. One protein, S, is soluble and the other, M, is rapidly bound to the microtubule. After 100 s of diffusion (assume that 1000 molecules are at one point and start diffusing at $t = 0$), what is the average displacement of protein S and of protein M? Remembering that the distribution is Gaussian, what are the average concentrations of the two proteins within the region from the origin to the average displacement point? Now calculate the average displacement and concentration, if 1000 molecules of soluble protein, S, are expressed with a signal sequence so that they are translocated into the ER lumen which is a tube 0.1 µm in diameter and hundreds of micrometers in length.

Answer: After 100 s, $\Delta X = \pm 7.6 \times 10^{-3}$ cm for S and $\pm 2.0 \times 10^{-4}$ cm for M. The volume for S is a cylinder 37 µm in radius and 0.2 µm thick ($2\pi r^2 l = 7.2 \times 10^{-9}$ cm^3), whereas for M the volume is a cylinder of the microtubule plus the protein ($2\pi r^2 l = 3.8 \times 10^{-15}$ cm^3). Because of the Gaussian distribution, 66% of the molecules will be within the volumes or 1.1×10^{-21} moles, giving concentrations of 1.6×10^{-13} M for S and 2.9×10^{-4} M for the microtubule-bound protein M. For S in the ER where the diffusion coefficient is threefold lower, $\delta X = \pm 3.3 \times 10^{-3}$ cm and volume is 2.6×10^{-13} cc. The concentration is 4×10^{-9} M.

2. A neuron (70 µm in diameter) is sprouting an axonal process and there are new proteins synthesized that are needed at the tip of the growing process. If a protein is synthesized (10,000 molecules) at the cell end of an axon that is 100 µm microns long (1 µm in diameter) and the protein has a diffusion coefficient of 2×10^{-7} cm^2/s, the average displacement of the protein molecules will equal the length of the axon after how many seconds? What will be the average protein concentration in the axon after that time? Now repeat the calculations for axons that are 1 mm and 1 m long (remember that you may have some axons 1 m in length). If there is a flow of cytoplasm into the axon as it is growing (the axon is elongating at 1 µm per minute typically), what will be the root-mean square average displacement of the 10,000 molecules of protein that are placed at the end of a 1-mm axon after 100 minutes?

Answer: Part A: If we assume 1D diffusion where $(\Delta X)^2 = 2Dt$ and $\Delta X = 10^{-2}$ cm and $D = 2 \times 10^{-7}$ cm^2/s, then $t = 250$ s. Part B: The concentration in the cell body was

1000 molecules (10,000 molecules/6×10^{23} molecules/mole = 1.6×10^{-20} moles) in a volume of $4/3\pi r^3$ where r = 3×10^{-3} cm. This gives 12.56×10^{-12} cm^3 or 12.56×10^{-15} l and the final concentration is then 1.3×10^{-6} Molar. Because the cell body is much larger than the axon, the concentration in the cell body will remain roughly constant and the concentration at the 0.01 cm will be half that, giving an average value for the concentration in the axon of about 0.9 μM. Part C: The diffusion times will be 2.5×10^4 s for 1 mm and 2.5×10^{10} s for 1 m. Part D: Because flow displacement is linear with time or 100 μm (10^{-2} cm) on average giving 10^{-4} cm^2 for the root-mean squared average displacement.

3. In real life, however, the protein in the neuron is not uniformly distributed. Because the protein is synthesized in the cell body, the concentration in the cell body will rise gradually with time. However, the concentration will plateau when the number of molecules moving from the cell body into the axon equals the number of molecules being synthesized per unit time. The linear transport rate of this protein down the axon is 5 mm/day and the axon has a cross-sectional area of 1 μm^2. If there are 1000 binding sites for the protein in 1 μm^3 of the axonal transport complex moving from the cell body to axon that have a dissociation constant of 1.3 micromolar (K_D = 1.3×10^{-6} M), then what is the plateau concentration of the protein in the cell body?

Answer: There are two parts to this problem. The first is the rate of movement of protein binding sites down the axon. The volume of 1 μm of axon is $\pi r^2 l$ or 3.12×0.25 μm^2 = 0.785 μm^3 or there will be 785 binding sites per micrometer of axon length. The displacement rate will be 5000 μm/day = 0.058 μm/s. In the second part of the question, we need to determine what fraction of the binding sites will be occupied. Because the concentration in the cell body is equal to the K_D, half of the sites will be occupied or 393 sites per micrometer. Thus the rate of loss of protein molecules from the cell body will be about 23 molecules/s, which must then be the synthesis rate at steady state.

4. Larger proteins in neurons diffuse at rates of about 3×10^{-8} cm^2/s. We are going to make the neuron more realistic and add a dendrite, which is 1 mm long and on the opposite side to the axon. If there are 10^7 molecules of protein injected into the cell body at time equals zero and those molecules can only diffuse into the axon or the dendrite, then how many seconds will it take until the width of the normal diffusion curve is ± 0.3 mm?

  Part B (quite hard): assuming that the cell body retains 5×10^6 molecules and the dendrite has a cross-sectional area of 4 μm^2 whereas the axon has a cross-sectional area of 1 μm^2, what will be the average concentration of protein in the first 0.5 mm of the axon and of the whole dendrite after the width of the normal diffusion curve has reached ± 0.5 mm?

Answer: Part A: Again, we assume that diffusion is one-dimensional. Solving the diffusion equation for D = 3×10^{-8} cm^2/s and $\Delta X = 3 \times 10^{-2}$ cm we calculate that t = 30,000 s. Part B: In this part, we assume that the number of molecules diffusing from the cell body will be proportional to the area of the process (four times more will go into the dendrite than the axon). Thus 4×10^6 will go into dendrite and 10^6 will go into axon and 64% of these will be within 0.5 mm of the cell. The concentration will be the same in both the axon and the dendrite (6.4×10^5 molecules/ 6.02×10^{23} molecules/mole = 1.05×10^{-18} moles contained within 500×10^{-15} l giving about 2×10^{-6} M.

5. A virus particle is a little over 100 nm in diameter and a bacterium is about 1000 nm. What are the diffusion coefficients of the virus and the bacterium at room temperature in water? If the virus and the bacterium invade a cell at 37°C, what will be the diffusion coefficient of each in the central portion of the cell that has a cytoskeleton network with a pore size of 300 nm (assume that small molecules, less than 3 nm, diffuse threefold slower in cytoplasm than in water alone)?

Answer: Because these are assumed to be spheres, we can use the standard equation Dsphere = kT/6πηr.

Plugging in the numbers for the virus: r = 50 nm, kT = 4.11×10^{-21} J = 4.11×10^{-14} erg (g cm^2/s^2), and η = 0.009 g/cm s. Thus $D_{virus} = 0.5 \times 10^{-8}$ cm^2/s and because the radius of the bacterium is 10-fold greater, $D_{bacterium} = 5 \times 10^{-10}$ cm^2/s.

After invading the cell, the increase in temperature from 25°C to 37°C will cause an increase in the diffusion coefficients by 4% (310/298 = 1.04), but the threefold higher viscosity of cytoplasm will decrease the diffusion coefficients by threefold. Thus, for the virus particle, the new diffusion coefficient will be $D_{virus} = 1.04 \times (0.5 \times 10^{-8}$ cm^2/s)/ $3 = 0.17 \times 10^{-8}$ cm^2/s. In the case of the bacterium, however, it exceeds the pore size of the cytoskeleton network and will not be able to diffuse significantly.

6. We want to compare two different ways to check for unpaired bases in terms of rate of scanning. The first involves the linear transport of a scanning protein complex along double-stranded DNA at the modest rate of 50 nucleotides per second. The second involves diffusion along double-stranded DNA with a diffusion coefficient of 10^{-9} cm^2/s (in both cases a single pass is needed). After one minute and one hour, how much DNA will be scanned by each method, on average?

Answer: After 1 minute, the linear transport system will have scanned 3000 nucleotides, whereas the diffusion scanner will cover $\Delta X^2 = 6 \times 10^{-8}$ cm^2 or $\Delta X = 2.44$ μm or $\Delta X = 7340$ nucleotides. After 1 hour, the linear transport system will have scanned 180,000 nucleotides, whereas the diffusion scanner will cover 56,855 nucleotides.

7. A gene gun is used to put DNA into brain slices to a depth of about 0.5 mm. The gun is so named because it uses air or blanks (in the old days) to propel particles coated with DNA into the tissue at high velocity. Assume that the particles need a Reynold's number in the tissue of greater than 100 initially to penetrate sufficiently far into the tissue. Because the particles are 3 μm in diameter and the density of the particles is 4 g/cm^3, how fast do they have to be traveling when they hit the surface of the brain tissue (effective viscosity of 10 times water)?

Answer: Reynolds number Re = $\rho v L/\mu$ where ρ is the density of the particle and μ is the viscosity of the tissue. Thus, $100 = 4$ g/cm$^3 \times$ v $\times 3 \times 10^{-4}$ cm/0.09 g/cm s = 0.0133 v or v = 7500 cm/s.

8. If we assume that a protein (D = 2×10^{-7} cm^2/s) is synthesized at a steady-state rate of 100 molecules per second and assembled into a larger complex at the same rate, then you can calculate the gradient of protein concentration in the cell from a site of synthesis separated by 20 μm from the site of assembly (assume that all the molecules are moving down a 20-μm diameter tube). If the concentration at the site of assembly is 10^{-9} M and the assembly site is at the pole of a spherical cell of 40 μm in diameter, what is the concentration at the synthesis site at the center of the cell and what is the concentration in the other half of the cell?

Answer: There is 1D diffusion down a 20-μm diameter tube with length L = 20 μm.
   The flux is J = (dN/dt)/S, cross-section S = πr^2 (r = 10 μm), rate of synthesis dN/dt = (100 molecules/s)/N_a (normalization to Avagadro number) or 5.29×10^{-17} moles/cm^2/s.
   The gradient can be calculated by:
   Jx = –D dC/dx or Jx/D = –dC/dx
   5.29×10^{-17} moles/cm^2/s /2×10^{-7} cm^2/s = 2.64×10^{-10} moles/cm^4 or 2.64×10^{-7} M/cm or 2.64×10^{-10} M/10 μm.
   Changing of concentration is ΔC = L*dC/dx, L – distance between sites of synthesis and assembly.
   Concentration at the center of cell: $C_{synthesis}$ = $C_{assembly}$ +ΔC. Thus, the gradient is from 1.5×10^{-9} M to 1.0×10^{-9} M from site of synthesis to assembly.

9. Many believe that the reduction in the number of dimensions of diffusion (e.g. from 3D solution phase to a 2D membrane phase or to a 1D filament phase) will result in an increase in the rate of transport. How will the problem change if a microtubule extends from the site of synthesis to the site of assembly and all of the protein binds to the microtubule surface and diffuses with the same D as in Problem 1?

Answer: The cross-section of the flux-tube decreases and the flux dramatically increases (if the radius of microtubule is ~10 nm, then the increase is proportional to the ratio of cross-sections: $S_{tube}/S_{microtubule} = 10^6$). This leads to an increased concentration gradient (also 10^6-fold higher) and the calculated concentration at the center of cell on the microtubule would be 2.64×10^{-4} M. However, in the cytoplasm, the concentration would be effectively zero. The decreasing of D by two orders of magnitude still leads to an increase in the concentration gradient (10^2-fold increase).

10. Axonal transport has been studied extensively by injecting the cell bodies with radioactive amino acids and then following the distribution of radioactivity in specific proteins along the axon (Nixon et al., 1994; Dillman et al., 1996; Jung et al. 2000). Originally the slow axonal transport of cytoskeletal proteins

(1–2 mm/day) was interpreted as the bulk movement of the assembled cytoskeleton down the axon. In the case of studies of the rat optic nerve (radioactive amino acids were injected into the eye and the optic nerves, 1.6 cm in length, were cut up at various times after injection), we can compare three different models of the transport process. (1) Active transport; in this case assume that the cytoskeletal proteins are carried in a complex that moves at the rate of 2 mm/day and all of the radioactive amino acids (10,000 counts per minute (cpm) total) were incorporated into protein within 1 hour. (2) Diffusive transport; assume that the protein has a diffusion coefficient of 10^{-7} cm^2/s. (3) Rapid transport and diffusion model; assume that the protein is part of a complex which moves at a fast axonal transport rate (200 mm/day) but for only 1% of the time. For the rest of the time the particles diffuse at a rate of 10^{-10} cm^2/s. Please calculate the number of cpms that would be expected in 2 mm slices of the axon for each of these cases after 1.5 days and one week.

Answer: The cpm can be considered as measures of the amount of protein and we can use cpms as the concentration in diffusion calculations. We presume that time constant of radioactive decay is much slower than time of experiment.

Active transport v = 2 mm/day, so in 1.5 days all labeled amino acids will be transported 3 mm (all 10,000 cpms in second 2 mm slice). In 7 days all cpms will be in the seventh slice.

In the case of diffusion there is one-dimensional diffusion of radiolabeled compound from the cell body down to axon. We have Gaussian distribution of compound along the axon (P(x)dx = (1/(4pDt)$^{1/2}$)*e$^{-x2/4Dt}$ dx) and you can calculate the amount of labeled aminoacids by using tables of normal (Gaussian) distribution or manually by using approximation of Gaussian integral: http://ece-www.colorado.edu/~bart/book/gaussian.htm

1.5 days or 1.3 × 10^5 s × 10^{-7} cm^2/s gives 0.013 cm^2 or ± 0.12 cm

7 days or 6.1 × 10^5 s × 10^{-7} cm^2/s gives 0.061 cm^2 or ± 0.25 cm.

In one standard deviation there is 66% and only 33% will go into the axon (the other half will go into the cell body). Assuming a constant concentration

$<\Delta X^2> = 2D t + (vt)^2$. The fact that protein is moved by fast axonal transport 1% of time with v = 200 mm/day is mathematically reflected in effective v = 2 mm/day. Diffusion would be negligible.

5.6 REFERENCES

Berg, H. 1993. *Random Walks in Biology*. Princeton, NJ: Princeton University Press.

Dayel, M.J., Hom, E.F. & Verkman, A.S. 1999. Diffusion of green fluorescent protein in the aqueous-phase lumen of endoplasmic reticulum. *Biophys J* 76: 2843–2851. www.biophysj.org/cgi/content/full/76/5/2843

Dillman, J.F., 3rd, Dabney, L.P. & Pfister, K.K. 1996. Cytoplasmic dynein is associated with slow axonal transport. *Proc Natl Acad Sci USA* 93: 141–144.

Fujiwara, T. & Ritchie, K. 2002. Phospholipids undergo hop diffusion in compartmentalized cell membrane. *J Cell Biol* 157: 1071–1081.

Jung, C., Yabe, J.T. & Shea, T.B. 2000. C-terminal phosphorylation of the high molecular weight neurofilament subunit correlates with decreased neurofilament axonal transport velocity. *Brain Res* 856: 12–19.

Luby-Phelps, K. 2000. Cytoarchitecture and physical properties of cytoplasm: volume, viscosity, diffusion, intracellular surface area. *Int Rev Cytol* 192: 189–221.

Luby-Phelps, K., Castle, P.E., Taylor, D.L. & Lanni, F. 1987. Hindered diffusion of inert tracer particles in the cytoplasm of mouse 3T3 cells. *Proc Natl Acad Sci USA* 84(14): 4910–4913.

Nixon, R.A., Lewis, S.E. , Mercken, M. & Sihag, R.K. 1994. [32P]orthophosphate and [35S]methionine label separate pools of neurofilaments with markedly different axonal transport kinetics in mouse retinal ganglion cells in vivo. *Neurochem Res* 19: 1445–1453.

Phillips, R., Kondon, J., Theriot, J. & Garcia, H. 2012. *Physical Biology of the Cell.* New York, NY: Garland Press.

Saxton, M.J. & Jacobson, K. 1997. Single-particle tracking: applications to membrane dynamics. *Annu Rev Biophys Biomol Struct* 26: 373–399.

Wang, Z. & Sheetz, M.P. 2000. The C-terminus of tubulin increases cytoplasmic dynein and kinesin processivity. *Biophys J* 78: 1955–1964.

5.7 FURTHER READING

Active transport of cargo: www.mechanobio.info/topics/cytoskeleton-dynamics/cytoskeleton-dependent-intracellular-transport/

Myosin ATPase activity: www.mechanobio.info/topics/cytoskeleton-dynamics/motor-activity/

PART II

Design and Operation of Complex Functions

6 Engineering Lipid Bilayers to Provide Fluid Boundaries and Mechanical Controls

Membranes form the boundaries of cells. In eukaryotic cells, membranes also create internal organelles that are critical for compartmentalizing cell functions and making the system more robust (note, from textbox below, that 95–98% of cell membrane mass is internal). As physical boundaries, they define the cell and organelle volumes and must be moved in cell motility and membrane trafficking. As chemical boundaries, they keep critical components in the cytoplasm and other factors outside the cytoplasm and possibly the cell. The hydrophobic barrier provided by lipid bilayers blocks movement of charged and hydrophilic substances into and out of the cell. An important robust property of cells is that plasma membranes can often be lysed multiple times before a cell will die. They serve a gateway function in that they need to facilitate entry of some components and the exit of others. Also, chemical and physical elements of the cellular microenvironment, including its architecture, need to be communicated to the cytoplasm. Cell shape is typically defined by the cytoskeleton for most mammalian cells, but membrane area to cell volume ratio affects the shape as well as the mechanical properties of the membrane. Membrane mechanics play a critical role in many cytoplasmic as well as cellular phenomena because cell membranes have a significant bending stiffness and resistance to stretch as well as a fluid nature.

Membrane Properties

Plasma membrane: 50% lipid and 50% protein by mass, lipid bilayer is fluid and will reseal if lysed, about 85% of membrane area is lipid bilayer (from erythrocyte membrane), most of membrane surface is covered by protein and carbohydrate (electron microscope images show plasma membranes to be 10 nm thick whereas lipid bilayer is about 5 nm and hydrophobic layer is 3 nm thick).

Area/lipid: 0.5 nm^2/phospholipid.

(*continued*)

Membrane elasticity: lipid bilayers have elastic modulus of 10 mN/m and 4% expansion before lysis.

Lateral diffusion in cell membranes: lipids typically 4–6×10^{-8} cm^2/s; proteins typically 5×10^{-9}–10^{-11} cm^2/s.

Plasma membrane dynamics: for a growing fibroblast the endocytosis rate is equal to the area of the plasma membrane every 60 min.

Fraction of membrane area in different compartments: plasma membrane (2–5%), endoplasmic reticulum (50%), Golgi (10–15%), nucleus (5%), endosomes (10–15%), mitochondria (5–10%), and lysosomes (2–4%) (based upon stereology).

6.1   Membranes as Compartment Boundaries

The basic role of membranes is to compartmentalize cells by blocking the free diffusion of water-soluble components. Biological membranes are organized around lipid bilayers that have a hydrophobic core and two hydrophilic surfaces. Hydrophobic molecules and surfaces cause an increase in the order of water by not forming hydrogen bonds and creating a surface like an air–water interface. Because order causes a decrease in entropy, the hydrocarbon layer of the bilayer separates readily from water by hiding in the center of the bilayer, making the bilayers quite stable. Breaking a bilayer and exposing the hydrophobic edges to water is energetically very unfavorable. For this reason, holes in bilayers reseal rapidly to cover the exposed hydrocarbon tails. Many membrane proteins extend across the bilayer and they have a hydrophobic surface that fits with the hydrophobic domain of the bilayer. The specific transport of ions and certain molecules by transmembrane proteins occurs through hydrophilic channels in those proteins (Chapter 8).

Historically, membranes were thought to have almost no proteins penetrating from one surface to the other because the electron microscope images of the protein in membranes showed two parallel layers separated by 3–5 nm. This was interpreted as meaning that the lipid bilayer separated cytoplasmic from extracellular proteins. This model was refuted through studies of cell–cell fusion by Michael Edidin that showed mixing of membrane proteins from one cell with the other and led to the Singer–Nicolson fluid mosaic model of membranes (Figure 6.1). The major modification of the model since it was formulated has been in the belief that membrane protein diffusion is very significantly inhibited by cytoskeletal interactions at the cytoplasm–membrane interface.

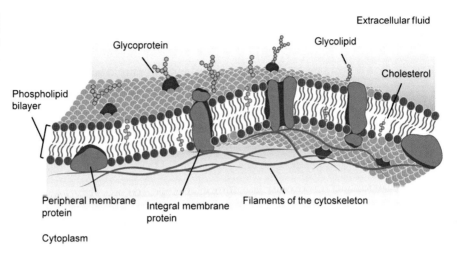

Figure 6.1

Organization of the lipids
and proteins.

6.2 Chemical Order of Membranes

The phospholipid bilayer has a hydrophilic surface and a hydrophobic core that is pierced by proteins whose hydrophobic domains are typically matched in length with the hydrocarbon region of the lipid. In different cells, there are a variety of different lipids with different head groups that interact somewhat specifically with proteins. Cholesterol is a major lipid, but it is primarily hydrophobic with only one hydroxyl group. Lipids in plasma membranes are asymmetrically distributed between the inside and outside halves of the bilayer, with the negatively charged lipids almost totally on the inside surface. One of the early signs of cell apoptosis is the movement of the negatively charged lipid, phosphatidyl serine, to the external plasma membrane surface, indicating that the lipid redistribution must be particularly sensitive to early changes in apoptosis. Lipids that contain carbohydrates, called glycolipids, are found on the external surface of the plasma membrane and the internal surfaces of the endoplasmic reticulum and Golgi where the carbohydrates are added (topologically the same surfaces, i.e. outside of the cytoplasm). Depending upon the lipid head group and the hydrophobic portion, lipids are more or less stable in bilayers. Major membrane properties, curvature, lysis, fusion, and resealing, can be strongly influenced by lipid composition. Proteins can alter the lateral organization of lipids through weak interactions with them, including the thickness of the protein hydrophobic domain that should be matched by the thickness of the bilayers hydrophobic domain (thickness is related to the fatty acid chain length of the lipid). In turn, changes in the lipids, particularly cholesterol depletion, can have a major influence on transmembrane proteins.

6.3   Lipid Composition

There are many different lipids in biological membranes both in terms of the hydrophilic head domains and the fatty acid length and degree of unsaturation (number of double bonds). The study of these lipids, and the various pathways and networks in which they function, is known as lipidomics, and this has been accelerated by mass spectrometry methods (Ivanova et al., 2009). The major phospholipids are illustrated in Figure 6.2. In each case, a glycerol backbone links two fatty acyl chains (typically 16–24 carbons in length) to the phosphate of the head group. In a bilayer, a typical phospholipid will occupy an area of about 0.5 nm^2 in one half of the bilayer. This means that there are about four million lipid molecules per μm^2 of bilayer. The majority of phospholipids are zwitterionic, with a negative phosphate and a positive amino group. The positive–negative dipole can be oriented by charge gradients but generally has a relatively low affinity for cytoplasmic proteins. There are no net positive phospholipids but several important negative lipids are commonly found. Phosphatidyl serine, which has a net negative charge of –1, is the most common negatively charged lipid. Lipids with higher charges, such as the multiply phosphorylated inositol lipids

Figure 6.2

Phospholipids.

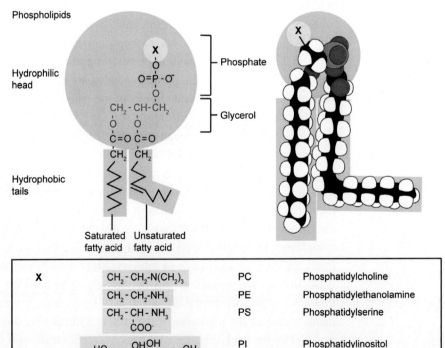

X		PC	Phosphatidylcholine
	$CH_2 - CH_2 - N(CH_2)_3$	PC	Phosphatidylcholine
	$CH_2 - CH_2 - NH_3$	PE	Phosphatidylethanolamine
	$CH_2 - CH - NH_3$ COO^-	PS	Phosphatidylserine
		PI	Phosphatidylinositol

(phosphatidyl inositol di- and triphosphates) have much higher charge densities that can strongly interact with charge pockets on proteins. Glycolipids are a class of lipids that have hydrophilic carbohydrate chains and are usually only found on the extracellular or intravesicular surface of membranes. They help to reduce the interaction of the lipid surface with extracellular proteins and other materials. Cholesterol is an uncharged lipid that is largely buried in the hydrophobic portion of the bilayer and complexes weakly with the sphingolipids (they differ from the phospholipids in that glycerol is replaced by sphingosine that has a long fatty acyl tail and one fatty acid is linked by an amide bond to sphingosine). In general, the diversity of lipids is to aid in the diverse functions of cells. Lipids are like most environmental factors in that they can significantly enhance or inhibit a given protein function yet they are generally not involved directly in protein functions. They do have significant signaling roles and head groups, fatty acids and diglycerides all are major second messengers in signaling pathways. In line with our discussion of localized functions and phases in chapter 4, lipid signals are relatively localized to the membrane surface and the relatively slow lateral diffusion restricts them to small (1–4 µm) regions of the cell membrane. Recent studies show that waves of lipid signals are propagated through membranes by positive feedback periods when signaling lipids are produced followed by negative feedback processes that degrade them.

6.4 Linking Proteins to Membranes

Proteins can be linked to membranes through acyl chains, and this is important for many functions. A surprising number of acyl transferase enzymes are present in cells that can add fatty acid chains to proteins. Once a protein is acylated (typically by myristic, palmitic, or geranylgeranyl groups) it has a very high affinity for a membrane surface. Acyl transferases can anchor a protein to a membrane in the region where the transferase is present. This will localize the protein and increase its effective concentration by over 100-fold. Thus, cytoplasmic proteins can be compartmentalized by acylation, indicating that compartments are often dynamic.

In thinking about the level of association of a protein with a membrane, it is useful to consider the fact that it requires about the energy of one ATP to pull a lipid out of a membrane. Similarly, diacylated proteins should require about the same energy to be pulled from membranes and mono-acylated proteins should require half of that energy. There are lipid transfer proteins that bind acyl chains and can facilitate the transfer of lipophilic groups from one membrane compartment through the cytoplasm to another membrane. These could aid in the specific transfer of acylated proteins from the site of synthesis to another site of function in the cell. Thus, the transfer of proteins from one area to another in the cell can occur through acylation–deacylation cycles or through specific transfer proteins.

Another aspect of lipid bilayers is that breaks in the membrane bilayers exposing the hydrophobic domains normally rapidly reseal. However, proteins can be designed to cover the edges of bilayers and such a designed protein has been used to create nanodiscs of lipid bilayers that are about 10 nm in diameter (Bayburt & Sligar, 2010). Nanodiscs are useful in the study of membrane proteins that are embedded in bilayers and become unstable when taken out of bilayers. As nanometer-level complexes they diffuse rapidly and enable studies where larger bilayers of membranes would potentially complicate the analysis of the protein structure or function.

6.5   Lipid Organization – Matching of Surface Areas, Phases, Charges

In an ideal planar lipid bilayer, the phospholipid head groups (the hydrophilic portion) will occupy the same lateral area as the hydrocarbon tails of the two fatty acids. However, the hydrophilic triphosphorylated inositol ring of phosphatidyl inositol 3,4,5 triphosphate (PIP3) is considerably larger than ethanolamine of phosphatidylethanolamine, while other phospholipid head groups are intermediate. There are also lyso-lipids where one of the two fatty acid chains has been removed, cutting the lateral area of the hydrophobic portion of the lipid in half. Furthermore, changes in the length of the fatty acid chains and their degree of unsaturation will affect the lateral area. The bilayer described so far represents a phase of lipid organization that is the predominant form of lipids in biological membranes. Additional phases have been studied extensively, such as the micellar phase of PIP3 (the pure lipid assembles into spherical micelles in water but is too dilute in membranes to produce that phase). However, there is a dearth of evidence that they play a significant role in biological phenomena. As noted, lipids that form into other phases are present in bilayers and it is possible that those lipids play significant roles in membrane activities such as membrane fusion and fission or curvature. For example, PIP3 at low concentrations in biological membranes can exert local effects on bilayer curvature that would encourage membrane fusion. Thus, the lipids of most biological membranes are heterogeneous such that they form bilayers; however, lateral diffusion can enable lipids to move to sites where they will affect membrane curvature locally for fusion or fission processes.

There is a general hypothesis that the relative lateral area occupied by the hydrophilic head group versus the hydrophobic fatty acyl chains determines the direction of membrane curvature. Bilayer curvatures have to be very severe to affect the relative areas occupied by the hydrocarbon and head groups. The typical bilayer

Figure 6.3

Bilayer curvature.

Bilayer Curvature

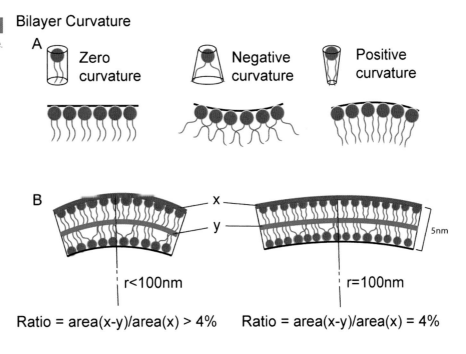

A

Zero curvature

Negative curvature

Positive curvature

B

x

y

5nm

r<100nm   r=100nm

Ratio = area(x-y)/area(x) > 4%   Ratio = area(x-y)/area(x) = 4%

thickness of 5 nm means that the membrane needs to have a curvature radius of 100 nm or less to see significant changes in surface area going through the bilayer. In Figure 6.3, we see that a radius of curvature of less than 100 nm causes at least a 4% change in the ratio of the lateral area at the surface versus the middle of the bilayer. Because that is the amount of bilayer expansion before lysis, this is likely to represent a significant change in the free energy of the bilayer lipids that could drive the sorting of lipids to curved versus flat regions. Because bilayers are fluid, lipids will preferentially distribute to curved versus flat regions based upon the area of their head region versus the hydrocarbon region. Many internal membranes (ER, Golgi, etc.) have curvature radii of 100 nm or less. At the edges of cells, the plasma membranes in the lamellipodia and filopodia possess a similar curvature. Thus, high membrane curvatures can cause changes in lipid distributions; however, proteins are usually responsible for producing highly curved membranes.

6.6   Fusion and Fission

Typically, an area of membrane equal to that of the plasma membrane is endocytosed and exocytosed every hour in actively growing eukaryotic cells. For such membrane dynamics to occur there must be rapid fusion and fission of membrane vesicles. Similarly, for the cell to form and maintain multiple membrane

Figure 6.4

Fusion and fission.

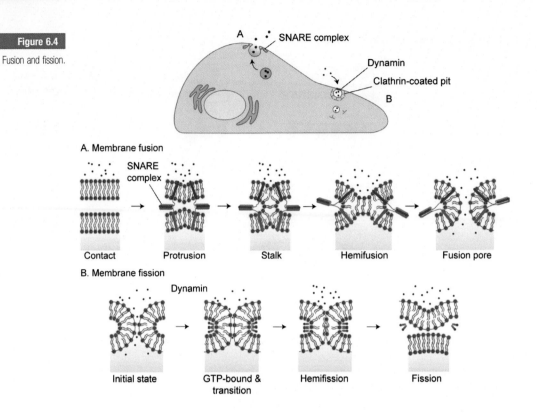

A. Membrane fusion

Contact   Protrusion   Stalk   Hemifusion   Fusion pore

B. Membrane fission

Initial state   GTP-bound & transition   Hemifission   Fission

compartments there must be exquisite control of membrane vesicle fusion and fission. During the fusion and fission processes there are energetically unfavorable steps (see stalk, hemifusion and fusion pore steps in Figure 6.4) that utilize proteins to catalyze the process. Going from the stalk to the hemifused state is favored by lipids with smaller head groups that would stabilize an inverted membrane with a large hydrophobic domain. In pure lipid systems, vesicle fusion will be catalyzed by tension within the lipid vesicles. In cells, however, proteins control the process and serve to regulate when fusion occurs and when it does not occur. Enzymatic modification of lipids to produce forms that can favor fusion is possible in some systems such as by hydrolysis of the charged head group. Thus, lipids that have different hydrophilic and hydrophobic areas are normally part of biological membranes, but they are a fraction of the total lipid and other lipids serve to stabilize the bilayer except when proteins drive deformation (fusion or fission).

6.7   Lipid Asymmetry and the Bilayer Couple

Because the human erythrocyte has only a plasma membrane with no internal membranes, it is a favorite system for the study of membrane behavior. Normal erythrocytes have a biconcave disc shape but show dramatic alterations in shape under a variety of conditions. When drugs are added to erythrocytes, they often have a dramatic effect on erythrocyte shape. Two major classes of shape change are observed after drug addition: either the cells become crenated with outward projections, or they become cup-shaped with inward projections. The drugs are amphiphilic in that they normally contain a hydrophobic domain and a hydrophilic domain. Most cause an increase in the surface area of a lipid monolayer when added to the water phase. Thus, a logical explanation for the behavior is that the drugs are binding to the membrane and causing the shape change. Because the drugs are inserting into the lipid head region, it is possible that asymmetric physical expansion of the bilayer is causing the shape change. Indeed, there is a clear difference between the drugs that cause crenation (negatively charged and membrane impermeable drugs) and those that cause cup formation (permeable, positively charged drugs). Logically, the hypothesis is that the drugs are preferentially distributing based upon lipid charge distribution. As a result of the negative charges on the cytoplasmic membrane surface, the positively charged drugs are attracted there and the negative ones repelled. Because drug binding causes an expansion of the lipid surface area and the inner and outer halves of the bilayer are always joined at the center, curvature of the membrane follows (see Figure 6.3). This is the bilayer couple hypothesis that has held up over time and explains many shape changes.

Lipid asymmetry was found through a number of different analyses. One of the earliest clues came from studies of the reactivity of the amine lipids [e.g. phosphatidyl ethanolamine (PE) and phosphatidyl serine (PS)] that would react with amine reagents (e.g. *N*-hydroxysuccinimide linked to a fluorophore), whereas the choline lipids (e.g. sphingomyelin and phosphatidylcholine) would not. When amine reagents were added to intact cells, they did not react with the amine lipids unless they could get into the cells after hypotonic lysis of the erythrocytes. This was a particularly surprising result for the human erythrocytes that only had a plasma membrane and the amine lipids could not be hidden on internal membranes. Thus, they suggested that the amine lipids were on the cytoplasmic surface of the plasma membrane, i.e. PE and PS were only on the cytoplasmic side of the plasma membrane. A similar asymmetry was found in the case of the phosphorylated inositol lipids. They rapidly incorporated radiolabeled phosphate in intact cells from cellular ATP, which meant that they were on the cytoplasmic surface as well. Another method was used to show that sphingomyelin was on the outside surface of the bilayer. The enzyme sphingomyelinase was active on intact cell plasma membranes and it caused invagination or cup formation of erythrocytes, indicating that sphingomyelin was on the outer surface. Thus, accessibility of head groups in intact versus lysed cells was used to establish the distribution of lipids

Figure 6.5

Bilayer couple with charge distribution and flippases.

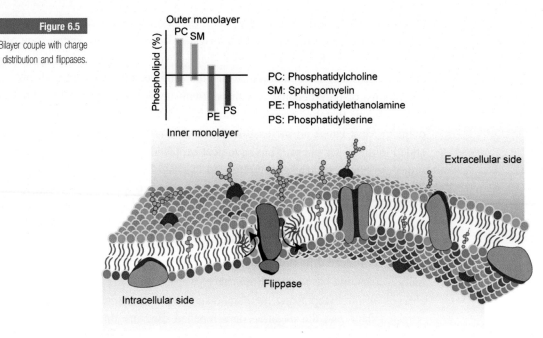

between the outer and cytoplasmic halves of the membrane bilayer. The majority of negatively charged lipids, PS and the phosphatidyl inositols were distributed on the cytoplasmic surface along with PE. This produced a significant charge imbalance across the bilayer with a neutral set of lipids on the outer surface (sphingomyelin and phosphatidyl choline) and the anionic lipids on the interior.

6.7.1 Transport across Membranes

Chemical transport of components across membranes is a critical aspect for the proper function of cells and membrane-bounded organelles. Highly charged or hydrophilic compounds need to be carried or actively transported through protein channels that span the membrane. These membrane transporters range from water channels to the flippases that transport phospholipids from one surface to another. The structure of these membrane proteins reflects the order of the lipid bilayer in several ways. Transmembrane proteins (those that integrate into the membrane) often have a string of positive amino acids on the cytoplasmic surface that can bind to and be stabilized by the negative charges of the cytoplasmic surface lipids (White, 2007). Glycoprotein groups are located on the extracellular or intravesicular compartments, and these groups lubricate the cell surfaces and inhibit nonspecific interactions with neighboring molecules. The transmembrane portions of the proteins are hydrophobic and are typically matched to fit the length of the hydrophobic portion of the lipid bilayer. There are some proteins like *N*-cadherin that have multiple forms with different cytoplasmic portions and a form that is lipid-linked (glycophosphoinositide, GPI) to only the external half of the bilayer with the same external domain. This enables the protein to elicit multiple responses with the same extracellular binding interactions. The GPI-anchored proteins are often associated with 'lipid raft' signaling complexes that were originally defined by detergent extraction in the cold. These complexes resisted extraction and sedimented as lipid–protein–detergent discs. At body temperature there are typically no rafts and the low-temperature detergent extractions caused the aggregation of those components. Still, there is evidence that differences in the thickness of the hydrophobic domain favor the lipid lateral association in the membrane particularly after the lateral aggregation of the proteins. When proteins with thick hydrophobic domains aggregate laterally in membranes, the lipids with longer hydrophobic tails will concentrate in those regions because they will have a lower energy configuration there. Thus, membrane proteins come in a variety of forms that can produce transmembrane communication by moving ions or molecules across the membrane or by aggregating to send signals to the cytoplasm (more about this when we discuss matrix–cell interactions in Chapter 11).

6.8 Physical Reality of Membranes

Membranes have often been likened to soap bubbles. Indeed, the endoplasmic reticulum membranes form tubes with triangular junctions that move as if they were fluid soap bubbles. However, plasma membranes are supported by the cytoskeleton most of the time and only occasionally form blebs (bubbles). Of the two important properties of membrane bilayers discussed below, only the fluidity is shared between bilayers and soap bubbles. In the case of elasticity, soap bubbles can change area dramatically as a result of changes in tension, while lipid bilayers are relatively inelastic, and can only expand by 4% before they will lyse.

6.8.1 (a) Membranes are Inelastic

Biological membranes are surprisingly inelastic. Tensions in the range of 10 mN/m are required to expand the membrane by 4%, yet those tensions are also sufficient to cause membrane lysis. For this reason, normal eukaryotic plasma membranes exist in an unstretched state, i.e. they are constant in area, with resting tensions that are 100- to 1000-fold lower than lytic tensions. Although membranes are inelastic, they are also fluid, and this means they have a number of interesting properties. The relative inelasticity means that they do not change thickness easily, which translates into a significant energy for the lateral aggregation of lipids and proteins that have similar lengths of their hydrophobic domains. Further, as cells change shape, the extension in one area must be matched by a retraction in another. This is particularly important during major cell shape changes or migration, as there must be significant flow of membrane from one area of the cell surface to another. The largest flows have been recorded during neurite outgrowth, where surface membrane flow may be a factor in the overall exchange of membrane proteins. Perhaps it is not surprising that membrane reservoirs exist in the cell to provide a buffer for cytoskeletal-dependent shape changes. Abrupt increases in cell volume can occur when the salt concentration of the medium decreases (water will move into the cell then because of the higher salt and small molecule concentration in the cell, a phenomenon called osmotic swelling). To prevent cell lysis in this situation, the membrane folds separate from the actin cytoskeleton, exocytosis increases the membrane surface area, and ion channels open to let out salts or other small molecules. The rapid release of small molecules can equalize the osmolyte concentrations inside and outside the cell. The channels that release these osmolytes are activated by tension in the membranes in bacteria and plants, but in eukaryotic cells the osmotic-sensitive channels are dependent upon the cytoskeleton. This has been interpreted to mean that the channels in eukaryotic cells are activated by membrane pulling away from the cytoskeleton, i.e. a vertical force on the membrane channel that is bound to the cytoskeleton. Tension in the plane of the membrane in eukaryotic cells is linked to several signaling systems (discussed below).

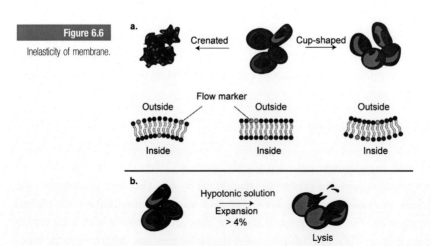

Figure 6.6

Inelasticity of membrane.

The inelasticity of membranes has additional effects on the cell, as it confers a significant resistance to bending of the cell membranes. In the case of human erythrocytes, the undisturbed shape is a biconcave disc that minimizes the bending energy of the membrane. For the cell surface to go from a flat to a highly curved surface, there must be relative change in the surface area at the inner and outer bilayer surfaces. Because there is no rigid cytoskeleton, it is the plasma membrane of mammalian erythrocytes that provides the basis for the change in shape that accompanies red blood cell aging and various hereditary abnormalities, such as sickle cell anemia and hereditary spherocytosis. The Bilayer Couple Hypothesis provides a simple way to understand how the differential expansion or contraction of the inner versus outer bilayer surfaces can cause dramatic changes in cell shape (Sheetz & Singer, 1974). A number of theoretical papers have confirmed that only small changes in the relative surface areas (0.1–0.2% relative surface area) can cause major changes in cell shape because the membrane is so inelastic. Although the same type of immediate shape changes are not seen in mammalian cells with a rigid cytoskeleton, it is expected that the tensions produced by relative changes in the inner and outer bilayer surface areas will alter cell morphology over time.

6.8.2 (b) Membranes are Fluid (Diffusion and Flow)

In the majority of cases when a cell changes shape, the membrane is a passive component that moves in response to the action of the underlying cytoskeleton. This is usually not a problem in the sense that there is often a membrane buffer where extension in one region is coupled with contraction in another. Rapid cell extensions or retractions are not coupled with changes in cell surface area but rather the flow of membrane into the region. The membrane must flow from some other area of the cell surface, but the areas are typically too small to measure reliably. The requirement for the maintenance of a constant area provides an important constraint for integrated models of cell migration. The subtle increase in tension caused by extension of the cell edge in one region can cause the inhibition of extensions in other regions. Because changes in membrane tension in one region of the cell can inhibit motility in other regions, cell asymmetry and directed migration can be reinforced. Recent studies find that cell polarization involves a gradient of tension in the membrane such that regions of higher tension have low levels of motility and regions of lower tension often activate actin-dependent membrane extension.

6.9   Plasma Membrane Protein Diffusion is Controlled by Cytoskeletal Corrals

In most cases, the movements of membrane proteins are due to diffusion within the membrane plane. Extensive analyses of membrane protein diffusion have shown that diffusion in the plasma membrane is significantly slower than in pure lipid bilayers. Diffusion coefficients are typically a 100-fold slower in biological membranes (0.01 μm^2/s in cell membranes versus nearly 1 μm^2/s in lipid bilayers). This means that membrane proteins move over 1 μm in about 2 minutes, or 5–6 μm in an hour, which is the time required for the exchange of the whole plasma membrane through exo- or endocytosis. However, if the membrane separates from the actin cytoskeleton in a bleb, then diffusion is much faster, almost reaching the level of diffusion in a pure lipid bilayer. This indicates that the actin cytoskeleton inhibits glycoprotein diffusion, which is further reinforced by the observation that actin filament depolymerization increases diffusion rates. The effect of the actin cytoskeleton on membrane protein diffusion is linked to the formation of lateral barriers, known as 'membrane corrals' or 'fences,' on the cytoplasmic surface of the membrane. Membrane proteins diffuse as rapidly in the actin corrals as in lipid bilayers, but moving from one corral to another requires a break in the corral or separation from the surface. Thus, diffusion of membrane proteins can be rapid very locally ($< 0.2\,\mu m$) but slow on scales of micrometers.

In historical experiments where migrating cells bound beads or other micrometer-sized objects, the beads invariably moved inward with the actin cytoskeleton. These observations prompted researchers to think that the whole membrane was being transported rearward. However, when the diffusion of individual membrane proteins or lipids was measured, there was no rearward transport. The individual membrane proteins diffused randomly and were unaffected by the rearward movement of the actin cytoskeleton. There is no good explanation for how the cytoskeleton can inhibit the diffusion of membrane proteins yet move

Figure 6.7

Cytoskeleton corrals.

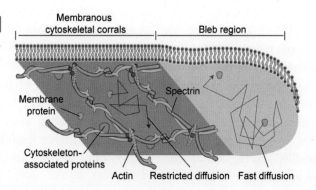

independently of those same proteins. It is reasonable to postulate that two different types of actin filaments are present: one set of filaments that is involved in restricting membrane protein diffusion by being closely apposed to the membrane, and another more extensive actin cytoskeleton that drives cell motility.

6.10 Membrane Exchange Rates are Sufficient to Produce Lateral Inhomogeneities

A consequence of the slow diffusion of membrane proteins is that the normal processes of exo- and endocytosis will produce significant diffusion gradients. A gradient can also be established as proteins gradually diffuse away from the sites where they might be localized (for example, integrins are concentrated at adhesions and adhesions disassemble). However, because a membrane protein can only diffuse about 5 μm per hour, these proteins will still be concentrated within 5 μm of the specific site after that time. Thus, in a 40 μm diameter cell that exchanges its plasma membrane area in an hour, local endocytosis can deplete an area of a receptor while exocytosis in another region can dramatically increase the concentration. This means that membrane dynamics can significantly change the local activities of the plasma membrane.

6.11 Membrane Curvature and Protein or Metabolite Concentration

Diffusion of membrane proteins can be dramatically influenced by the physical properties of the membrane such as membrane curvature. Certain transmembrane proteins and membrane-bound proteins can diffuse to and concentrate in curved membranes because of the curvature alone. Thus, curvature has a number of physiological consequences for the cell. For example, membrane curvature impacts the concentration of the BAR (Bin–Amphiphysin–Rvs) domain containing proteins at membrane surfaces. High curvatures during endocytosis can cause BAR protein binding and their presence can facilitate membrane vesicle formation. In another example, membrane proteins that interact with new matrix sites as the cell extends to new areas are localized to curved membrane regions at the leading edges of lamellipodia. This is because cellular projections that function in these cellular processes, such as filopodia or lamellipodia, are defined by the cytoskeleton, and possess highly curved regions at their outer edges. Indeed, the major matrix binding protein, integrin, preferentially distributes to such curved regions. Thus, the preferential distribution of certain membrane proteins to highly curved regions is important for a variety of localized functions.

Figure 6.8

BAR domains.

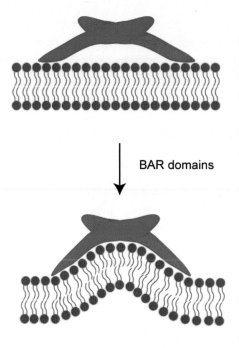

BAR domains

Another interesting consequence of curvature in cell projections and thin processes is that the ratio of membrane to cytoplasm is very high. For example, axons are often only 1 μm in diameter, yet centimeters long, and this translates to a large membrane surface encasing a relatively low volume of cytoplasm. Membrane-bound enzymes that create soluble signals that are broken down by cytoplasmic enzymes can produce higher concentrations of the signals in regions of high membrane curvature. This is due to the reduced ratio of cytoplasm to membrane, and subsequent alteration in the metabolization rate compared to regions of low curvature, where the cytoplasm/membrane ratio is higher. This is observed in the production of cyclic AMP from the membrane-bound adenylyl cyclase. In that case, highly curved regions of the cell respond differently to hormones that activate cyclic AMP production, compared to regions of low curvature. As there are a number of cases where signals are created by membrane-bound enzymes and broken down by cytoplasmic enzymes, this phenomenon may impact a wide range of complex functions.

6.12   Concentration at Membrane Surfaces Enables More Dynamic Interactions

A very important aspect of membranes is that they are essentially planar and, as 2D surfaces, they can concentrate components and reduce diffusion-binding times. This is perhaps best illustrated by the realistic case of a cytoplasmic protein having a fatty acyl chain added that now causes it to bind to a membrane. If we assume that there are 1000 copies of the protein per μm^2 at the membrane surface and it reaches only 20 nm from the surface, then the concentration in our hypothetical normal fibroblast will be 2×10^{-17} liters for 1.6×10^{-21} Moles, which is about 80 μmolar at the membrane surface. In contrast, the same number of proteins in cytoplasm would have a concentration of about 1 μmolar (2×10^6 molecules in 3000 μm^3 vs. 2000 μm^2 × 0.02 μm = 40 μm^3 at the membrane or a 75-fold lower concentration). Thus, by simply moving from the cytoplasm to the plasma membrane, the effective protein concentration would increase by about 100-fold if its major binding partners were membrane-bound.

Equilibrium Binding and Dynamics

For many protein–protein binding events, the interactions are readily reversible based upon the low binding constants and the diffusion limit to the binding rate. This enables the cell to rapidly reverse interactions and adapt to changing conditions or to move on to the next step in a process. If we consider a normal equilibrium binding system in which protein A binds to protein B with a binding constant of 10^6 M^{-1} (this is a reasonable binding constant for many protein interactions and concentrations of proteins are often in this range), then we can write the following equation:

$$A + B = AB$$

$$\text{then } [AB]/[A][B] = 10^6 \text{ M}^{-1} = K_A$$

Alternatively, we can describe the Association Constant as the association rate, k_a, divided by the dissociation rate, k_d

$$K_A = k_a/k_d$$

Now, the association rate constant times the concentration of B will give the rate of change in A. In a water medium, the diffusion limited binding rate is maximally 2×10^6 M^{-1} s^{-1} (this estimate has been experimentally tested). With this number, it is easy to calculate dissociation times

$$d[A]/dt = k_a \, [B] \text{ and diffusion limits } k_a \text{ to about } 2 \times 10^6 \text{ M}^{-1}\text{s}^{-1}$$

$$k_d = d[AB]/dt = k_a/K_A = 2 \times 10^6 \text{ M}^{-1}\text{s}^{-1}/K_A = 2\text{s}^{-1} \text{ for } K_A = 10^6 \text{ M}^{-1}$$

$$\text{Weaker } K_A \text{ of } 10^4\text{M}^{-1} \text{ will give a rapid } k_d = 200 \text{ s}^{-1}(\text{bound time} \sim 5 \text{ msec})$$

$$\text{Stronger } K_A \text{ of } 10^9\text{M}^{-1} \text{ will give a slow } k_d = 0.002 \text{ s}^{-1}(\text{bound time} \sim 500 \text{ s})$$

In designing a robust system, it is important to have rapid processing and release of components. The high concentrations of membrane-attached proteins and ligands at membrane surfaces enable components to bind and release rapidly for efficient processing (see text box for the quantitative treatment of the changes in off rates). From the text box we see that a weak interaction has a fast dissociation rate or short bound lifetime (5 ms for $K_A = 10^4$ M^{-1}), which means that it can rapidly interact with and modify many weak binding molecules. On the other hand, a strong interaction with high affinity will have a slow dissociation rate with a long bound lifetime (500 s for $K_A = 10^9$ M^{-1}). The caveat in this argument is that if the concentration of the components is low, $< 10^{-6}$ M, then less than 1% of the components will be in a complex if $K_A = 10^4 \, M^{-1}$, which means that the reaction will proceed slowly despite the rapid off rate. However, if the local concentration is about $10^{-4} \, M$ as in the case of the concentrated membrane protein, then nearly half of the components will be in the complex and rapid modification and release can occur. In contrast, with the strong binding constant, the complex stays for 500 s no matter what the surrounding concentration might be. Thus, the way to get rapid processing is to have weak binding constants and concentrate components in fluid compartments, such as at the surface of a membrane.

In this view, the membrane surface effectively forms a separate compartment where the processing of membrane-attached components is ~100-fold faster than cytoplasmic forms of the same components. This provides a rationale for the many different fatty acid transferase enzymes that can attach fatty acids to soluble proteins to attach them to the membrane. This increases the local concentration of these proteins at the membrane surface and that can dramatically increase the rate of product formation at regions near the membrane surface. For example, many phosphorylation/dephosphorylation systems have a membrane-bound kinase and a soluble phosphatase that balance each other. However, when the membrane kinase is activated in one region of the cell, its much greater local activity will create a regionally high level of phosphorylated product. Overall levels in the cell may not change significantly, but local levels will. Over time, the kinase will become inactivated and the balance will be restored, but there can be a memory of the event locally through the downstream actions of the phosphorylated proteins. Thus, the reduced dimensionality of membranes can be an important aspect of many signaling systems (for example, Src family kinases, inositol lipid kinases, many small G-proteins and protein kinase C).

For the inositol lipids that have major roles in signaling processes, the concentrations at the cytoplasmic surface of the plasma membrane are impressively high. For example, the apparent concentrations of a minor lipid, namely phosphatidyl insitol 4,5 diphosphate (PIP2), which is involved in many protein functions, are in the millimolar range. PIP2 is present in many cells at a level of about 0.5–1% of the total lipid. It is concentrated in the cytoplasmic half of the lipid bilayer and thus constitutes 1–2% of the lipid in that surface. Because there are 2 million lipid molecules per μm^2 of membrane bilayer surface, there should be 18,000–36,000 PIP2 molecules at the cytoplasmic surface of a membrane per μm^2 (lipid areas are 80–90% of plasma membrane areas). Within 10 nm of the plasma membrane, the concentration of PIP2 is 16,000–36,000 molecules per 0.01 femtoliter or 3–6 mM and the off rate for binding constants in this range are in the 1–10 ms range, meaning that the binding is very reversible. In contrast, the concentration of PIP2 in the whole cell is about 100-fold lower, which means that cytoplasmic proteins have to bind much more tightly to PIP2 to be able to compete with membrane proteins for PIP2 binding. Thus, the high effective concentrations of inositol lipids at membrane surfaces means that cells can rely on relatively low-affinity interactions and get rapid signal propagation through binding reactions.

Another advantage of the binding to membrane surfaces is that the diffusion to a site in two dimensions is more efficient than going in three dimensions. Thus, the ability of membranes to restrict the dimensionality of movement has a great benefit for processes where a newly synthesized component must diffuse to a site for a function. There is a penalty in that membrane diffusion is slower than cytoplasmic diffusion, but that is compensated by restricting diffusion to two dimensions.

6.13   Lipid Asymmetry and Charge Effects

For many cell functions, the physical aspects of lipid asymmetry are important and the inositol lipids contribute a large component of the high negative charge density at the cytoplasmic membrane surface. A variety of proteins use charge interactions as well as fatty acid tails to concentrate at the cytoplasmic surface of the membrane. Rapid hydrolysis of the inositol lipid phosphates and the increase in protein negative charge by phosphorylation can reverse the positive protein binding to the negative lipids. This provides another way of controlling the amount of bound protein at the membrane surface.

There are other physical aspects of the charge and lipid asymmetries that affect cell shape and ion transport. With the normal lipid composition of a plasma membrane, the outer surface lipids are neutral except for the glycolipids that have charged sugars at the outer ends of the carbohydrate chains. Here, the carbohydrate charges are separated from the membrane surface by the length of the carbohydrate molecules. In contrast, at the cytoplasmic membrane surface, where over 90% of the phosphatidyl serine and inositol lipids reside, those lipids constitute 12–20% of the total phospholipids and 24–40% of the cytoplasmic half of the bilayer. Because those lipids are negatively charged, there is a very high density of negative charges on the cytoplasmic surface (0.5–0.8 charge/nm^2). There are many cationic proteins and/or cationic domains of proteins that are localized to the cytoplasmic surface of the plasma membrane and these may neutralize a significant fraction of the lipid charges. However, the negative charges create a surface potential that will attract positive ions and repel negative ions (see text box about charge pairs in solution to better understand how ions partition in water near charged surfaces). Indeed, the charged lipid in the cytoplasmic face of the plasma membrane can affect the concentration of ions in the region as well as the pH. If we assume that the concentration of anionic lipids is 33% of the total lipids on that surface, then the surface potential in 0.1 M NaCl will be about –50 mV.

According to the Boltzmann distribution, the concentration of a cation at the surface (c_s) will be greater than the concentration in cytoplasm (c_c) because of a lower energy due to interaction with the surface charge

$$c_s = c_c \, e^{+ze\,0.05/kT} = c_c \, e^{+8\times10^{-21}/\left(1.38\times10^{-23}\times300°\right)} = c_c\,e^{1.9} \qquad (6.1)$$

This applies for all singly charged ions and corresponds to roughly a sevenfold increase in the concentration. For a divalent cation, the concentration would be about 50-fold greater ($e^{3.8}$). These are theoretical values and the large size of hydrated ions in solution will change the observed values dramatically.

The Debye length is the distance from the surface at which the surface potential drops to 1/e of the original potential. Debye length increases with increasing temperature and with decreasing ion concentration. In 0.1 M salt the Debye length is about 1 nm and the potential drops by 1/e in about 1 nm. Thus, there will be a drop in the potential from –50 mV at the surface to –0.3 mV at 5 nm from the surface. Regional changes in the distribution of anionic lipids or the binding of proteins to the surface through cationic domains would result in major inhomogeneities in the surface potential (Peitzsch et al., 1995; Murray et al., 1999).

However, we should remember that this is a general property of ions in water and applies to proteins and carbohydrates as well as membranes. Because the charge decays rapidly in physiological solutions with potentials dropping by over twofold in one nanometer, charge effects occur at a single protein level and can affect conformation changes. Still, the charge density at the cytoplasmic surface can cause a decrease in the apparent pH by a full unit at the membrane and will cause a potential gradient across the membrane that can orient charged dipoles. This all can be modulated by inositol lipid metabolism.

6.14 Inositol Lipids and Changes in Level of Charge

The most highly charged lipids are the most dynamic in cells. Phosphatidyl inositol mono-, di-, and triphosphates rapidly turn over in cells and have been linked to many signaling pathways. Hormonal signaling (calcium release), cell chemotaxis, actin assembly, and secretion are all linked to the hydrolysis or synthesis of phosphorylated inositols. The basic metabolic events are noted below.

PtIns + ATP ⟶ PtIns,4P + ATP ⟶ PtIns,4,5P2 + ATP ⟶ PtIns,3,4,5P3

PI-4 kinase   PI-5 kinase   PI-3 kinase

Major roles have been proposed for PI-4,5P2 in many cell functions. It is the most abundant phosphorylated inositol and; when hydrolyzed by phospholipase C, it is converted to IP3 (1,4,5 triphosphoinositol) and diacylglycerol. IP3 causes calcium release from cytoplasmic (ER) stores and diacylglycerol activates protein kinase C (PKC) to phosphorylate serine groups on membrane-associated proteins.

Recent studies have linked the production and perhaps the dynamics of PtIns-3,4P2 and PtIns-3,4,5P3 with chemotaxis and cell asymmetry generation. PI-3 kinase is a major enzyme in chemotactic pathways, but how kinase activity is linked to the asymmetric assembly of actin and migration is not clear. Furthermore, PI-3 kinases are linked to many cancers as are the phosphatases that degrade them. Thus, the phosphorylated inositols appear to be involved in many signaling pathways, and that could be aided by the fact that their local synthesis or degradation would result in dramatic changes in their local concentrations.

6.15   Flippases and the Loss of Asymmetry

To produce and maintain lipid asymmetry requires considerable energy and the loss of energy results in dissipation of the asymmetry. For the lipid asymmetry to be established, there must be selective transport of the choline lipids (phosphatidyl choline and sphingomyelin) from the site of synthesis on the cytoplasmic surface of the ER to the lumen of the ER by flippases that carry them across the membrane. Similarly, in the plasma membrane there are flippases that move phosphatidyl serine from the outer half of the plasma membrane to the inner half. When lysophosphatidyl serine (lyso lipids have only one fatty acid and thus can be carried by serum albumin to cells) is added to the outside surface of cells, it will be transported to the cytoplasmic surface, which will expand that surface rapidly and cause invagination immediately. Over longer timescales, these enzymes are sensitive to the differential in the internal versus external surface pressure that gives rise to shape changes in erythrocytes through the bilayer couple and will work to establish a balance of the surface pressures. In addition to flippases, there are proteins that will bind lipids and carry them from the plasma membrane to other membrane surfaces, or vice versa. This can also restore balance in the bilayer couple. Although we do not fully understand the phenomena that rely upon lipid asymmetry, it is a major factor in controlling membrane protein functions. For example, PIP2 is needed for the function of many ion channels and other membrane proteins. The loss of lipid asymmetry is the hallmark of cell apoptosis and proteins that bind phosphatidyl serine on the outside surface of cells are markers of that process and also activators of other signaling pathways that may aid removal of dying cells. It is logical to speculate that the membranes of cytoplasmic organelles are also asymmetric.

Charge-pairs in Solution

The physical properties of charged ions in a water environment have important effects on the behavior of charged nucleic acids, proteins, and lipids. Take the case of a charge-pair such as a sodium cation and a chloride anion dissolved in water. These ions will diffuse independently, yet will be drawn together by a charge–charge attraction. A balance will be struck between the tendency to diffuse away from each other and the attractive electrostatic force. The mathematical description of this balance is provided by the Debye–Huckel model. We can assume that the distribution of counterions around a point charge will follow the Boltzmann relation

$$n_i = n_0\, e^{-U/kT} \tag{6.2}$$

where n_i is the number density (concentration) at the ith point, which is different in energy from the 0th point by U, T is the temperature, and k is the Boltzmann constant (k = 1.38×10^{-23} $J^\circ K^{-1}$). U is only electrostatic energy that is dependent upon the potential (Ψ_i) at the ith point

$$U = ze_0\Psi_i \tag{6.3}$$

where z is the charge on the counterion and e_o is a unit charge ($e_o = 1.6 \times 10^{-19}$ C). The potential will decrease rapidly with distance from the first charge (independent of whether it is a salt or a protein or a lipid surface), particularly in a concentrated salt solution.

To understand the relationship between the salt concentration and the change in potential with distance, a useful parameter is the Debye length, which corresponds to the effective distance between the two counterions. Debye length is calculated from the equations above using a number of approximations and derivations for Ψ_i.

$$\kappa^{-1} = L_D = \left(\varepsilon_o\varepsilon kT/2e_0^2 N_A I\right)^{1/2} \tag{6.4}$$

where ε_o is the dielectric constant, ε_o is the permittivity of free space ($\varepsilon_o = 8.854 \times 10^{-12}$ C^2 $N^{-1}m^{-2}$), N_A is Avagadro's number ($N_A = 6.023 \times 10^{23}$ molecules/mole), and I is the ionic strength (I = $1/2\Sigma c_i z_i^2$, where c_i is the concentration of the ith ion and z_i is the charge).

Debye length for	0.1 M NaCl	L_D = 0.96 nm
	0.01 M NaCl	L_D = 3.04 nm
	0.01 M MgCl$_2$	L_D = 1.75 nm

6.16 SUMMARY   The hydrophobic lipid core of biological membranes presents a reliable barrier to the movement of hydrophilic molecules into or out of cells. Furthermore, the rapid resealing of plasma membranes through the rezipping of hydrophobic edges enables cells to survive transient lytic events. The lipids of biological membranes are important cofactors in many biological functions and the bulk of internal membranes are involved in compartmentalizing the cytoplasm and making cellular functions more robust. By concentrating components on membrane surfaces, nearly a 100-fold increase in concentration is achieved. Many enzymes move from cytoplasm to membranes as a critical step in their activation. Lipid asymmetry (anionic lipids are cytoplasmic) is aligned with protein asymmetry and signaling systems use the phosphorylated inositol lipids for coordinating growth and motility processes, whereas phosphatidyl serine flipping is a hallmark of apoptosis. Mechanical properties of the lipid bilayers are also critical in that the bilayers are not stretched and tension transients can signal many motile processes. The bilayer couple and other mechanical aspects of bilayer bending contribute to the processes of membrane trafficking as well as cell morphology. In addition to compartmentalizing the cell, the membrane barriers are critical for many processes including ionic and charge movements that underlie neuronal processes.

6.17 PROBLEMS

1. If the bilayer has an elastic modulus such that it expands by 4% when there is an expansive tension of 10 mN/m and we assume that half of the bilayer has half of that elastic modulus, then the addition of an amphiphilic compound that expands one half of the bilayer by 0.5% will cause an expansive tension in the other half of the bilayer of how much, assuming that the bilayer does not change curvature?

Answer: For an increase in area by 0.5%, the membrane bilayer will be under a tension of 1.25 mN/m. One half of the bilayer should then have a tension of 0.6125 mN/m.

2. The rate of an enzymatic reaction is dependent upon the concentration of the enzyme. The rate of lipid kinases and hydrolases will depend upon whether they are in contact with the membrane surface. Can you think of a way to use lipid kinases or hydrolases to sense a force on the membrane (an object pushing on the membrane)? Remember that the cytoskeleton is close to the membrane but often only indirectly in contact with it.

Answer: If the enzyme is kept from the membrane by a compressible protein, then a vertical inward force would activate the enzyme by forcing the membrane onto the enzyme through the compression of the protein. Alternatively, a membrane channel that is linked to the cytoskeleton would be stretched by a vertical outward force (as in osmotic swelling).

3. A membrane protein has a diffusion coefficient of 10^{-10} cm^2/s and yet it travels 1.5 μm in 25 s toward the leading edge. What is the probability that the protein would move in this way from a simple consideration of one-dimensional diffusion (remember the Gaussian (Normal) distribution curve)?

Answer: For one-dimensional diffusion, the displacement of a membrane protein with a diffusion coefficient of 10^{-10} cm^2/s after 25s would be $X^2 = 2Dt = 50 \times 10^{-10}$ cm^2 or $X = 7.1 \times 10^{-5}$ cm, which is 0.71 μm. If the protein actually traveled 1.5 μm or a little over 2 standard deviations, then assuming a Gaussian distribution about 95.5% of the diffusing proteins would be within 1.5 μm of the initial sight. If we assume equal probabilities of moving toward and away from the edge, then the probability of moving 2 standard deviations toward the edge would be less than 2.25%.

4. A membrane channel has a large cytoplasmic domain that covers a circular area 4 nm in diameter (the pore is in the center) at the membrane surface and neutralizes the negative lipid charges beneath it. If the potential at the uncovered membrane surface is –50 mV, and the solution contains 0.1 M of monovalent salt, what is the pH difference at the mouth of the pore vs the edge of the protein at the bilayer?

Answer: In the chapter, the concentration for a monovalent cation at a –50 mV surface was sevenfold greater than away from the membrane. Because the pore in the center of the protein is two Debye lengths from the lipid surface, the potential will be decreased by $e^2 = (2.718)^2 = 7.4$-fold to a value of about –7 mV, which will give a value of about 30% higher proton concentration or a drop in pH by 0.11 pH units.

5. We want to understand the importance of the cytoplasmic surface potential to the transmembrane potential. Transmembrane potentials are typically –100 mV (negative inside the cell). If the cytoplasmic surface has a charge of –50 mV, what is potential gradient across the 5 nm of the bilayer? Dielectric breakdown of biological membrane occurs at about 1 V across the membrane. What is the fraction of the potential gradient across the membrane at dielectric breakdown that is contributed by the cytoplasmic surface potential given above?

Answer: The potential gradient across the bilayer will be the total potential across the bilayer of –150 mV divided by 5 nm or 300,000 volts/cm. At dielectric breakdown of 1 V, the –50 mV from the surface potential is 5% of the potential gradient.

6. Membrane elasticity is important for understanding how cells change shape. Cells like lymphocytes can swell in hypotonic solutions from an apparent diameter of 6 μm to an apparent diameter of 7.5 μm without the addition of any internal membranes to the plasma membrane. What is the apparent increase in the plasma membrane area? Is this due to stretching of the plasma membrane or to smoothing of folds in the membrane, such as around filopodia or lamellipodia?

Answer: 113 μm^2 to 177 μm^2 or a 64 μm^2 area increase, which corresponds to a 57% increase from the original apparent area of the sphere. This is much greater than the elastic limit of membrane stretch (4%) and thus due to the smoothing of folds in the membrane around filopodia or lamellipodia.

7. Mitochondria constitute the main energy production compartment inside animal cells. The energy is generated by the F_0/F_1 ATPase, which is powered by a proton-motive force across the mitochondria inner membrane. Consider a rat liver cell with high respiration rate, the pH in the mitochondria matrix is about 8.25, whereas the pH in cytosol is about 7.37. Mitochondrial membrane potential is about 118 mV. Can you calculate the net proton-motive force across the mitochondria inner membrane?

    (http://bionumbers.hms.harvard.edu//bionumber.aspx?id=101103&ver=5; Mollica et al., 1998, p. 216, table 2.)

Answer: The concentration of protons is $10^{0.88} = 7.55$-fold higher in the cytoplasm.

8. There is a dramatic increase in the concentration of enzymes at the membrane surface when they are modified with fatty acids that cause them to go from cytoplasm to the membrane. Assuming that the cytoplasm of our normal cell has a volume of 4000 μm^3 and that membrane surface area is 2000 μm^2, what will be the final concentration of an enzyme at the membrane surface if all of the molecules have two palmitate chains added that cause them to be within 5 nm of the membrane surface? The cytoplasmic concentration is initially 20 nM.

Answer: There are two ways to answer this. One is to calculate the relative volume of the membrane versus the cytoplasmic compartment and the second is to calculate the number of molecules in the cytoplasm and divide that by the volume of the membrane compartment. In the first case, the volume of the membrane compartment is $2000 \times 0.005 = 10$ μm^3. Thus, going from 4000 to 10 represents a 400-fold increase in concentration or going from 20 nM to 8000 nM or 8 μM. In the second, there are 2×10^{-8} moles/l $\times 4 \times 10^{-12}$ l $\times 6.023 \times 10^{23}$ molecules/mole = 48,000 molecules. With 48,000 molecules in 10 μm^3, the concentration is (48,000 molecules)/(6.023 $\times 10^{23}$ molecules/mole) = 8×10^{-20} moles in a volume of 1×10^{-14} l or 8 μM.

9. When a tubulosicular ER strand of 10 μm in length and 0.2 μm in diameter is swollen by hypotonic media, the tube of membrane goes to several spherical beads on a string. If we assume that the volume increase is from 0.314 μm^3 to 0.4 μm^3, then why can't the membrane just stretch to accommodate the larger volume (the overall length is 10 μm, i.e. 6.28 μm^2 of membrane and a volume of 0.314 μm^3)?

Answer: The maximal stretch of the membrane before lysis is 4% and that will produce an increase of about 8% in the volume, which would not accommodate the over 25% increase in volume.

6.18   REFERENCES

Bayburt, T.H. & Sligar, S.G. 2010. Membrane protein assembly into Nanodiscs. *FEBS Lett* 584: 1721–1727.

Ivanova, P.T., Milne, S.B., Myers, D.S. & Brown, H.A. 2009. Lipidomics: a mass spectrometry based systems level analysis of cellular lipids. *Curr Opin Chem Biol* 13: 526–531.

Mollica, M.P., Iossa, S., Liverini, G. & Soboll, S. (1998). Steady state changes in mitochondrial electrical potential and proton gradient in perfused liver from rats fed a high fat diet. *Mol Cell Biochem* 178(1–2): 213–217.

Murray, D., Arbuzova, A., Hangyas-Mihalyne, G., et al. 1999. Electrostatic properties of membranes containing acidic lipids and adsorbed basic peptides: theory and experiment. *Biophys J* 77: 3176–3188.

Peitzsch, R.M., Eisenberg, M., Sharp, K.A. & McLaughlin, S. 1995. Calculations of the electrostatic potential adjacent to model phospholipid bilayers. *Biophys J* 68: 729–738.

Sheetz, M.P. & Singer, S.J. 1974. Biological membranes as bilayer couples. A molecular mechanism of drug–erythrocyte interactions. *Proc Natl Acad Sci USA* 71: 4457–4461.

White, S.H. 2007. Membrane protein insertion: the biology-physics nexus. *J Gen Physiol* 129: 363–369.

6.19   FURTHER READING

Membrane: www.mechanobio.info/topics/cellular-organization/membrane/

7 Membrane Trafficking: Flow and Barriers Create Asymmetries

Membrane Dynamics

Plasma Membrane Turnover: About 2% of the plasma membrane per minute is endocytosed.

Endocytic Recycling: About 95% of endocytosed membrane is recycled.

Early Endosomes: Near membrane, pH is dropping to ~6.

Late Endosomes; Movement from early to late is microtubule-dependent, lower pH of ~5.

Lysosomes: low pH 4.5 and hydrolases.

Endoplasmic Reticulum: Synthesis of lipids, membrane proteins, and secreted proteins. Flow of synthesized proteins to exit sites is driven by association.

Golgi: Glycosylation of ER proteins and lipids as well as some reglycosylation of plasma membrane proteins, transit time about 10 min.

Exocytic Vesicles: Can be rapid and microtubule-dependent or local through diffusion.

There are two distinct systems involved in the establishment of membrane-compartmentalized functions in cells. Synthesis and secretion are involved in producing new materials, whereas endocytosis and recycling are involved in metabolism and turnover. Both are involved in the sorting of components for maintaining the compartments. In the synthetic pathway, it is critical for cells to move membranes from the site of protein and lipid synthesis to the correct membrane compartments. In the endocytic pathway, endocytosis of the plasma membrane brings in components from the environment that are sorted and processed for cell growth and activity. These functions rely upon two major mechanisms to move membranes from one compartment to another, either vesicle fission/fusion or fusion of membranes and flow followed by fission. The material to be transported is selected by coat proteins or other filters. The endocytic and synthetic pathways involve bidirectional flows of material that sort components for recycling or forward transport. In the endocytic pathway, most endocytosed components are rapidly recycled, while some are processed before recycling, and an even smaller fraction are degraded. In the synthetic pathway, the ER synthesizes the whole volume of the cellular membranes in a day for a typical growth rate (higher rates are found for specialized secretory cells). Newly synthesized membrane proteins are typically glycosylated in the ER and moved to the Golgi for further glycosylation before transport to either secretory vesicles, lysosomes, or peroxisomes. Flow through these pathways can be likened to liquid flow between a series of pools in that the inhibition of the inflow will cause shrinkage and of the outflow will cause expansion of the pool. Membrane bending stiffness, bilayer asymmetry and tension all play significant roles in membrane trafficking and the proteins involved are designed to sense and respond to each one of these membrane factors. In terms of robust systems engineering, the dynamic membrane system is designed to create linked compartments for sequential processing of proteins either as part of synthesis or metabolism.

Membrane dynamics is critical for many functions, including secretion, renewal, nutrient supply, and signaling. Cells have the capacity to transport large amounts of membrane and perform sorting operations that confine many components to a compartment while selectively removing others for further processing. Normally the transport mechanisms are robust, and this enables tasks to be performed, albeit at a slower rate, even when one or another protein sorting system is lost. This robustness will often enable the *de novo* establishment of new compartments from available components, after which functions will resume. Although compartment self-assembly is robust, there are no hard rules for how components are sorted, and in most cases the sequence of sorting events will involve overlapping compartments with soft boundaries. Along with biochemical parameters, physical properties of the membrane, including bilayer curvature, thickness and surface charges, play significant roles in these sorting events. What is missing from our understanding of these systems is a complete description of the pathways involved, as well as a model system defined in one mammalian cell type with good quantitation that can serve as a paradigm for other cell types.

The synthetic system relies on cycles of vesicle fission and fusion to move materials, whereas in the endocytic system, membrane flow facilitates the movement of large volumes of membranes that can be driven by small differences in membrane tension. In terms of their primary functions, the synthetic pathway is critical in the production of the various components that the cell is programmed to secrete, as well as to replenish the cell components needed during growth. For example, in lactation the epithelial cells need to rapidly produce large amounts of milk and the process is highly efficient. In contrast, the endocytic pathways facilitate an intake of materials for nutrition, remove inactive proteins from the plasma membrane, and recycle activated receptors. The transferrin pathway is a good example where the circulating transferrin complex with iron binds to the cell surface and is taken into the cell where the acidic environment causes unloading of iron and the recycling of the transferrin.

In the synthetic pathway, secreted proteins such as hormones, matrix, etc., are secreted without lipid. This is because the vesicles carrying the components are integrated into the plasma membrane while their contents are released. Although lipid synthesis will produce enough lipid to sustain cell growth, it is insufficient to keep pace with the requirements of the secretion system, which sees a large number of vesicles created for transport purposes. For these reasons, the recycling of lipids, which occurs through endocytic pathways, is essential in maintaining a quasi-steady state. Here, vesicles are transported in the opposite direction, from the plasma membrane or intravesicular membranes, to the ER or Golgi. In considering the whole process of synthesis, it is important to consider both the forward (exocytosis) and the rearward (endocytosis) transport of vesicles.

Although the Golgi complex is centrally located in the cell, it plays a major roll in both the secretory and the endocytic pathways and the mechanisms that drive the two systems are considerably different. Observations using light microscopy show that in the secretory system, a flurry of vesicular activity occurs in the Golgi region. Directional movements of vesicles are driven by the transport of motor proteins along microtubules in the cytoskeleton (see Chapters 9–11). Surprisingly, microtubule depolymerization does not block Golgi function, as evidenced by the appearance of fully glycosylated proteins outside the cell, at a roughly normal rate, despite the absence of microtubules. In the endocytic pathway there are multiple mechanisms including phagocytosis, macropinocytosis, clathrin-dependent, and clathrin-independent endocytosis. In each case, there are detailed differences in the mechanisms, but all have the effect of decreasing the plasma membrane area and as a result must work against the expansive tension in the membrane. If the tension is very high, the rate of endocytosis decreases. A dramatic example is observed during mitosis, where the membrane tension is about 10-fold greater than in interphase cells. In addition, the tension sets the relative rates of endocytosis between the different mechanisms because they all will have different sensitivity to membrane tension. This chapter is organized around the general mechanisms of fission/fusion and sorting because they are critical for the secretory pathway. For the endocytic pathway, the flow of membrane in tubes and physical methods of sorting appear to be more important. There are numerous examples of flow in the secretory system and vesiculation or sorting in the endocytic system, but the loose generalities can help to organize the chapter.

ENDOCYTOSIS	EXOCYTOSIS
LARGE VOLUME INITIALLY	Synthesis on ER and move inside
Recycling of most membrane (> 95%)	Processing and exit to Golgi
Drop in pH, move to Golgi complex	Golgi processing and 3 exits
Movement to lysosomes (degradation)	PM, Lysosome, or Secretion

7.1   Secretory Pathway

The secretory pathway accounts for well over two-thirds of the volume of membrane in the cell. About 50% of the cell's membrane is in the ER and another 15% in the Golgi. In addition, the nuclear membrane, which is connected to the ER and the secretory vesicles, accounts for another 5%. Thus, it is important to understand how this membrane is organized and how dynamics occur in it. In fibroblasts and many epithelial cells, the ER forms a dynamic tubulovesicular network by moving on microtubules and extending throughout most of the endoplasm. Like microtubules, this network is excluded from the cortical actin meshwork. The Golgi typically forms around the microtubule-organizing center or centrosome because specific microtubule motors carry the complex there. Surprisingly, the Golgi and ER are self-forming compartments that can be mixed and will reform. For example, a drug, brefeldin A, will cause the rapid fusion and mixing of the Golgi into the ER, but upon removal of the drug, the Golgi reforms in less than 45 minutes. Similarly, the depolymerization of microtubules results in the dispersion of the large central Golgi complex into multiple, small peripheral Golgi complexes. Furthermore, nuclear membrane, ER and Golgi fragment into small vesicles during mitosis so that both daughter cells will receive about half of each compartment. The compartments will then reform in the daughter cells. Tubular membranes in the ER are formed and stabilized by the DP1/Yop1p proteins (similar to BAR proteins in that they induce membrane curvature) that bind to the ER cytoplasmic surface and curve the membrane (flat regions are defined by other ER proteins). Thus, the different compartments in the secretory system are very robust, and the system appears to be able to reform the compartments rapidly from dispersed or newly synthesized components.

In cells that are secreting large volumes of material such as in lactating epithelia, the rate of secretion is strongly dependent on microtubule-dependent transport

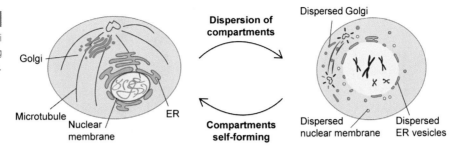

Figure 7.1

Membrane dynamics (Golgi and ER: self-forming compartments).

from the Golgi complex to the plasma membrane. However, in the case of slower secretion processes such as viral glycoprotein secretion, early studies found that the rate of viral secretion was not significantly inhibited by microtubule depolymerization when assayed after 2–3 hours. However, more recent studies with higher time resolution determined that depolymerization of microtubules caused a delay in secretion of viral glycoproteins for about 45 minutes and during that period the central Golgi complex dispersed into many peripheral Golgi complexes close to the plasma membrane. Secretion resumed from the reformed Golgi complexes because the secretory vesicles were now close to the plasma membrane surface. Thus, although microtubule transport from the Golgi to the plasma membrane is the normal secretory pathway, there is an alternative pathway for secretion in the absence of microtubules that requires a redistribution of the Golgi complex. This is an excellent example of how a critical function must have multiple mechanisms to provide backup.

Membrane Trafficking as a Series of If/Then Decisions

EXOCYTOSIS	ENDOCYTOSIS
If synthesized with signal sequence then imported into ER	If binding to clathrin adaptors then endocytosed
If processed in ER, folding, sugars, then exported to Golgi	If not moved into recycling vesicles then moved to TGN
If glycosylated in proximal Golgi then moved to distal	If to be degraded because denatured then moved to lysosome

Movement from the ER to Golgi and from Golgi to the plasma membrane can be modeled as a first-order process where the probability of a protein exiting the ER or the Golgi is constant over time (2% per minute for a viral glycoprotein moving out of the ER). This rules out a number of possible mechanisms of

Figure 7.2

Endocytosis and exocytosis.

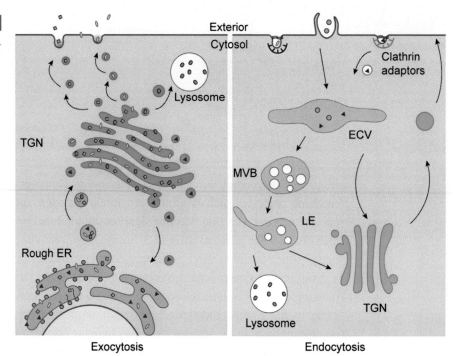

Exocytosis   Endocytosis

transport, such as a processing-driven mechanism in which the processing takes a constant time. In that case, the inhibition of further synthesis would result in all of the material exiting the ER after the period required for processing. However, the actual exit rate decreases more gradually, and is normally defined as a constant fraction of the material in the ER. In some respects the trafficking of lipids and proteins through the exocytic pathway can be likened to a decorative fountain where the water flows between pools with the added feature that some flow is recycled back to the previous pool and is not just coming from synthesis at the source in the ER. There is a finite amount of lipid in the system and it can build up in a given pool if the outlet is partially blocked. Similarly, it can be depleted if there is an increase in the outflow. Movement between cellular compartments may be blocked in a temperature-sensitive manner, and this is characterized by an expansion of the compartment before the movement is blocked and a depletion of the compartment after the block. Typically, the synthesis of new membrane lipids and proteins occurs at a much slower rate than the flow between the compartments. For example, the cell doubling time is 24 hours, but the outflow of a GFP-tagged protein from the ER occurs with a half-time of 30–40 minutes (Hirschberg et al., 1998). It is important for the cell to maintain quality control to be sure that processed proteins and lipids move to the next compartment. Unfolded proteins will accumulate in the ER and be subjected to autophagy (a mechanical process).

For most of these steps, multiple criteria for making the cellular decision to move forward will exist, and typically, the reverse traffic will take improperly processed or sorted material back for proper sorting. Thus, in the case of most secreted proteins that must compete with other newly synthesized components for exit vesicles, a constant fraction of the protein in the ER will exit per minute.

First-order Processes (Radioactive Decay, Exit from ER, Exit from Golgi)

If the change in X with a unit time is a constant fraction of X, then the loss of X will be described by an exponential decay $[X_t] = [X_0] e^{-kt}$ where X_t is the concentration of X at time t (minutes), X_0 is the concentration of X at $t = 0$ and k is the fraction of X that is lost in a minute. This is common for many reactions where the process depends upon a single rate that is independent of common factors in the system.

7.2 Fission–Fusion Mechanisms

For much of the internal trafficking to occur, there must be rapid and controlled fission and fusion of membranes. In the case of the ER and Golgi, they can form extensive tubulovesicular networks that can fuse with homologous membranes. These will move along microtubules but will not fuse with heterologous membranes even as tubulovesicular strands. The basis for allowing fusion with self but not with the other compartment is not understood; however, it is necessary to maintain the integrity of the compartments. For example, Brefeldin A inhibits Arf-dependent coat assembly on the Golgi and causes in the fusion of the Golgi with the ER, i.e. it indicates that the coats are important to keep the compartments from fusing. Because the Golgi can reform after fusing with the ER, it is clear that Golgi proteins in the ER will be sequestered and incorporated into vesicles that can then fuse with other Golgi protein-containing vesicles to reform the Golgi complex. What constitutes a critical mass of Golgi proteins to enable the assembly of the fission machinery is not clear. There is a 10–100-fold concentration of the proteins into ER exit vesicles and this depends upon a number of sequence features that cause presumed interactions between the coats, which then cause fission. Fission in general requires more energy than fusion and the steps are similar to those in clathrin-dependent endocytosis except that the invagination is not required (see Chapter 3 for description of clathrin-dependent endocytosis). For example, fission of clathrin-coated internal vesicles that bud from the Golgi, and fission of many other internal membranes, is reliant on a protein known as dynamin.

Fission of Vesicles

Important elements

1. Assembly of coat or other protein complex to curve the membrane and assembly often involves accumulation of PIP2.
2. Pinching proteins such as dynamin. These proteins assemble around the neck region of the vesicle-larger membrane and physically constrict the neck (both membrane curvature and tension forces inhibit the constriction).
3. Lipids that destabilize the bilayer. In the clathrin system, inositol lipid hydrolysis to remove the head group and produce the diacylglycerol is a critical step.
4. After fission, then disassembly should be rapid.

See the detailed treatment in Liu et al. (2009) that considers the balance of forces and how the chemical changes may influence them.

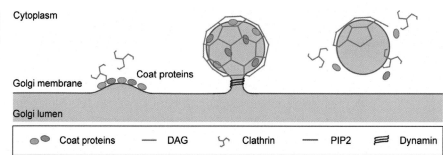

Figure 7.3

Fission of vesicles from Golgi.

On the other hand, fusion should occur in a specific manner due to the function of vesicle SNARE proteins, which form complexes with the proper target SNARE proteins on the acceptor compartment. Fusion of vesicles with cytoplasmic compartments often occurs as soon as they contact each other. However, there is a physical feedback mechanism whereby blocking the exit from one compartment causes the build up of membrane in that compartment and eventually the inhibition of further entry of new vesicles. Exocytic vesicles, however, fuse upon specific stimuli that include mechanical tension in the plasma membrane. Estimates of the amount of trafficking, both internal and external, are very high with the turnover of the ER in 20–30 minutes and the plasma membrane in 30–60 minutes. This means that the components involved stay at fission and fusion sites for only short periods of time. Thus, working out the mechanisms of fission and fusion are complicated by the rapid dynamics and the potential for alternative pathways of transport.

Fusion of Vesicles

The formation of SNAP–SNARE complexes is believed to catalyze secretion.

1. SNAP–SNARE forms tetrameric intertwined helices (specific interaction of the vesicle SNARE with the target SNARE is required for docking and fusion to the correct site).
2. There are roles for the lipids and for several other proteins in the fusion complex.
3. Fusion is stimulated by Ca^{2+} in neural synapses and by membrane tension or other factors in other systems.

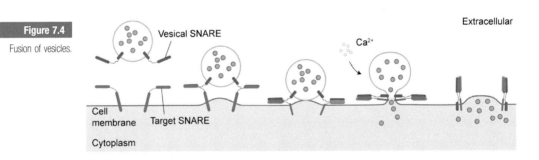

Figure 7.4

Fusion of vesicles.

7.3 Coatamers in Vesicular Exocytic Pathway

Exocytosis is a complex function involving numerous steps that start with the formation of the trafficking or exocytic vesicle. The coatamer proteins, such as clathrin or COPI, ensure this occurs by packaging the processed proteins for budding and subsequent fusion with the next compartment (Zanetti et al., 2012). A number of interesting and important principles are associated with coatamer assembly that confer this functionality. First, the shape of the assembled coat is usually spherical and the size of the assembled spheres is often, but not always, correlated with the size of the transport vesicle. For the process to work effectively, the assembly of the coat must be linked to the assembly and packing of the membrane proteins under the coat. Although not fully understood, proper carbohydrate processing as well as proper folding and sulfhydryl state of the transported protein is required for assembly and packing in the exit vesicle. For movement to the next step in exocytosis, the properly processed proteins should be selectively chosen and those improper proteins should be recycled back from the next compartment.

The whole exocytic process can be considered as a series of If/Then decisions that result in the formation of the transport vesicle completely loaded with cargo.

Figure 7.5

Coatamers.

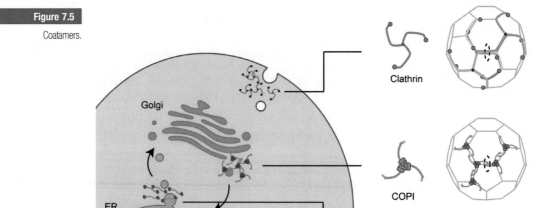

Coat proteins aggregation   Multiple weak interactions

It is important that the correct components are modified selectively at each step with high fidelity and reasonable efficiency. There are several coat protein complexes that aid in both the concentration of cargo and the formation of a transport vesicle. In addition to clathrin, the COPI and COPII coats can help in sorting, packaging or delivery steps (Szul & Sztul, 2011; Zanetti et al., 2012). In all cases, there are binding steps, and steps that regulate concentration in the coated membrane bud, further coat assembly and pinching off of the coat. Before fusion to the targeted membrane compartment can occur, the coats must be broken apart. The dynamic nature of these processes requires that cycles of phosphorylation/dephosphorylation, folding/unfolding, or other post-translational modifications must occur in order for the fission/fusion and recycling steps to proceed through to completion. Because no single sequence has been identified that encodes for ER exit or Golgi exit, the interaction must be weak and rely upon multiplicity (oligomerization) to enable sorting to occur with high fidelity.

Importantly, the specificity required in vesicular packaging and transport can come from multiple weak binding interactions between the proteins to be transported and the coat proteins, which together produce a strong interaction (even a weak interaction between the transported proteins can greatly strengthen the overall binding affinity of the proteins for the coated membrane bud). The basis for this comes from the fact that the free energy of each binding interaction is additive if a single protein has multiple binding interactions. This is advantageous to the cell because each interaction can be readily dissociated due to the low binding affinity. Furthermore, the loss of one binding interaction due to phosphorylation or other signal can cause the rapid dissociation of other interactions. If, on the other hand, the cell relied upon a single strong interaction, the rate of dissociation would naturally be quite slow, and this would slow down the overall process of membrane trafficking, and its related signaling events. Thus, the assembly of multiprotein complexes such as the coats that rely upon multiple weak interactions is critical. We explore this in greater depth in the text box below, because it is a very important concept that will be repeated often, in the context of many different cellular functions.

Cooperative Binding To Coat Complexes in Sorting

There is considerable evidence that coat complexes are critical for sorting components moving from ER to Golgi and moving between other compartments. Why is this a good mechanism for sorting and what is the basis? There is evidence for a di-lysine motif that is involved, but that seems insufficient to provide the specificity that is needed. One common principle is that there are cooperative interactions involved. In the case of coats, only a weak interaction with the coat and a weak interaction with neighboring proteins is needed for the proteins to concentrate in the coated regions.

Consider the free energy of binding

$$RT \ln K_A = -\Delta G_A \text{(Note that favorable association requires a negative free energy of binding.)}$$

This free energy describes all of the chemical and energetic factors involved in the association reaction. It is extremely useful to break this term down into two opposing energies, the bond energy (bond) and the change in entropy (s).

$$\Delta G_A = \Delta G_{bond} + \Delta G_s$$

In the case of sorting, the bond formation between the coat and the protein is weak, as is bond formation between the proteins to be transported, and the entropy term is positive, i.e. there is an entropy penalty for binding. Because the binding sites in the coat are closely spaced, the binding to those sites enables self-association (a second bond energy) without the entropy penalty for binding. Thus,

$$\Delta G_A = \Delta G_{bond1} + \Delta G_{bond2} + \Delta G_s$$

This means that a strong bond can form because of the cooperative nature of the complex, whereas individual bonds would be very unstable. When the coat disassembles after the vesicle pinches off, the first bond energy is lost and thus the aggregate of transported proteins can rapidly disperse into the new compartment upon fusion.

A number of tools are commonly used to study the trafficking of membrane proteins from the ER to the plasma membrane. These include temperature blocks, viral glycoprotein mutants, and drugs such as brefeldin A, which blocks Golgi to plasma membrane exocytosis. The most commonly used viral glycoprotein mutant is the temperature-sensitive vesicular stomatitis viral G-protein (VSVG-ts045). The movement of this glycoprotein mutant is delayed at physiological temperatures (37°C). It was originally thought that VSVG-ts045 was aggregated in the ER at the non-permissive temperature; however, more recent studies show that delays in its movement to the Golgi complex are due to the decreased binding interactions to other components at the ER exit sites, and not aggregation (Malchus & Weiss, 2010). This indicates that interactions with the coatamers or luminal binding proteins may be the key determinants in the temperature-dependent retention of VSVG-ts045. This has the broader implication that clustering, and interactions with multiple binding sites, is the important step in the concentration of proteins for export. An additional implication of the first-order transport of the VSVG-ts045 at the permissive temperature is that the coated vesicles are transporting a relatively constant fraction of the proteins in the ER per unit time (~2% per minute). Presumably, at steady state, the vesicles are being filled with a variety of different proteins that are to be transported to the Golgi and then on to the plasma membrane. Thus, diffusive transport to the exit site drives ER exit for the properly folded proteins.

7.4 Golgi Processing Involves a Series of Steps with Ordered Enzymes

The ability of the Golgi complex to reassemble after dispersion into the ER following brefeldin A treatment indicates that the important interactions involved in organizing the Golgi are readily reversible. At present, we don't know the ordering principle, but it is likely to involve some of the associated cytoplasmic proteins that are found with the planar stacks of the Golgi complex. As there are both vesicles to carry proteins from the ER to the Golgi and vesicles to recover ER proteins from the Golgi, it is easy to understand how the Golgi proteins would be separated from the ER proteins. Because the Golgi proteins would be selected against for movement on to the plasma membrane, they would concentrate in a forming Golgi. If their concentration reached a certain level, then they could bring cytoplasmic proteins to the surface and catalyze a self-assembly process. Although it is easy to imagine how the ER and Golgi complex components could separate, it is more difficult to imagine how *cis* and *trans* Golgi proteins could separate unless there was a similar vesicle transport system that would carry out *trans* Golgi proteins and recover *cis* to a *cis* compartment. The complexity required to assemble many compartments is very high and there is debate over whether the *cis*, medial and *trans* Golgi compartments are true compartments.

Proteins that are processed by the Golgi complex move through it in a progressive manner, entering from the ER at the *cis*-end and exiting in vesicles from the *trans*-end. The initial carbohydrate transfer enzymes in this process are needed to modify the proteins in the *cis*-Golgi, as well as organize the early *cis*-Golgi compartment. This compartment requires a site for fusion of the ER vesicles, as this will ensure initial contact is made between the Golgi enzymes and the proteins to be secreted. As the proteins progress through the Golgi, they will move on to the second compartment (the medial-Golgi) where they will be further processed. Final processing in the *trans*-Golgi will prepare them for transport to the extracellular space. The carbohydrate modifications are often far from the membrane surface of membrane proteins and similar modifications are made to soluble secreted proteins. Thus, it seems likely that luminal proteins would interact with the carbohydrates to help in concentrating them for transport to the next compartment.

Recent studies have developed a model to describe Golgi function that relies heavily on the retrograde trafficking of enzymes. Initially, the fusion of several COPII-coated vesicles from the ER was studied. The ER components involved were normally transported back to the ER via vesicles coated in COPI, leaving the

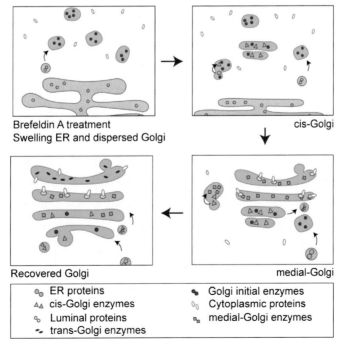

Figure 7.6

Golgi are reassembled with ordered enzymes.

Brefeldin A treatment
Swelling ER and dispersed Golgi

cis-Golgi

Recovered Golgi

medial-Golgi

⚬⚬ ER proteins		⚭ Golgi initial enzymes	
△△ cis-Golgi enzymes		₀₀ Cytoplasmic proteins	
⚬ Luminal proteins		⚭ medial-Golgi enzymes	
⚬- trans-Golgi enzymes			

Golgi enzymes that would become concentrated. If the *cis*-Golgi enzymes in this new compartment were activated by concentration, then they could modify the proteins there. Modification of the proteins in the *cis*-Golgi compartment could signal that the proteins should be moved to the next Golgi compartment, the medial, leaving behind the *cis*-Golgi enzymes and moving forward the medial-Golgi enzymes. When the concentration of medial-Golgi enzymes was high enough, they would be activated to further modify the proteins to be secreted. Such a combination of forward movement of enzymes involved in later modifications and retrograde movement of the enzymes involved in earlier modifications would ensure that the components are sequentially modified. Furthermore, the continual addition of ER vesicles would create the flow. Other schemes are possible, but the important elements are that there needs to be a mechanism for sequentially activating the enzymes for the next step so that the flow of membrane to the exit site is directional and sorts the earlier enzymes to the beginning of the flow and later enzymes to later compartments.

7.5 Endocytosis is Designed to Send Back Good and Degrade Bad Proteins

There are multiple reasons for endocytosing membrane, including the uptake of nutrients, clearance of damaged proteins and recycling of receptors. After endocytosis, however, most ($> 95\%$) of the membrane is recycled back to the plasma membrane for reuse. In the endocytic pathway, the endocytosed material, both plasma membrane and other proteins, is tested both mechanically and chemically to determine whether the endocytosed material should be returned to the plasma membrane or passed on to the next compartment. If the proteins continue to the next compartment, they may face denaturation due to the changes in vesicle conditions, including a lower pH. However, if they do not denature and aggregate, they may flow into early recycling tubulovesicles. Some native membrane proteins will remain in the low pH of the endosome and they will get transported to the *trans*-Golgi network (TGN), where they will be further stressed and some will be enzymatically repaired before being transported back to the plasma membrane. If they should aggregate or appear damaged, they will be moved on to the lysosomes, where they will be degraded. Those that move to late endosomes are transported to the TGN on microtubule tracks. Carbohydrates can be added to proteins in the TGN prior to their return to the plasma membrane. Thus, endocytosed components can move into the secretory pathway for further processing. The pathways for recycling to the plasma membrane are primarily dependent upon the biochemistry and the continued native configuration of the proteins. In contrast, denaturation and protein aggregation will activate proteins for degradation based upon fairly non-specific aspects that are primarily physical in nature. Similarly, autophagy of aggregated or improperly folded proteins relies upon the physical properties of those aggregates rather than the specific biochemical nature of the components.

7.6 Sorting Mechanisms

A number of different mechanisms potentially exist to specifically move proteins from one compartment to another, or to retain them in their current compartment. Using viral proteins that target different membrane areas, many studies have tried to define the sequence of amino acids present in the proteins that determine how they are sorted. Although some general rules were found, there was no single sequence motif or other chemical factor that proved critical for protein sorting (Dancourt & Barlowe, 2010). In the case of ER retention, however, one sequence tag (KDEL) was described as a reliable marker for retention. In this case, KDEL binds to a specific receptor that carries the tagged proteins back to the ER. Physical factors such as partitioning into curved versus flat or thick versus thin membrane regions will also contribute to sorting. However, greater specificity comes from associations with coat proteins where the proteins to be secreted are often concentrated by an order of magnitude or two into the coated regions. Although there are no general sequence rules for sorting, it is clear that sorting factors are important and they are specific for specific proteins.

7.7 Physical versus Biochemical Sorting

There are two different mechanisms of sorting. One mechanism is dependent on the physical properties of the protein in the lipid bilayer while the other is based upon protein binding interactions. In many cases, there is synergy between the physical and biochemical properties in the sorting process. Again, it is important to keep in mind that the relative priorities that determine sorting will be different in different cells and may even dynamically change under different conditions. Two physical properties have been observed as determining how proteins are sorted in the Golgi; specifically, membrane curvature and size-related barriers. In the case of curvature, the membrane proteins will partition preferentially with either flat or curved surfaces because the protein has either a curved or a flat membrane-binding region. Integrin $\beta1$, for example, prefers curved regions, as does the early membrane antigen in neurons (Sheetz et al., 1990; Schmidt et al., 1993). *In vitro* studies have established that membrane proteins and even some lipids can segregate based on bilayer curvature (Aimon et al., 2014). Size-related barriers, on the other hand, prevent proteins and even lipids from penetrating or moving further. The best example of this is the axonal hillock, which is the region in the transition from the cell body to the axon of hippocampal neurons (Kobayashi et al., 1992; Winckler et al., 1999). It has been postulated that similar barriers could exist within internal membranes that could filter components based upon size. A similar concept could potentially work for the plasma membrane as well, where lipid diffusion is less inhibited than protein diffusion; this will not be the case for all proteins, however, as observed for glycoprotein diffusion, where the rate-limiting step has only a weak size dependence (Kucik et al., 1999). Physical restraint of diffusion is a generally accepted mechanism to sort or keep membrane proteins in certain regions, but again, we don't understand many of the detailed mechanisms. The robust nature of the membrane protein diffusion measurements indicate that the barriers are always coming and going but with cell type variance.

Physical Sorting	Biochemical Sorting
Membrane curvature	Coat proteins on membranes
Size or aggregation state	Multivalent binding
Tension	Fission and fusion
Thickness of bilayer	Biochemical aggregation
Flow	Flipping
Bilayer couple	
Sieve	

Membrane curvature has an important role in many cell functions and can clearly provide a simple means for concentrating membrane proteins. Barriers and sieves for membrane proteins and lipids are seen in plasma membranes in epithelial cells and neurons. Some of the models of Golgi sorting include membrane proteins organized by coat proteins to act as sieves. Thus, physical sorting mechanisms can clearly play a role in the creation of membrane domains and the concentration of membrane proteins.

Biochemical sorting mechanisms, on the other hand, are linked to the coat protein binding, and the creation of vesicles that then fuse with the correct compartment (Dancourt & Barlowe, 2010). In some cases, there can be a combination of the physical and biochemical sorting mechanisms. Most of the

Figure 7.7

Axon hillock.

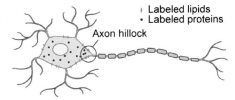

biochemical mechanisms do not involve high-affinity binding of specific sequences to transport receptors. Instead, there are weak interactions with the coat proteins that enable a wide spectrum of proteins to be transported based partially upon steric factors as well as biochemical interactions. Quantification of the dynamics is not sufficiently accurate to really understand the exact mechanism of selection.

7.8 SUMMARY Membrane trafficking within the cell is involved in the synthesis of new proteins and lipids as well as endocytosis to process or degrade many of the membrane-bound proteins. Similarly, internal membrane trafficking is involved with both synthesis and autophagy or clearance of damaged proteins. Although the movement of membrane by flow in lipid tubes is faster than vesicle fission and fusion, the control and specificity afforded by the latter processes make them preferred for many functions. Endocytic and recycling pathways appear to rely more on flow than the synthetic pathways. Cooperativity in the concentration of components by vesicle coats during synthesis and processing makes the vesicular process more efficient and more selective. The system is robust in its ability to resort compartments after mixing and the bidirectional nature of the traffic enables correction of errors as well as the ability to maintain reasonable separation of components between compartments (e.g. ER and Golgi). Both lipid and protein modifications are used to make the fission and fusion processes directional and there are mechanisms for the reverse traffic as well as the forward traffic, but with different specificities. Over 10 specific fission and fusion processes are needed to explain the observed trafficking. Again, specificity and functional issues win out over efficiency. However, the membrane dynamics are critical for quality control and maintenance at the intracellular level both during times of growth and times of stress.

7.9 PROBLEMS

1. We will use a 14°C temperature block to hold a GFP-tagged membrane glycoprotein in the ER until a significant amount is synthesized. When the temperature is raised to 37°C, the protein will be released to transit to the Golgi and on to the plasma membrane. If further synthesis of the protein is blocked and we use the constants defined in the Hirschberg et al. (1998) study, then about how long will it

take before 10% of the protein reaches the plasma membrane? From Hirschberg et al. (1998), for ER to Golgi transport the mean rate constant (i.e. the fraction of VSVG-GFP moved per unit of time) was 2.8%/min, for Golgi to plasma membrane transport it was 3.0%/min, and for transport from the plasma membrane to a degradative site it was 0.25%/min.

Answer: You can do this by solving the integral for sequential first-order reactions A to B to C with the k_1 as the rate constant of A to B and k_2 of B to C. The number of C $= A_0 - A - B = A_0[1 + 1/(k_1 - k_2)(k_2 e^{-k1t} - k_1 e^{-k2t})]$. If $k_1 = 0.028$ and $k_2 = 0.03$, then $C/A0 = 0.1 = 1 - 1/0.002(0.03\ e^{-0.028t} - 0.028\ e^{-0.03t}) = 1 - 15e^{-0.028t} + 14e^{-0.03t}$ or $15e^{-0.028t} - 14e^{-0.03t} = 0.9$. As there is not an exact solution, we tried several values and found that 19 minutes was the best fit. If we calculate the approximate answer then after 2 minutes there will be 5.6% in the Golgi and 3% of that (0.17% of the total) will transfer to the plasma membrane in the next minute. In the fourth minute, $0.94 \times 0.028 = 0.026$, thus $8.2 \times 0.03 = 0.25\%$ will transfer. In the fifth, $10.8 \times 0.03 = 0.32\%$ will transfer. In the sixth, $13.3 \times 0.03 = 0.40\%$. In the seventh, $15.7 \times 0.03 = 0.47\%$. In the eighth, $18 \times 0.03 = 0.54\%$. In the ninth, $20.3 \times 0.03 = 0.61\%$. In the tenth, $22.5 \times 0.03 = 0.68\%$. In the eleventh, $24.7 \times 0.03 = 0.74\%$ (total of 4.01 %). In the twelfth, $26.8 \times 0.03 = 0.80\%$. In the thirteenth, $28.9 \times 0.03 = 0.87\%$. In the fourteenth, $30.9 \times 0.03 = 0.93\%$ (total of 6.61%). In the fifteenth, $32.8 \times 0.03 = 0.98\%$. In the sixteenth, $34.7 \times 0.03 = 1.04\%$. In the seventeenth, $36.5 \times 0.03 = 1.09\%$ (total of 9.72). Thus, the percent in the plasma membrane will exceed 10% in the eighteenth minute.

2. If endocytosis is randomly sampling the surface and 4% of the surface is endocytosed every minute, then how long will it take to endocytose 80% of a membrane protein, assuming that it is no longer synthesized? If the protein is at steady state and the protein on the surface is damaged, then how long will it take before 50% of the damaged protein is replaced by newly synthesized protein?

Answer: Use the first-order reaction $[MP] = [MP_0]e^{-kt}$ where k is the rate constant of 0.04/s. Then solve for the concentration of the membrane protein [MP] relative to the original concentration, $[MP_0]$, is 0.2, i.e. $[MP]/[MP_0] = 0.2 = e^{-0.04t}$. Solving for t gives 40 s.

3. Many receptors are recycled after endocytosis, but a fraction often moves on to the lysosome where it is degraded. If the endocytosis rate is the same as in problem 2 but 75% of the endocytosed protein is recycled, then what is the half-time for the degradation of the protein, assuming again that there is no further synthesis of the protein?

Answer: The effective k for the degradation is 0.01 and the equation becomes $[MP]/[MP_0] = 0.5 = e^{-0.01t}$. Solving for t, we get 69 s.

4. Plasma membrane glycoproteins are processed in the ER and passed on to the Golgi before being moved to the plasma membrane. We assume that there are

100,000 viral glycoproteins retained in the ER because they have a temperature-sensitive defect in processing, and then the temperature is changed to allow them to move on to the Golgi and to the plasma membrane. If 5% of the glycoproteins in the ER move to the Golgi per minute and 10% of those in the Golgi move to the plasma membrane per minute, then how many glycoproteins are going to be in the plasma membrane after 15 minutes?

Answer: Using the equation from problem 1, we can calculate $C = 100,000[1 + 1/(k_1 - k_2)(k_2 e^{-k_1 t} - k_1 e^{-k_2 t})]$ where $k_1 = 0.05$ and $k_2 = 0.1$ and $t = 15$. $C = 100,000[1 - 20(0.1 \times 0.47 - 0.05 \times 0.223)] = 38,000$ in the plasma membrane after 15 minutes.

5. In the case of viral membrane proteins, they need to be processed through the ER and Golgi before reaching the plasma membrane. If the membrane protein is synthesized at a rate of 500 molecules per minute in the ER and exits the ER at 4% per minute, then how long will it take for 10,000 molecules to reach the Golgi?

Answer: Basically 4% of the material that has accumulated will be transferred in the next minute. Again using an approximation approach, it will take about 10 minutes for 1000 molecules to reach the Golgi.

6. Studies of the concentrations of a secreted protein in coated vesicles that move from the ER to Golgi indicate that they are 400-fold greater than in the ER. Assuming that this is due to equilibrium binding, what is the energy difference for the secreted protein in the bud of the coated region of the ER versus the rest of the ER?

Answer: Using $\Delta G = RT \ln K$ where K is 0.0025, we get a value of $-6RT$.

7. Rab proteins function in guiding of selective membrane targeting. Please hypothesize a scheme whereby a Rab protein could target a vesicle from the ER for transport to the Golgi. Remember that the Rab proteins can be in either a GTP or GDP form.

Answer: The postulate is that the Rab protein is processed to the GTP form on the ER membrane and then it associates with elements of the coat complex that forms the ER transport vesicle. The vesicle will then diffuse or be transported on microtubules to the Golgi. At the Golgi surface, the Rab-GTP will bind to a receptor protein complex and cause the fusion of the vesicle with the Golgi. In the process of fusion, the Rab-GTP will encounter a GAP (GTPase-activating protein) that will convert it to Rab-GDP. The Rab-GDP will then transit back to the ER where it will go back to Rab-GTP.

8. Because about 2% of the plasma membrane protein is endocytosed every minute and the cell divides once every day, there must be recycling of the membrane. What is the fraction of membrane protein that is recycled if the average lifetime of a membrane protein is 8 hours?

Answer: In 24 hours, 28.8 plasma membrane equivalents will be endocytosed. During the same period, three plasma membrane equivalents will be degraded and one will be synthesized as a result of growth of the cell. Thus, on average $1 - 4/28.8 = 86\%$ will be recycled.

9. If the binding of an ER protein to the coat protein, COP2, at ER exit sites has an apparent affinity constant of $K_A = 10^4$ M^{-1} and forms a bond with other proteins at the exit site that has an enthalpy (ΔH) of -3 kcal/mol and the change in entropy (ΔS) upon binding can be ignored, what is the apparent affinity constant of the ER protein for the exit site that has COP2 and the other exit proteins? (Assume that R has the value 2 cal/deg.mol, and T is the absolute temperature = 300 degrees for 27°C. Also, remember that $\Delta G = \Delta H - T\Delta S = -RT \ln K_A$.)

Answer: We know that the free energies of binding interactions are additive. Thus, the free energies of the two interactions should be added. Protein–COP2 binding has a free energy of $\Delta G = -RT \ln K_A = -600$ cal/mol ln 10^4 $M^{-1} = -5.5$ kcal/mol. If we add the free energy of the second interaction $\Delta G = \Delta H - T\Delta S = -3$ kcal/mol, then the total binding has a $\Delta G = -8.5$ kcal/mol $= -RT \ln K_A$. Solving for $K_A = 1.42 \times 10^6$ M^{-1} which is a significant increase in affinity.

10. We want to know the relative concentration of KDEL proteins in the ER versus the Golgi. Assume that the proximal Golgi has an area of 1000 μm^2 and that it accepts vesicles from the ER that have an area of 2% of the ER every minute (about 400 μm^2). KDEL containing proteins leak out of the ER to the Golgi at a rate of 1000 per minute. If the KDEL receptor binds KDEL with an affinity of 10^6 M^{-1} and there are about 1100 KDEL receptors moving back to the ER per minute, then what would be the apparent concentration of the KDEL proteins in the Golgi at steady state? (Assume that (1) there is 300 μm^2 of membrane moving on from the Golgi and 100 μm^2 moving back to the ER per minute with KDEL receptors that are only in the recycling membranes, and (2) the volume occupied by the KDEL proteins and the receptors is 20 nm from the Golgi membrane surface.)

Answer: For the reaction A + B going to AB we have 100 free receptors (A) and 1000 liganded receptors (AB). This gives the equation that the concentration of B $= 10^{-6}$ M (AB/A) $= 10^{-5}$ M.

7.10   REFERENCES

Aimon, S., Callan-Jones, A., Berthaud, A., et al. 2014. Membrane shape modulates transmembrane protein distribution. *Devel Cell* 28: 212–218.

Dancourt, J. & Barlowe, C. 2010. Protein sorting receptors in the early secretory pathway. *Annu Rev Biochem* 79: 777–802.

Hirschberg, K., Miller, C.M., Ellenberg, J. et al. 1998. Kinetic analysis of secretory protein traffic and characterization of Golgi to plasma membrane transport intermediates in living cells. *J Cell Biol* 143: 1485–1503.

Kobayashi, T., Storrie, B., Simons, K. & Dotti, C.G. 1992. A functional barrier to movement of lipids in polarized neurons. *Nature* 359: 647–650.

Kucik, D.F., Elson, E.L. & Sheetz, M.P. 1999. Weak dependence of mobility of membrane protein aggregates on aggregate size supports a viscous model of retardation of diffusion. *Biophys J* 76: 314–322.

Liu, J., Sun, Y., Drubin, D.G. & Oster, G.F. 2009. The mechanochemistry of endocytosis. *PLoS Biol* 7(9): e1000204.

Malchus, N. & Weiss, M. 2010. Anomalous diffusion reports on the interaction of misfolded proteins with the quality control machinery in the endoplasmic reticulum. *Biophys J* 99: 1321–1328.

Schmidt, C.E., Horwitz, A.F., Lauffenburger, D.A. & Sheetz, M.P. 1993. Integrin–cytoskeletal interactions in migrating fibroblasts are dynamic, asymmetric, and regulated. *J Cell Biol* 123: 977–991.

Sheetz, M.P., Baumrind, N.L., Wayne, D.B. & Pearlman, A.L. 1990. Concentration of membrane antigens by forward transport and trapping in neuronal growth cones. *Cell* 61: 231–241.

Szul, T. & Sztul, E. 2011. COPII and COPI traffic at the ER–Golgi interface. *Physiology (Bethesda)* 26: 348–364.

Winckler, B., Forscher, P. & Mellman, I. 1999. A diffusion barrier maintains distribution of membrane proteins in polarized neurons. *Nature* 397: 698–701.

Zanetti, G., Pahuja, K.B., Studer, S., Shim, S. & Schekman, R. 2012. COPII and the regulation of protein sorting in mammals. *Nature Cell Biol* 14: 20–28.

7.11 FURTHER READING

Caveolar endocytosis: www.mechanobio.info/topics/cellular-organization/membrane/membrane-trafficking/caveolar-endocytosis/

Clathrin-independent endocytosis: www.mechanobio.info/topics/cellular-organization/membrane/membrane-trafficking/small-scale-clathrin-independent-endocytic-pathways/

Clathrin-mediated endocytosis: www.mechanobio.info/topics/cellular-organization/membrane/membrane-trafficking/clathrin-mediated-endocytosis/

CLIC/GEEC endocytosis pathway: www.mechanobio.info/topics/cellular-organization/membrane/membrane-trafficking/clicgeec-endocytosis-pathway/

Exocytosis and vesicle secretion: www.mechanobio.info/topics/cellular-organization/membrane/membrane-trafficking/exocytosis/

Membrane trafficking: www.mechanobio.info/topics/cellular-organization/membrane/membrane-trafficking/

8 Signaling and Cell Volume Control through Ion Transport and Volume Regulators

Many cellular functions depend heavily upon the ionic environment within cells and upon the gradients of ions across membranes. In terms of robust devices, the transmembrane potential and ion movements provide a simple means of coordinately regulating multiple functions. Not only are the levels of calcium, magnesium and protons critical, but trace ions like zinc and iron are also needed for many specialized reactions. Gradients of protons and ions across membranes drive neuronal and muscular signaling, ATP production, and even cell death signals. Because ions cannot freely pass through membranes, there are a large number of proteins that have been created with specialized functions to move ions under the right conditions. In general, the cytoplasm of mammalian cells is high in potassium and low in sodium with several millimolar of magnesium, submicromolar of free calcium, and a pH of about 7.0. Outside of the cell, the fluid is high in sodium and low in potassium with several millimolar of both magnesium and calcium (pH 7.4). Total solute concentrations (designated as osmolality or the sum of the concentrations of negative ions, positive ions, proteins and small molecules) are matched inside and out in resting cells. During most cell functions, proteins act as pores, ion-selective channels, or exchangers to control ion movements. Due to the small volume in cells, only a small number of ions need to cross the membrane to produce major changes in concentration, pH, and particularly transmembrane voltage. These parameters control chemical- and electrical-based signal transduction and, therefore, rapid ion movements across membranes must be carefully orchestrated in a robust system. Neuronal signaling and muscle contractility, for example, are controlled by rapid and reliable electrical signaling, which occurs through cell depolarization. Imbalances or mistiming of ion movements in these and other physiological cases are linked to cell damage and disease.

To respond to changes in ionic strength due to dehydration, or rapid intake of large amounts of water, cells possess intrinsic mechanisms that regulate their volume. These rely heavily upon osmosensing, whereby the movement of water out of or in to the cell changes the cell volume, causing the compression of the membrane onto the cytoskeleton or pulling of the membrane away from the cytoskeleton, respectively. In response, the cell will open channels that will allow the entry of salt or the exit of metabolites like taurine. Over a longer timescale or when certain tissue repair mechanisms are stimulated (e.g. in liver regeneration), the cell can increase its volume or size permanently. This is observed, for example, when cellular hypertrophy is stimulated by mTOR signaling. Thus, the critical aspects of transmembrane gradients of ions are rigidly maintained through metabolic or membrane transport activities which restore ion homeostasis following perturbations to either signaling or volume.

8.1 Intracellular and Extracellular Ionic Environments Differ

There are a multitude of reasons why a mammalian cell wants to maintain a different composition of ions from the surrounding medium. Fortunately, the plasma membrane provides an ideal permeability barrier that maintains defined ion gradients between the intracellular and extracellular spaces. Calcium signaling, as well as depolarization-dependent signaling, which are both critical in neuron and muscle function, rely heavily on the use of ion gradients. In addition, many metabolites (nucleotides and some amino acids) are anionic, and these can be released with potassium during hypotonic swelling to decrease the concentration of osmolytes in the cytoplasm. Somewhat surprisingly, the cell maintains a low concentration of chloride in the cytoplasm, and *in vitro* motility with kinesin and other motors is dramatically disrupted by high chloride concentrations. This indicates that some enzymes are sensitive to the substitution of chloride for organic anions. To create the transmembrane gradients, cells have transport

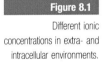

Figure 8.1

Different ionic concentrations in extra- and intracellular environments.

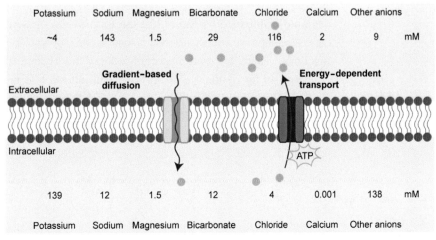

proteins that utilize energy derived from ATP hydrolysis to move ions against a concentration gradient. Once the concentration gradients are created, the energy of a diffusive gradient can be used by ion-selective channel proteins to rapidly move ions during signaling processes. Ion gradients across the plasma membrane are critical physical factors that enable a variety of biophysical functions in cells. Thus, ion gradients provide a robust means of supplying energy to many complex functions in the cell as well as a way to signal across the length of the cell very rapidly.

8.2   Transmembrane Gradients Store Useful Energy

In order to carry out many functions, the cell must actively create gradients of ions or metabolites across membranes and the energy stored in those gradients can be used for signaling and other processes. For functions to work reliably the cell must maintain an intracellular ion and metabolite environment that is relatively constant even when there are rapid fluctuations in metabolites, ions, hydration, or other factors. At the cellular level this involves balancing the uptake of nutrients with the output of waste products plus the maintenance of the ionic environment while signaling events are being carried out. In the case of calcium, there are proteins like calmodulin that rapidly bind free calcium ions when they enter the cytoplasm, thus buffering the concentration of free calcium before the calcium pumps can transport calcium out of the cytoplasm.

In culture, cells divide about once a day. This means they need to take in an amount of amino acids, carbohydrates, and nucleic acid precursors equivalent to that already in the cell within that day. Trace metals like zinc and iron, which are needed for many enzymes, must also be absorbed. To power all of this transport, as well as to facilitate rapid signaling, energy is required. Energy stores are therefore maintained by the cell in the form of ion concentration gradients, and these also serve as energy reservoirs that can be tapped to power a wide variety of functions. There are two important energy reservoirs, specifically the Na–K gradients set up by the Na–K ATPase, and the calcium gradients set up by the Ca^{2+} ATPase transporter. The Na–K ATPase is responsible for transporting sodium out of, and potassium into, the cytoplasm. This will create ion concentration gradients of about 20-fold, which can then power the movement of nutrients and other ions from one compartment to another or across the plasma membrane. Because the plasma membrane is normally permeable to potassium ions, Na–K ATPase activity creates a potassium-based membrane potential of about –70 mV (cytoplasm negative). Although the calcium concentration gradient is often 1000-fold (with more calcium outside the cell than inside), calcium is rarely used as an energy source to power the transport of other components. Instead, transient increases in the cytoplasmic concentration of calcium trigger many functions including muscle contraction and immune responses. Thus, the transmembrane potentials and ion gradients provide the simplest mechanism for coordinately activating a variety of cell functions simultaneously and also providing an energy source for some of those functions.

Cells can be described as storage devices for electrical energy in that the gradients of ions are coupled to ion-specific channels that let ions move through the membrane until the transmembrane potential balances the energy of the concentration gradient. The Na–K ATPase enzyme is primarily responsible for giving cells their electrical qualities by creating the gradients of Na^+ and K^+. The ultimate consequence for the cell is the ability to rapidly signal through changes in the ion specificity of the channels as well as to use the energy of the gradients to concentrate amino acids, glucose and other metabolites through cotransporter enzymes. The gradients created generate the cell potential, and form the basis of action potentials and nerve transmission. In terms of the energy required to generate this gradient, one ATP will be hydrolyzed for every 2–3 Na^+ (moving out) and K^+ (moving in) ions transported. With approximately 100 mM of Na^+ and K^+ being transported across the membrane, about 50 mM of ATP is required to generate the original gradient. In addition, the uptake of glucose, which cells use as an energy source, is also driven by sodium movement into the cell (symporter). There are ion transporters that use the movement of sodium into the cell (down a concentration gradient) to carry calcium out of the cell (called anti-porters, and another anti-porter carries protons out). The range of enzymes that use the transmembrane ion gradients is even larger if we include those that rely upon the transmembrane potential as well as the sodium gradient. Thus, the ion concentration gradients are generally needed for a variety of cell functions through special sym- or anti-port systems.

Figure 8.2

Na–K-ATPase and Na-glucose cotransporter.

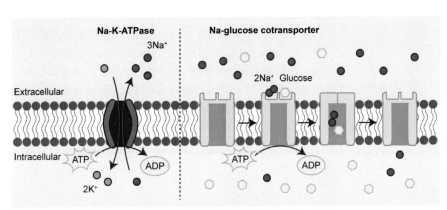

8.3   Ion Transport by Membrane Transport Proteins

For a cell to function, there must be transport proteins to circumvent the hydrophobic barrier of the plasma membrane so that cells can absorb nutrients, release waste, and control intracellular ion concentrations, ions and other water-soluble molecules. Membrane transport mechanisms have therefore had to evolve hand-in-hand with the appearance of the membrane. In some cases, the rate of movement of the substance across the bilayer is sufficient and no transporters are needed, whereas in other tissues transport must be facilitated. For example, water can move in to and out of most cells rapidly enough so that water channels are not needed; however, in the kidney, the water must be moved very rapidly for kidney function, and water channels are needed. There are two basic mechanisms that enable membrane permeability of small molecules (including ions); first, the formation of pores, which allow the passage of either all molecules smaller than the pore diameter or specific ions in the case of ion-specific pores (without direct energy dependence); and second, the specific transport of ions and molecules by transport proteins (usually energy-dependent).

8.3.1   Permeability through Membrane Pores

In some cases, membrane pores are non-specific, meaning that once they are open they will let many different molecules pass across the lipid bilayer, down a concentration gradient. However, due to the lack of specificity, membrane pores cannot be open for long periods. In bacteria, pores open to prevent membrane tension-mediated lysis; however, in mammalian cells, most pores are regulated by the cytoskeleton and could be opened by swelling-induced forces at the plasma membrane–cytoskeleton interface. Some smaller pores have 'ion specificity', meaning they preferentially allow the transport of only one ion. These pores are often called channel proteins, and are also normally gated. Ion specificity is related to the strength of ion hydration and the organization of the amino acid charges in the pore channel.

Pores range in size from the nuclear pores, that can pass 20 kDa proteins through simple diffusion, to the various ion channels that pass only specific ions one at a time. Nuclear pores must traverse the two bilayers of the nuclear membrane. Although they are not gated, they do have a fibrous cloud around them that can actively transport large mRNAs or ribosomes, yet passively impede the transport of many proteins. This allows specific proteins to be concentrated in the nucleus while others are excluded. However, ions and small molecules freely pass through the nuclear pores. Because diffusion through nuclear pores is rapid enough to support the dynamics of small molecules, ATP production and ion regulation are performed in the cytoplasm.

Figure 8.3

The regulation of mitochondrial pore proteins during apoptosis.

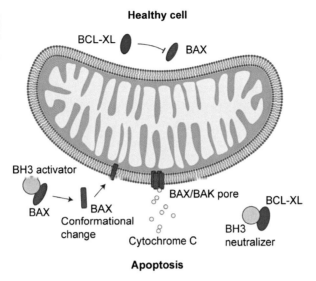

In the case of ion channels, there is greater interaction between the protein pore and the ion that is diffusing across the membrane compared to the larger non-specific pores and that interaction is responsible for the ion specificity. Specific channels are often open for only short periods to only allow specific solutes (usually inorganic ions of appropriate size and charge) to pass through them for the needed time. A major exception is the K^+ channel in mammalian cell membranes that is normally open and only closes transiently during cell depolarization. When ion channels are open, about 10^6–10^8 ions/second (turnover number) can move across the membrane. In comparison, most enzymes have a turnover number of 10^2–10^4/s, and carbonic anhydrase has the largest turnover number (6×10^5/s). Ion channels afford both specificity and high transport rates, but rely upon a concentration gradient, which is essential for processes like action potentials, where the cell wants to move specific ions rapidly across the membrane.

Ion channels can be coupled to mechanical work in the case of the bacterial flagellar rotor or to the production of ATP from ADP and phosphate in the case of the F_0/F_1 ATPase that is concentrated in mitochondria for aerobic metabolism and ATP production. The aerobic metabolism in the mitochondria yields much more useful energy than anaerobic. Metabolic energy from aerobic metabolism of sugars and fats produces electrons that create a proton gradient across the inner membrane of the mitochondria. The proton gradient is converted into ATP by the F_0/F_1 ATP synthase enzyme, which is a member of the AAA+ family of ATPases and possesses a sixfold symmetry. The rotation of a protein core, which occurs due to proton movement across the membrane, causes conformational changes in the three dimeric subunits, and this results in the synthesis of ATP. Pores exist in the mitochondrial outer membranes that allow diffusion of ATP out of, and ADP into, the gap between the inner and outer membranes. The whole system is particularly efficient in that the ATP synthase converts about 99% of the energy stored in the proton gradient into ATP, which can then passively diffuse through the pores.

Normally the pores in the outer membrane only pass small molecules like ATP, but there are pore-forming proteins in cells that can create larger pores leading to cell death or apoptosis. The whole process of apoptosis is critical in development and many disease processes where the targeted or abnormal cells are meant to die. In apoptosis, the outer mitochondrial membrane pores are enlarged, allowing cytochrome C to be released into the cytoplasm, and that activates proteolytic and DNA hydrolysis pathways. There is a strong belief that mitochondria originated from the adaptation of symbiotic bacteria in eukaryotic cells and in concert with this view, the mitochondrial pore proteins that cause apoptosis are analogs of the bacterial porins. The delicate balance between life and apoptotic cell death is highly dependent upon the concentration of pore-forming proteins, BAK and Bax, which oligomerize in the outer mitochondrial membrane and increase the permeability of the mitochondrial membrane. Studies indicate that the size of the pores can grow to facilitate the passage of proteins 100 kDa in mass. This is achieved through the gradual addition BAK to the apoptotic pore. In contrast, normal pores should enable metabolites and ATP to freely pass the membrane but not proteins. Thus, the cascade of enzymes that cause cell death is activated downstream of the formation of large pores in the outer mitochondrial membrane. This is again a cooperative phenomenon that is highly sensitive to small changes in concentration of the pore-forming proteins. Furthermore, other factors such as the lipid composition of the mitochondrial membrane will affect pore formation. This is an excellent example of how the control of pore formation was adapted to the complex but important process of apoptosis. Thus, we suggest that the major pathway of apoptosis is controlled by a highly cooperative assembly of pore-forming proteins on the surface of the mitochondria.

8.3.2   Membrane Transport Proteins (Na^+/K^+ Pump, Ca^{2+} Pump and Na^+, H^+ Pump)

The energy reservoir aspect of the cellular ion gradients requires that energy be put into the system by producing ion gradients. The sodium and potassium gradients produced by the Na^+/K^+ ATPase are the most important. It is critical for the Na^+/K^+ ATPase to be robust and reliably create a transmembrane gradient of these ions. There was a brief period when the role of membrane transporters in producing the ion gradients was challenged by the hypothesis that equilibrium binding of potassium in the cytoplasm was responsible for the high concentration there. However, it was proven that active transport was needed and the potassium was in solution so that it can produce a transmembrane potential. Transporters, such as the sodium–potassium, sodium–hydrogen and calcium ATPases undergo a series of steps to convert ATP energy into the movement of ions across the membrane. These three transporters share many structural and mechanistic similarities and contain a phosphorylation site that is similar to the ATPases associated with diverse activities (AAA) motif (however, they are not multimeric like the AAAs). The binding of ions to specific sites is needed to induce conformational changes in the transporter leading to the movement of the ion from one side of the membrane to the other, and those conformational changes are coupled in a unidirectional fashion with ATP hydrolysis. Upon release of ADP, Pi and the transported ions, the process will start over again. In all three transporters a conserved Asp residue located at the cytoplasmic surface (the P-domain) becomes phosphorylated by ATP (complexed with Mg^{2+}) when the ions to be transported bind to the transmembrane region of the channel. This region can be accessed from the cytoplasm as, prior to ion binding, the cytoplasmic gate is open and the extracellular gate is closed. Phosphorylation causes the cytoplasmic gate to close, sequestering the bound ions until a large conformational change opens the extracellular gate. The conformational change also lowers the affinity for bound ions, allowing them to exit the channel to the extracellular space. Release of the ions into the extracellular space enables the binding of extracellular ions. The extracellular gate closes following extracellular ion binding to adjacent sites, and this causes dephosphorylation of the same Asp residue that was phosphorylated earlier in the cycle (Figure 8.5). Another large conformational change back to the original state weakens the binding of the extracellular ions and opens the cytoplasmic gate so that they will exit, thereby completing the cycle. This transport cycle is unidirectional in that it couples ATP hydrolysis to the movement of typically two sodium ions outward and three potassium ions inward.

Transmembrane Potential Gradients are Large

Because the lipid bilayer is only 5 nm thick, an intracellular potential of –70 mV can produce a voltage gradient of 140,000 V/cm across the bilayer. Further, the anionic lipids on the cytoplasmic surface of the plasma membrane will generate a surface potential from the incomplete neutralization by counterions of about –25 mV as calculated from the Guoy–Chapman theory. This will add another 50,000 V/cm to the transmembrane potential gradient that is experienced by charges in the bilayer. Although these are very high gradients, the dielectric breakdown of membranes only occurs at transmembrane potentials of about 1000 mV. In electroporation procedures where holes are produced in membranes by high voltages, the actual voltages at the cell membrane are only in the range of 1 V.

(continued)

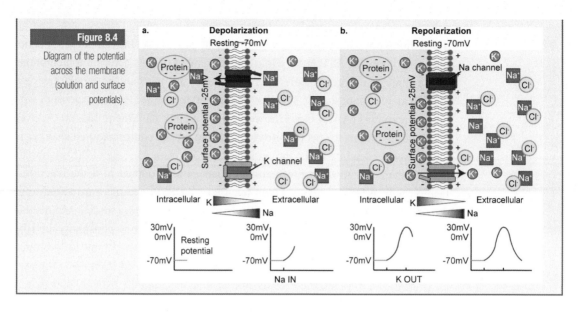

Figure 8.4

Diagram of the potential across the membrane (solution and surface potentials).

Metabolites like glucose need to be transported into the cell and a glucose–Na^+ anti-porter is responsible for concentrating glucose (similar to the reverse transport of Na^+ and K^+ in that Na^+ is moved out and glucose is moved in to the cell). The movement of a single glucose molecule again requires the hydrolysis of a single ATP molecule, and based upon free-energy calculations, this system can result in a 30,000-fold increase in glucose concentration (www.ncbi.nlm.nih.gov/books/NBK21687/). Similar cotransport systems are known for amino acids, inositol, and other metabolites.

8.4   Electrochemical Gradients

With a concentration gradient established from the active unidirectional transport of ions across membranes, the cell is able to transmit electrical signals through depolarization, and to do work by coupling the concentration gradient to other transport events. As mentioned earlier, ion channels allow specific ions to cross the membrane passively ('downhill'), in a process called *passive transport*, or facilitated diffusion. Many ion channels bind to the positive ions through oxygen atoms in the amino acids and replace water molecules that are normally strongly bound to the ions. Ions move past a series of these oxygen atoms that line the pore in the protein to go from one side of the bilayer to the other. Movement through the channels is by passive diffusion and net movement is driven by the difference in their concentration on the two sides of the membrane (the concentration gradient). If the solute is charged, however, both the concentration gradient and the electrical potential difference across the membrane (the membrane potential) influence transport. In most

Figure 8.5

Transmembrane ion pumps.

a. Ion channel: single gate

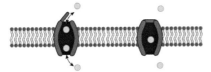

b. Ion pump: alternating gates

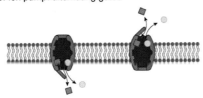

c. Ion pump: alternating gates and occluded states

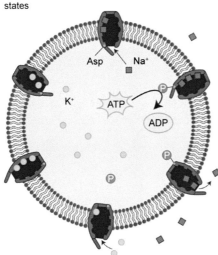

Extracellular Ions

K^+ for the Na^+/K^+ ATPase and H^+/K^+ ATPase

H^+ for the Ca^{++} ATPase

eukaryotic cells, the K^+ channel is open and the higher concentration of K^+ in the cell will result in an outward flow of K^+ that will create a negative potential inside the cell relative to outside. The negative potential will drive K^+ ions into the cell and counteract the effect of the higher cytoplasmic K^+ concentration. Because the number of K^+ ions that have left the cell is small when an electrical potential is large enough to balance the concentration gradient, we don't have to worry about changes in ion concentrations due to the channel. The concentration gradient for each permeable ion and the electrical gradient combine to create a net driving force, known as the *electrochemical gradient*, for each charged solute. Although an open K^+ channel will cause the cytoplasm to have a negative potential of about −70 mV normally, the opening of other channels instead of the K^+ channel will create different potentials (opening a Na^+ channel will produce about a +70 mV potential).

Defining the electrochemical gradient of each solute requires us to calculate the free energy of the ions in the two environments. The Nernst equation was derived for this purpose, and hence will give the electrical potential of a given solute. The Nernst equation provides a formula that relates the numerical values of the *concentration gradient* to the *electric gradient* that balances it. For example, if a concentration gradient was established by dissolving different amounts of KCl in two chambers separated by a membrane permeable to K^+ ions, then K^+ ions would diffuse from the more concentrated to less concentrated chamber. Assuming that Cl^- ions cannot pass the membrane, K^+ movement would create a potential across the membrane with the high concentration negative. The potential will increase until the flux of K^+ ions moving down the concentration gradient equals the flux that moves in the opposite direction due to the electrical potential.

$$\text{Nernst Equation} : E = V1 - V2 = \frac{RT}{zF} \ln \frac{[K]_2}{[K]_1} = 2.303 \frac{RT}{zF} \log_{10} \frac{[K]_2}{[K]_1}$$

$$R = \text{universal gas constant}$$
$$T = \text{absolute temperature}$$
$$F = \text{Faraday constant}$$
$$z = \text{electrical charge(valence) of ion}$$

$$\text{At } 20°C, 2.303RT/F = 58.17\text{mV}, z = +1 \text{ for } K^+ \qquad E = +58.17\text{mV} \times \log_{10} \frac{[K]_2}{[K]_1}$$

$$= +58.17 \times \log_{10} \frac{1\text{mM}}{10\text{mM}}$$

$$= -58.17\text{mV}$$

To correspond to the physiological convention in which all membrane potential is measured inside minus outside, we define side 1 as inside (intracellular), side 2 as outside (extracellular)

$$E_K = \frac{RT}{F} \ln \frac{[K]_o}{[K]_i} ; \qquad E_{Ca} = \frac{RT}{2F} \ln \frac{[Ca]_o}{[Ca]_i} ;$$

$$E_{Na} = \frac{RT}{F} \ln \frac{[Na]_o}{[Na]_i} ; \qquad E_{Cl} = \frac{RT}{-F} \ln \frac{[Cl]_o}{[Cl]_i} = \frac{RT}{F} \ln \frac{[Cl]_i}{[Cl]_o}$$

In terms of the energetics of the system, the transport of solutes across the membrane *against* their electrochemical gradient ('uphill') is called *active transport*. The energy for 'uphill' transport comes from ATP hydrolysis or a gradient of another ion. Conversely, the transport of solutes down their electrochemical gradient can be passive and require no expense of energy. Passive diffusion can also be coupled to the active transport of other components and this is observed in the case of another glucose transporter that couples the movement of two Na^+ ions into the cell with the transport of one glucose molecule into the cell in what is called a symport mechanism (please note that ATP is not needed here, whereas the anti-porter that carried two Na^+ ions outward for each glucose inward did require ATP, as discussed in Wright et al., 2011). Thus, the ion gradients are exploited in some cases for the uptake of metabolites but often multiple transport pathways are available for the uptake of critical components.

8.5 Ion Channels Fluctuate between Closed and Open Confirmations

In addition to ion selectivity, the other important property that distinguishes ion channels from simple aqueous pores is that ion channels do not remain open all the time. Instead, they are *gated*, which allows them to open and close in a cyclical fashion. The timing of opening and closing transitions are random, but the fraction of time they remain in the open state can be controlled. The random nature of opening and closing is not a property unique to only ion channels as it is inherent in the stochastic behavior of other *single molecules*, and ion channels are essentially single molecules. From patch clamp measurements it is clear that channels open stochastically, but the average behavior follows predictable rules and responds to drugs and other factors often as predicted. This serves to reinforce the idea that biological systems are driven by diffusion, but the averaging over time or over large numbers of molecules gives rise to predictable behaviors.

Patch Clamp Technique for Measuring Single Ion Channels

A major advance in the ion channel field came from the discovery that small glass electrodes commonly used to penetrate cell membranes could pick up a small portion of the plasma membrane covering the tip, a patch of membrane. Because the tips were sometimes 1 μm or less in diameter, the patch covering the tip would have few and sometimes single channels. This enabled the measurement of currents due to single channels in response to agonists, potential differences or other factors through voltage clamping of the patches. This patch clamp technique revolutionized our ability to measure the functional characteristics of single ion channels and the discovery by Erwin Neher and Bernd Sackmann was rewarded with a Noble prize.

Although the timing by which the ion channel opens and closes is stochastic, the relative occupancy of the open state (open probability, Po) can be controlled by different stimuli. In particular, changes in the voltage across the membrane will control voltage-gated ion channels, a mechanical stress will control mechanosensitive ion channels, and the binding of an extracellular or intracellular ligand will control ligand-gated ion channels. When an ion-specific channel opens, ion movement through the channel down the concentration gradient will rapidly create a potential equal to its Nernst potential. Relatively few ions need to move through the channel to create an electrical potential even though the membrane has a relatively high electrical capacitance. For example, the movement of approximately one micromolar of K^+ ions ($\sim 10^7$ ions), which is just 1/100,000th of the K^+ ions in the cell, will create a normal cell potential. As only a few ions need to move through a channel to create the electrical potential, it can be rapidly adjusted as per the requirements of the cell. In terms of the consequences of changes in membrane potential, the important issue is that the transmembrane electrical gradients are very large and can drive dramatic conformation changes in membrane proteins, as will be discussed in the next section on action potentials. Thus, we suggest that only a few channels are needed for a dramatic change in the membrane potential and that those will follow stochastic opening and closing events that overall fit with the predicted behaviors.

8.6 Equilibrium Potentials and Action Potentials

The ion and electrical gradients that span the plasma membrane make it possible to propagate an action potential along a narrow process like a neuron. Electrical signaling along neurons is one of the fundamental processes that allows organisms to sense their environment, and act in response to a given stimulus, whether reflexively or following further processing (reason).

Because neurons are only about 1 μm in diameter, the closing of a potassium channel and the opening of a sodium channel can cause the cytoplasm to go from a

Figure 8.6

Neuron ion channels.

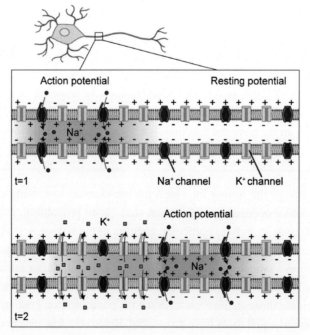

negative to positive charge in a fraction of a millisecond (the small volume means that less than 10^4 ions need to move and ion channels have permeabilities up to 10^8 ions per second). This will cause a dramatic change in the transmembrane potential of over 10^5 V/cm that will change the forces on membrane protein charges and cause the opening or closing of a protein channel. In some cases, the local depolarization of the axon will trigger the activation of neighboring channels, causing them to open and thus propagating the electrical signal like a wave along the axon. As noted in the text box below, the physical properties of neurons are particularly suited for the propagation of the depolarization waves along the axons to the synapse, where a neurotransmitter is subsequently released from docked vesicles. This will then cause depolarization in the next neuron. Propagation of depolarization waves provides a rapid means of signaling from one part of an organism to another.

8.7 Important Aspects of Depolarization Waves

Neuronal signaling in mammals has many refinements through evolution that make depolarization waves a rapid and effective method of signal transmission. First, the small diameter, and hence volume, of axons ensures that changes in transmembrane potential occur very rapidly. Myelin sheaths of the axons also increase the speed by which potential gradients change by limiting the volume of the extracellular space. Although the small diameter of axons confers a fragility that makes it difficult to maintain continuity over long distances, the benefits of a rapid signaling system outweigh the risks that damage could occur. Another feature is the high sensitivity of the voltage-gated ion channels that propagate the wave along the axon. This sensitivity is due to the presence of conserved sequences that include multiple charged amino acids, which would experience high forces from transmembrane potentials. Furthermore, the transport of sodium and potassium is critical for restoring the gradients of these ions across the membrane and both ATP concentration as well as Na^+/K^+ ATPase activity need to be very high all along the axon. Furthermore, neuronal activity is very energy-intensive and mitochondria are positioned all along the axons to provide enough ATP to drive repeated depolarizations. Thus, the design of neurons is excellent and they have many refinements at both a structural and biochemical level to enable the neurons to rapidly and repeatedly transmit signals throughout the organism.

Physical Parameters of Cellular Action Potentials

- The lipid bilayer of the biological membrane is an electrical capacitor with an extremely thin insulating layer. The cell membrane has specific capacitance (conductance/cm^2) near 1.0 μF/cm^2 (0.01 pF/μm^2).

(continued)

- Suppose that the membrane contains K^+-channels with 20 pS (20×10^{-12} siemens)/channel of electrical conductance (conductance g = 1/R). If an average of 0.5 channel/μm^2 is open, the specific membrane conductance is 10 pS/μm^2.
- Then specific membrane resistance $R_m = 1/g_m = 1000$ $\Omega.cm^2$ and membrane time constant for $C_M = 1\mu F/cm^2$ would be $\tau_M = R_M C_M = 1$ ms.
- Suppose that the concentration ratio of K across the membrane is 52:1 (inside:outside) so that E_K is 58.2 log(1/52) = –100 mV.
- Now what happens right after the salt solutions are introduced and K^+ ions start to diffuse?
- The voltmeter reports a membrane potential changing from 0 mV to –100 mV along an exponential time course with a time constant of 1 mS (τ_M): E = [1 – exp(–t/1ms)] $\times$ (–100 mV).
- After a few ms, the system reaches equilibrium and an excess charge of $Q = EC_M = 10^{-7}$ C/cm^2 has been separated across the membrane. This equals to an outward movement of $Q/F = 10^{-12}$ mol of K^+ ion per cm^2 of membrane (a very tiny amount, ~0.004% of K^+ ions inside a cell with a diameter of 10 μm).

8.8   Muscle Cells Propagate Depolarization Waves

Like neurons, muscle cells also undergo depolarization. This is most evident with the propagation of depolarization waves from one cell to another in cardiac muscle, which is critical for the coordination of cardiac contractions. Similarly, effective contraction of long skeletal muscle fibers (cm in length) requires the propagation of the action potential through a depolarization wave. If the activation of contraction occurred within an innervated region, but without an almost simultaneous depolarization in adjacent regions, uncoordinated muscle contraction would result. In all skeletal muscle cells, it is the entry of calcium with depolarization that activates muscle contraction. Much of the calcium in muscle actually comes from the sarcoplasmic reticulum, which is a differentiated endoplasmic reticulum that has a high density of calcium release and calcium transport proteins. Muscle fibers can be large in diameter (100 μm is common). To ensure the depolarization-dependent calcium release wave reaches the center of the muscle cell, invaginations of the plasma membrane penetrate deep into the fiber. These invaginations are known as transverse tubules, and are located at each of the actin anchoring z-lines of the sarcomere and they abut the sarcoplasmic reticulum so that they can stimulate the release of calcium in the center of the cell at the same time as at the periphery.

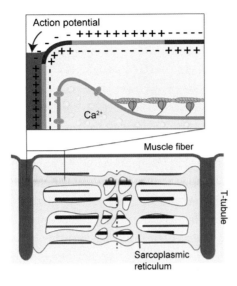

Figure 8.7

Muscle cells propagate depolarization wave.

Again, the important aspect of ion-dependent signaling by depolarization is that it can occur very rapidly and propagate over long distances whether it involves calcium, sodium, or other ions.

Although hormonal and other chemical signaling processes can spread through the circulatory system and can act on the second to minute timescale, depolarization of trans-membrane potentials is the only process that can propagate signals on a subsecond timescale over meters.

8.8.1 Broader Implications of Depolarization Waves

The selective control of ion movements, and the ability of the cell to generate waves of depolarization, have immense implications for both the cell and its host organism. For mammals and other large organisms, the ability to rapidly propagate signals from one limb or organ to another forms the basis of rapid movements and thought. These capabilities provide a tremendous selective advantage in terms of competition for food and other resources. Again, the basic concept of depolarization has been refined over many generations to the current neuronal networks in our brain that are responsible for complex thoughts. The basic physics and the tools involved are similar, but refinements do make the neuronal signaling more robust and resistant to many environmental perturbations.

8.9 Calcium Signals can be Different

One may think that the presence of calcium alone is sufficient to activate the various processes it controls. However, there are several properties of calcium that make it possible for multiple control processes, and multiple responses to changes in its cytoplasmic concentration. For example, non-muscle cells typically have millimolar concentrations of the calcium-binding protein calmodulin. The entry of low levels of calcium from a channel in the plasma membrane would result in a local rise in calcium levels, but calmodulin would rapidly bind the free calcium and prevent it from diffusing throughout the cell. Thus, slow calcium leaks at the plasma membrane can produce quite different effects than the major calcium release events from the endoplasmic or sarcoplasmic reticulum. Furthermore, the calmodulin–calcium complex can also interact with other proteins such as the calmodulin-dependent kinase, which needs to bind 12 calcium–calmodulins to become fully active. The small amounts of calcium that enter the cytoplasm with each depolarization would be unlikely to saturate the kinase, meaning that many depolarizations are required for full activation. The activity of the kinase could therefore be a way to measure the integrated depolarizing activity over time. This would have to occur in a short period, however, as there would also be a gradual loss of calcium–calmodulin complexes due to the normally low calcium levels in cytoplasm. Thus, it is perhaps not surprising that this kinase plays an important role in learning and memory where multiple depolarizations of a given neural pathway are needed to cause reinforcement of synapses in that pathway, i.e. memory. This example serves to highlight how calcium can have multiple signaling roles through multiple entry pathways and multiple high-affinity binding events.

8.10 Resting Membrane Potential Formation and Relevant Diseases

A possible 'weak link' in the cellular control of depolarization waves is that the active transport of ions is needed to restore the normal ion gradients, and this is certainly the greatest energy-requiring step. In the heart, where the depolarizations are occurring about every second of the life of the organism, there is a major need for rapid adjustments of homeostasis. For example, alterations in kidney function resulting in high K^+ levels in blood will lead to arrhythmias in the heart that are life-threatening. Another weak point is the transport of calcium, and the levels of calcium during contraction and relaxation need to be maintained within tight limits. These restrictions are maintained through the dynamic regulation of both the calcium release and calcium transport systems. Changes in the distribution of pumps and channels can cause regional differences in ion levels that can seriously affect contractility. Furthermore, there must be rapid feedback mechanisms that adjust the level of calcium release to the demands of the heart such as in the startle reflex. For some animals, the startle or stress response can be so strong that they will suffer heart failure and die because of the stress. A slight imbalance in the activation of either the release system, or the transport system, can result in a major disruption of cardiac function. An ion imbalance need only occur for a short period to cause death.

In this system, there are many steps that are occurring within subsecond periods to cause the proper release, and recapture, of calcium. Often, many of the problems associated with cardiovascular disease arise with age and are the result of hormonal changes that are part of the aging process. If steps in the response to neuronal activation are slow, then there will be a greater chance of imbalance and a negative cycle that will damage the tissue. Furthermore, dehydration or altered kidney function in disease will change the ion concentrations in the extracellular medium and alter cardiac function. Thus, many of the changes in cardiac function in disease are related to alterations in the handling of ion movements.

8.11 Mechanosensitive (MS) Ion Channels Initiate Cell Volume Regulatory Responses

The maintenance of cell volume is another important factor in the effectiveness of cellular functions. Because water crosses membrane bilayers faster than salts, cells will swell or shrink in response to hypotonic or hypertonic solutions, respectively. Osmotic pressures can be very high (see text box) and could easily lyse the plasma membrane. In mammalian cells, the cytoplasmic pressure is greater than the surrounding medium, but that is because of cytoskeleton contractions that create an inward force on the membrane. These contractions correspond to an increased concentration of cytoplasmic ions of about 5 micromolar, which is only a small fraction of the total 300 mM of ions in cells and the surrounding medium (this is from totaling the concentrations all of the different ions in the medium, i.e. the osmolarity). In order to avoid lysis when severe hypotonic changes occur, mammalian cells use mechanosensitive ion channels to release small molecules (osmolytes) but not proteins. Osmotic changes in the environment are particularly problematic for bacteria, especially as rain can dilute salts that are concentrated by long dry periods. To overcome these problems, bacteria have several similar channels that open when the lateral tension in the membrane becomes high (i.e. during swelling). In mammalian cells that have excess membrane area in membrane folds and invaginations, the links between the membrane and cytoskeleton provide a mechanical tension that is perpendicular to the membrane and can open larger pores to release large numbers of ions rapidly. Thus, the very high forces that are produced by osmotic pressures have caused the evolution of mechanosensitive channels to rapidly restore osmotic balance across the membrane.

The osmotic pressure Π of an ideal solution with a low concentration can be approximated using the Morse equation (named after Harmon Northrop Morse):

$$\Pi = iMRT,$$

where i is the dimensionless van't Hoff factor that is 2 for NaCl, M is the molarity, $R = .08205746$ L atm K^{-1} mol^{-1} is the gas constant, and T is the absolute temperature. The osmolarity is the sum of the concentrations of all of the separate species, e.g. a 150 mM solution of NaCl has an i of 2 because there is 150 mM of both Na$^+$ and Cl$^-$ giving an osmolarity of 300 mOsm. Thus, the pressure from a 150 mM NaCl gradient would be 7.4 atm.

The differences between the bacterial and the mammalian cells highlight the differences in the two models of mechanosensitive channels, which are known as the *bilayer model* and the *tethered model*, respectively, and explain how mechanosensitive channels are activated by mechanical forces. In the bilayer model, as the bilayer is stretched or bent, the forces at the channel–lipid interface will change. These changes will eventually trigger channel conformational changes. However, in the tethered model, the channel is connected directly to either extracellular or cytoskeletal proteins. This tether may either pull on the channel or act as a gating spring. In the former scenario the channel will be displaced from the bilayer and this causes conformational changes similar to the bilayer model. In the latter scenario, when the gating spring is stretched, it favors the open state of the channel. Of course, the bilayer model and tethered model are not mutually exclusive and could coexist for the same channel. Up to now, the channels known to solely use the bilayer model are bacterial MscL and MscS channels, and maybe the NMDA receptor channel. The mechano-response of purified MscL reconstituted in bilayer has been demonstrated. Most other MS channels seem to use the tethered model or both models.

Perhaps the best-understood tethered channel is the tip link involved in hearing and resides near the tips of the hair cell stereocilia. In the current model, the thin link between neighboring stereocilia is anchored on one side by a channel complex and there is a motor complex that can adjust the tension on the link. When sound causes the stereocilia to vibrate, tension is generated on the link and this opens the channel and creates a depolarization signal that is transmitted to the brain, where it is interpreted as a sound. The sophisticated architecture of the mechanosensing system in the ear is fascinating and highlights how selective pressures develop very efficient sensory systems that help in the survival of the organism.

Mechanosensitive channels are involved in many other sensory functions, including touch, the detection of blood pressure changes, and the response to

Figure 8.8

Bilayer model and tethered model.

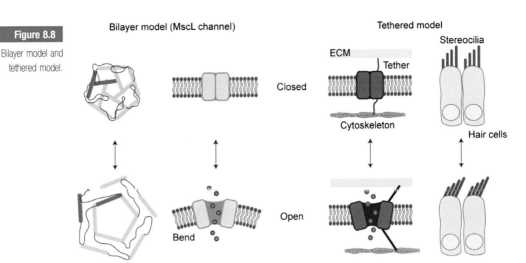

changes in tissue volume or contractions. Many different forces impinge on a tissue and the response of the tissue should be tailored to the cause of the force. Although there are many different types of MS channels, only a small fraction have been analyzed at a cellular level. There is controversy over the exact roles that many of these channels play in cell functions. However, there are many mechanical processes where channels could play critical roles. Again, it is important for our further understanding of where ion movements are used in cellular functions that the assays of those functions enable the rapid monitoring of activities along with ion movements.

8.12   Cell Volume Control over Longer Timescale or in Tissue Repair via Cellular Hypertrophy

Although cells regulate volume via local and short-term ion signaling and cellular responses to ion homeostasis, over longer timescales of hours and days, cells can regulate their volumes more stably via mTOR and/or IIS signaling pathways in a process called compensatory cellular hypertrophy (CCH). Upon tissue damage or identification of malignant cells, a void in tissue space is generated by apoptosis, necrosis, or active extrusion. The neighboring cells respond to mechanical and/or chemical signals to alter sizes and/or numbers to fill the void. Similar scenarios are also found in exercise-induced hypertrophy or compression-induced size regulation. The signals to grow cell size in response to injury involve a plethora of chemical signals known to be involved in inflammation signaling and wound healing. More recently, mechanical signals such as those in epithelial cell sheets and *Drosophila* embryos have been shown to be early signals as well. The decisions to switch between growing into larger cells or activate cell proliferation to fill in the void in tissue space caused by damage has been a quantitative one. In liver regeneration, it is known that up to 30% liver hepatectomy activates mostly CCH, while larger injury of 70% activates a mixture of CCH and cell proliferation responses. Thus, it is very likely that transient and short timescale volume control and signaling via ion regulation can be integrated into longer timescale decisions on cell volume or size regulation via CCH.

8.13 SUMMARY

Cells create ion gradients between cytoplasm and the surrounding medium and these have proven to be essential for many cellular functions. In mammals, the rapid depolarization of cells provides the basis of neuromuscular signaling as well as brain activity. This relies upon the rapid movement of sodium ions through passive channels into the cell to reverse the resting potential that was created by a gradient of potassium ions. With repeated depolarization of cells, transport proteins must work rapidly to restore the normal ion gradients and this is a weak point that often causes problems for the organism when functions are compromised by disease. Calcium acts as a signal for many cell processes because the normal cytoplasmic levels are below micromolar, whereas the surrounding serum contains calcium in the millimolar range. The high level of calcium-binding proteins in the cytoplasm can restrict the diffusion of calcium, thereby limiting the effects of an increased concentration to the local region. Mechanosensitive channels play sensory roles that detect and control acute changes in cell volume or detect and transduce signals that originate from stimuli in the larger environment, including sound. These are transduced as neurological signals to our brains, where further processing takes place. Thus, ion movements are critical components for many basic as well as higher-order functions of cells.

8.14 PROBLEMS

1. We talked about cell depolarization as the change from being permeable to potassium to being permeable to sodium. If the only ions that move during the depolarization of a cell are sodium ions, then how many ions will move in a cell with a membrane area of 4000 μm^2 and a volume of 4000 μm^3 upon depolarization (assume the concentrations of sodium and potassium given for cytoplasm (12 and 138 mM, respectively) and serum (143 and 4 mM, respectively)?

Answer: From the Nernst equation, we know that the potential is initially 58.17 mV $\times$ $\log_{10}$ $[K_o]/[K_i]$ = –90 mV and that the depolarization goes to the Na potential

58.17 mV $\times \log_{10}$ [Na$_o$]/[Na$_i$] = 63 mV. Thus, the total change in the intracellular potential is 153 mV and we can go to the text box where the charge movement for 100 mV change was calculated to be 10^{-12} moles of charge per cm^2 of membrane. For a 153 mV change, there should be 1.53×10^{-12} moles of Na$^+$ moving inward per cm^2 of membrane. With 4.0×10^{-5} cm^2 of membrane, there would be 6.12×10^{-17} moles of Na$^+$ or 3.7×10^7 atoms of Na$^+$ moving inward (15.3 micromolar or 0.1% of total cytoplasmic Na$^+$).

2. If we assume that 99 mM of anionic charges are immobilized on DNA in the nucleus, then what will be the potential developed by a Donnan equilibrium from the inside to the outside of the nucleus with a cytoplasmic concentration of 100 mM NaCl? Remember that a diffusion potential can be generated by either positive or negative charges.

Answer: The 100 mM of anionic charges are balanced by 100 mM of Na$^+$ but the chloride in the nucleus will be only about 1 mM. Using the Nernst equation $-58.17 \times \log_{10}$ [Cl$_o$]/[Cl$_i$] = -116.3 mV.

3. The Debye Huckel theory describes the distribution of counterions around charges and the distance over which potentials are neutralized. If we assume that a membrane has a surface potential of -100 mV and concentration of Ca^{2+} is 1 micromolar in cytoplasm, then what will be the Ca^{2+} concentration at the surface of the membrane?

Answer: This is going to be defined by the Nernst Equation for a charge of two, because the plus ions will concentrate due to the high negative charge. If we modify the Nernst equation for z = 2, we get $-29.08 \times \log_{10}$[Ca$_m$]/[Ca$_c$] = -100 mV, where [Ca$_c$] is the calcium concentration in the cytoplasm and [Ca$_m$] is the concentration at the membrane surface. Solving for $\log_{10}$[Ca$_m$]/[Ca$_c$] = 3.43 and the taking the anti-log [Ca$_m$]/[Ca$_c$] = 2750 or [Ca$_m$] = 2.75 mM.

4. The Donnan potentials can arise without a membrane as the result of high concentrations of counterions for fixed charges (for example, in a polymer bead) that then are diffusing away from the bead into a potentially lower concentration of that ion in the surrounding medium. Something similar happens when the membrane is made permeable to both cations and anions. For example, let's take the simple case of a cell membrane, which has pores that allow both potassium and chloride ions through the membrane (assume the values for potassium are 140 mM inside and 4 mM outside, and for chloride are 110 mM outside and 10 mM inside). What will be the final potential?

Answer: Using the Nernst equation, we have $-58.17 \times \log_{10}$[Cl$_o$]/[Cl$_i$] $+58.17 \times \log_{10}$[K$_o$]/[K$_i$] = $-58.17 \times \log_{10}$[110]/[10] + 58.17 $\times \log_{10}$[4]/[140] = $-60.6 - 89.8 = -150.4$ mV.

5. A cytoplasmic protein that binds to membranes has a charge of +10 on the surface that faces the membrane and is neutral over the rest of its surface. It has an alpha

helix of 25 amino acids that contains the plus charges with rough dimensions of 2×7 nm. If the bilayer surface has 33% phosphatidyl serine (-1 charge) and 4% PIP2 (phosphatidyl inositol 4,5 diphosphate with a -3 charge), then what will be the overall charge density in the region of the membrane containing the peptide (assume that concentration of lipids in the peptide region is the same as the average and that the area per lipid is 0.5 nm^2)? If the potential of the surface scales with the charge density and the potential on the bare membrane surface is -52 mV, then what is the potential over the surface of the peptide. Assume that the proton (H^+) concentration in the cell is 10^{-7} M (pH 7), then what is the pH at the bare membrane surface and at the surface of the peptide?

Answer: The peptide covers an area of 14 nm^2 or 28 lipid molecules. Because this lipid surface is 33% PS, there will be 9.3 PS molecules on average under the peptide as well as 1.12 PIP2 molecules on average. Thus, the average charge in the region of the peptide will be partially neutralized by the 10+ charges of the peptide from -12.62 to -2.62 on average. Assuming that the potential scales with the average charge density, then the potential will be $-52 \times 2.62/12.62 = -10.8$ mV. Using the Nernst equation $-58.17 \times \log_{10}[H_m]/[H_c] = -52$ mV, then $[H_m]/[H_c] = 7.8$ or $pH_m = 7 - 0.89 = 6.11$ at the bare membrane surface and $[H_m]/[H_c] = 1.53$ or $pH_m = 7 - 0.19 = 6.81$ at the peptide surface.

6. The Donnan equilibrium relies upon the establishment of potentials based upon the fixed charges that cannot diffuse from a cell or any small porous solid. When you consider a permeabilized nucleus, the fixed phosphate charges (equivalent of over 200 mM anionic carboxyl groups) contribute to an increase in the total concentration of ions in the nucleus above the concentration outside of the nucleus. If the concentration of salt outside is 10 mM KCl and we keep the concentration of fixed anionic charges inside the nucleus at 200 mM (K^+ is the cation in the nucleus), what will be the difference in total salt concentration between the nucleus and the surrounding solution? In addition, how far do you have to move from the permeabilized nucleus into the surrounding 10 mM KCl medium before the Donnan potential drops by 10-fold ?

Answer: The total salt concentration in the nucleus is 200 mM because the amount of Cl in the nucleus will be very small whereas the total ion concentration in the surrounding medium will be 20 mM or the sum of the K and Cl. For the second part, the Debye length provided in Chapter 6 for 10 mM salt is 3.04 nm. If we assume that the potential drops as $\psi_d/\psi_o = 1/e = e^{-d/LD}$ at the Debye length (where ψ_d is the potential at that distance d and LD is the Debye length). Solving for 1/10 the original potential gives 2.3 times the Debye length.

Extra credit: Assume that the nucleus above is coated with a semipermeable membrane that will let water in but will not let salts cross. What will be the osmotic pressure across that membrane at 25°C (use the gas constant of 0.082 liter-atmospheres/°K/mole)?

Answer: Using the equation for osmotic pressure and the fact that 0.18 M more salt will be in the nucleus, we calculate that the osmotic pressure will be 4.4 atmospheres.

7. In the case of the Nernst potential, the diffusion of an ion down a concentration gradient is balanced by the electrophoresis of the ion up the gradient. For most cells the membrane is permeable to potassium (4 mM outside and 140 mM inside), but in some cases calcium channels can open. Assuming that the interior concentration of calcium is 1 micromolar and the serum level is 2 millimolar, what is the change in potential upon closing the potassium channel and opening the calcium channel?

Answer: Initially the potential is $+58.17 \times \log_{10}[K_o]/[K_i] = 58.17 \times \log_{10}[4]/[140] = -89.8$ mVolts. When the calcium channel opens, the Nernst potential becomes $+29.08 \times \log_{10}[Ca_o]/[Ca_i] = 29.08 \times \log_{10}[2mM]/[0.001mM] = +96.0$ mV. Thus, the change in potential is 185.8 mV.

8. The mitochondrion is the main energy production compartment inside animal cells. The energy is generated by the F_0/F_1 ATPase, which utilizes energy from the proton-motive force across the mitochondrion inner membrane. Consider a rat liver cell with high respiring rate; the pH inside the mitochondrion is about 8.25, whereas the pH of cytosol is about 7.37. The mitochondrial membrane potential is about 118 mV inside positive. Can you calculate the net proton-motive force across the mitochondria inner membrane?
(http://bionumbers.hms.harvard.edu//bionumber.aspx?id=101103&ver=5; Mollica et al., 1998, p. 216, table 2).

Answer: The net potential for a proton is the membrane potential plus the Nernst potential for protons $-58.17 \times \log_{10}[H_m]/[H_c] = -58.17 \times (7.37 - 8.25) = 51.2$ mV. Thus, the total potential is 169.2 mV.

8.15 REFERENCES

Mollica, M.P., Iossa, S., Liverini, G. & Soboll, S. (1998). Steady state changes in mitochondrial electrical potential and proton gradient in perfused liver from rats fed a high fat diet. *Mol Cell Biochem* 178(1–2): 213–217.

Wright, E.M., Loo, D.D., et al. 2011. Biology of human sodium glucose transporters. *Physiol Rev* 91(2): 733–794.

8.16 FURTHER READING

Nuclear Pore Complex: www.mechanobio.info/topics/cellular-organization/nucleus/ and www.mechanobio.info/topics/mechanosignaling/ran-gtpases/

Plasma membrane and membrane proteins: www.mechanobio.info/topics/cellular-organization/membrane/

9 Structuring a Cell by Cytoskeletal Filaments

Despite a lack of exoskeleton, isolated animal cells normally exhibit stable, non-spherical shapes with excess membrane area. To maintain cell shape, the plasma membrane conforms to a cytoskeleton located inside the cell, but the membrane surface area to cell volume ratio determines how extensively the membrane can conform. In a few cases, such as endothelial cells in capillary tubes, the cytoskeleton alone can support the cell shape. In most cases, however, the cytoskeleton is dynamic, and defines the cell shape in concert with the cell volume, and the effect of contracting against external contacts, or adhesions. Modulation of these adhesions, whether they are between adjacent cells, or cells and matrices, provide the stimuli to initiate cytoskeleton dynamics. These dynamics, which include filament assembly and disassembly, typically determine any changes in the shape of the cell. Actin filaments are often assembled from adhesion sites or sites of membrane extension, both of which lie primarily at the cell periphery. They are then pulled inward and disassemble in a process known as 'actin treadmilling.' Actin filaments are contracted by myosin and this creates tension in the cytoskeleton and is responsible for tissue organization and the cohesive property of cells. Paradoxically, the rapid loss of contractile force increases filament dynamics, and cytoskeleton assembly. Cohesion of cells over longer periods is stabilized by intermediate filaments. These appear to be the safety net that prevents acute fracture of cells in epithelia under tension. Microtubules, on the other hand, help stabilize and define cell polarity and structure. Microtubules originate from the microtubule-organizing center (MTOC) and provide directional pathways on which motors will transport membranous organelles and other cellular components. During mitosis, the dynamic microtubules of the mitotic spindle play critical roles in the transport of chromosomes to the daughter cells. Thus, the cytoskeleton is a dynamic, mechanically active component of the cell that is responsible for shaping the cell and its environment as well as for organizing the intracellular organelles.

The cytoskeleton is comprised of three long filament systems; specifically, microtubules, actin filaments, and intermediate filaments. These filaments are crosslinked by motor proteins or multivalent binding proteins, which serve to stabilize the cytoskeleton, and provide it with dynamic qualities including the ability to disassemble, reassemble, and contract. As evidenced by photobleaching recovery experiments, filament subunit exchange occurs on the timescale of minutes for actin and microtubules. Such dynamics are critical to the structural functions of the cell, as well as the motility of both the whole cell, and several types of cellular protrusions. Cells use motile functions to produce a wide variety of cellular and tissue morphologies in nature.

Cytoskeletal Filament Systems

Microtubules: tubes of about 25 nm in diameter, composed of 13–16 protofilaments that are polar, linear arrays of tubulin αβ dimers, GTP in dimers is hydrolyzed after assembly into microtubules, typically microtubules polymerize from one microtubule organizing center that anchors minus ends. Molecular weight 50 kDa/subunit, cellular concentration 1–2 mg/ml (10–20 μM), fraction polymer ~50%.

Actin filaments: filaments have two strands in a helix (4–7 nm in width, 34 nm per turn), ATP in monomers is hydrolyzed in filaments, assembled from multiple sites at adhesions or membrane extension sites from barbed end (defined by myosin head decoration), molecular weight 42 kDa, cellular concentration 5–10 mg/ml (100–250 μM), fraction polymer ~50%.

Intermediate filaments: made from overlapping tetrameric complexes that have anti-parallel subunits (overall diameter about 10 nm). Subunits are transported by microtubule motors, assemble in periphery, and move inward.

For further information see Alberts et al. (2014) and Pollard et al. (2016).

To understand how cells achieve the observed variety in morphology, one can consider the macroscopic problem people face when needing to build different structures for different functions. In this analogy, a number of standard best practices and tools have been developed for building, whether it be a small wooden hut or a huge wooden hotel.

Importantly, the surrounding environment of the structure will always be taken into consideration when deciding which tools and practices are required. Building a hut on soft sand will require a different set of tools and procedures than building the same hut on a city street. Similarly, in the cell, a standard set of tools and procedures exist that are used to determine the overall shape, and functional state of the cell. These tools, which are in the form of protein complexes, support many functions and are used for specific tasks. Because the cytoskeleton determines the

Figure 9.1

Cytoskeleton in cell.

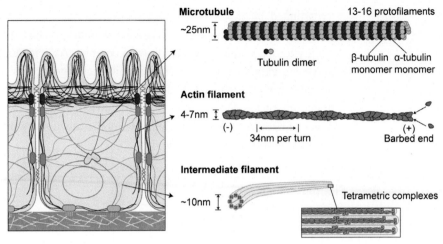

overall shape of the cell, organizes the cellular organelles and enables its motile functions, we should break down the systems of cytoskeletal assembly, disassembly, and contraction and examine the tools and procedures involved. Again, the dynamics are surprisingly rapid and force us to think about the molecular bases of functions in different ways than we would expect from our understanding of the macroscopic world.

9.1   The Case for a Dynamic Cytoskeleton

For the cell, the primary 'building materials' that determine its size and shape are the cytoskeleton and the plasma membrane. Although the name 'cytoskeleton' may conjure up images of the human skeleton or prehistoric animal skeletons, the cytoskeleton is significantly different. Rather than being extremely long-lasting and unchanging like skeletons, the cytoskeleton is highly dynamic and turns over at a rate that is faster than the cell changes shape. Turnover rates of most cytoskeletal components are on the order of a few minutes and yet cell shapes are preserved for hours to days. Thus, unlike the static framework of buildings, or even the human skeleton, the structural elements of cells are constantly exchanging. This dynamic exchange often occurs in response to environmental changes. For example, cells must quickly accommodate relatively large deformations from external forces imposed on them. Body movements (impacts), pumping of blood or fluid movements in tissues can all cause 5–20% or even greater strains in one or another dimension. Furthermore, the cytoskeleton can also contract and relax, thereby maintaining contractile stresses in both compressed as well as expanded states. Thus, the standard mechanisms by which cells are shaped are based upon dynamic tensions rather than static tension-bearing structures.

9.2   What Are Major Cytoskeleton Functions?

Because the form of every organism is ultimately constructed by the cytoskeleton and the cell surface area to volume ratio, the cytoskeleton plays a critical role in life. There are various properties of the cytoskeleton's filament networks that give rise to specific functions. These properties often stem from the structure and chemistry of the basic filament subunits, which include actin monomers, tubulin dimers, and intermediate filament tetramer–octamers. Actin filaments and microtubules are, for example, polarized and can therefore support motor movements. Intermediate filaments, on the other hand, are non-polarized and there are no motor proteins known to move along them.

One of the primary functions of the cytoskeleton is to develop the correct cellular force at the correct sites that is needed to properly shape a tissue. This will shape the organism and enable motility. For this function, filaments must assemble in the correct location with the proper orientation and be anchored to the correct sites. Therefore, filament assembly is a critical process, especially as filaments are continually turning over and must be constantly reassembled.

A lot of attention will be paid to the factors that control local filament assembly and disassembly, as those functions are at the heart of cytoskeletal functions. The response of filaments to imposed force and force generated by filament assembly are critical components of cytoskeleton dynamics with fluctuations in force stimulating dramatic changes in cytoskeleton filament turnover. For example, when force is applied to the cell from adhesive contacts, fluid pressures or the immediate mechanical environment, there will be an initial elastic response followed by an active response. In that active response, force is generated within the filament network by motor movements on polarized filaments, or by filament assembly. Transport of material such as RNA, membrane vesicles, and ribosomes is driven by appropriate motor systems on microtubules, particularly in animal cells, and on actin filament bundles in plant cells. Again, the sites of filament polymerization, filament dynamics and the filament anchoring sites are all critical parameters that determine whether components are carried to their intended site. Unlike in small bacteria where the transport of material predominately occurs through diffusion, higher-order organisms of larger dimensions rely heavily on motor systems to transport crucial material. This is particularly evident in long axons, where only motor transport can supply material to distant regions.

Understanding how cytoskeletal filaments are organized is therefore critical for defining the functions of larger organisms.

9.3   Polymerization in Cellular Systems

The assembly of actin filaments, microtubules and intermediate filaments all require polymerization systems that involve the non-covalent assembly of their subunit proteins into polymeric filaments. Each assembly process shares some common characteristics, and these reveal important properties of both the polymer systems and cytoskeletal filaments. In general, it is important for the cell to be able to control the site of filament assembly as well as the polarity of the filament. Furthermore, most of the common filament networks have the property of being composed of dimeric or multimeric filaments, and there are advantages for the cell to have multimeric rather than monomeric filaments, as will be discussed. The physical chemistry of filament assembly will also be considered below.

9.4   Monomeric Filaments

Although rarely found in the cytoskeletal system, it is instructive to consider the properties of monomeric filaments as those properties will differ in important ways from the filament types that usually make up the cytoskeleton. We suppose that monomeric filaments of protein A can form as the result of the polymerization of homogeneous protein subunits. In this model, we can assume that opposite sides of the protein have complementary binding sites and that subunit–subunit bonds have the same free energy of binding.

$$A + A = A_2 \qquad K_D = X_1$$
$$A_2 + A = A_3 \qquad K_D = X_2$$
$$\vdots \qquad\qquad \vdots$$
$$\vdots \qquad\qquad \vdots$$
$$A_N + A = A_{N+1} \qquad K_D = X_N$$

In this case, we suggest that $K_D = X_1 = X_2 = \ldots\ldots = X_N$

There are a number of important consequences for the properties of the linear polymer. First, the probability of forming a dimer is equal to that of forming a trimer or other multimer. This means that the polymer is likely to form anywhere in the cell, if the concentration is high enough, which makes it difficult for the cell to control the localization of polymer assembly. Second, the probability of dissociation of single subunits from the ends of the filaments will be the same as the probability of breaking bonds anywhere within the filaments. Thus, there will be no order to the disassembly of the filaments. These two factors differ in dimeric and multimeric filaments.

9.5 Dimeric Filaments

The classic case of a dimeric filament is actin, which is the most abundant protein in most eukaryotic cells. Actin filaments are homopolymers that consist of actin subunits assembled into helical, two-stranded filaments (Figure 9.2).

An important aspect of these filaments is that they will preferentially polymerize from seeds that can be either a short actin filament or actin-related protein complexes, which resemble the barbed end of an actin filament. This comes directly from the fact that the rate of dissociation of a monomer from a dimer is much more rapid than the rate of dissociation of a monomer from the end of a filament. There are several consequences of seed formation: first, when starting with pure monomer, the rate of polymerization often has a lag phase which corresponds to the time needed to form seeds; second, the addition of monomers to polymeric seeds removes the lag phase; third, at the cellular level, the site of polymerization can be controlled by the site of seed formation; and finally, a solution of pure actin monomer or tubulin dimer at the critical concentration (the concentration needed to form polymer from pure monomer) will only polymerize onto crosslinked seeds. The critical concentration (K_c) is defined by rate of polymerization and depolymerization at the ends of the filaments.

$$Kc = k_{off}/k_{on} = K_D$$

A further implication is that the release of subunits from the middle of a filament is unlikely. This can be understood by considering the number of bonds between subunits. In the case of actin, a monomer binding to the end of the filament forms bonds with two neighbors (one in the same protofilament and one between protofilaments), whereas a monomer in the filament has bonds with four neighbors (two in the

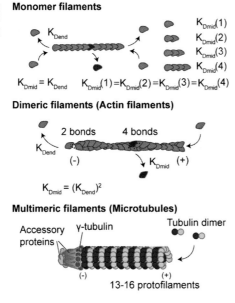

Monomer filaments

K_{Dend}

$K_{Dmid}(1)$
$K_{Dmid}(2)$
$K_{Dmid}(3)$
$K_{Dmid}(4)$

$K_{Dmid} = K_{Dend}$ $K_{Dmid}(1) = K_{Dmid}(2) = K_{Dmid}(3) = K_{Dmid}(4)$

Dimeric filaments (Actin filaments)

2 bonds   4 bonds

K_{Dend} (-) K_{Dmid} (+)

$K_{Dmid} = (K_{Dend})^2$

Multimeric filaments (Microtubules)

Accessory proteins   γ-tubulin   Tubulin dimer

(-)   (+)
13-16 protofilaments

same protofilament and two in the adjacent protofilament). Thus, the free-energy penalty for a subunit leaving the side of a filament is twice the free energy of leaving the end of the filament. This means that the K_D for leaving the side is equal to the K_D of the end squared (e.g. K_D for actin at the filament ends $= 10^{-7}$ M, which means that at the side $K_D = 10^{-14}$ M, giving a rate of dissociation of about 10^7 s or about 4 months). Thus, there are major advantages for the cell to have dimeric actin filaments that will assemble and disassemble from the filament ends, primarily at sites of seeding.

9.6   Multimeric Filaments

Microtubules constitute the major multimeric filaments in cells. They are hollow tubes, about 25 nm in diameter, and are composed of 13–16 protofilaments each that extend roughly parallel to the long axis of the cell. Microtubule seeds, from which tubulin will polymerize, are primarily gamma tubulin circles. Seeds stabilize the end of the filament, and therefore the probability of forming a complex without a seed is very low. Thus, the vast majority of tubules will form on active seeds. These seeds are typically found only in very specific regions of the cell. The fact that microtubules only form in specific sites confers on them the ability to organize cell architecture. Furthermore, as microtubules are polarized, cell polarity can be defined by stabilization of microtubules pointing in a given direction.

9.7   The Consequences of Filament Polarity

Actin filaments and microtubules have a polarity that can be defined from various perspectives. First, the subunit structures are asymmetric. Second, the rate of polymerization is greater at one end than the other. Third, individual motor proteins will only move in one direction along the filaments. Finally, the ends are dramatically different in terms of the components that bind to them. An important aspect of the polymerization process is that it requires energy and prior to their polymerization into actin filaments and microtubules, the actin and tubulin subunits are bound to ATP and GTP, respectively. The structures of individual actin monomers and tubulin dimers are known and show a clear asymmetry in the structures at the two sides that show complementary docking of the subunits. This asymmetry means that different energy barriers must be overcome during polymerization, causing different rates of assembly at the two ends. Typically, filament seeds block the slow growing end and catalyze polymerization from the

Figure 9.3

Polarity, polymerization, and depolymerization.

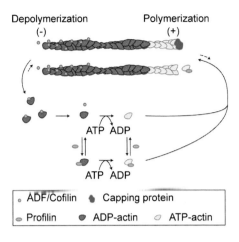

fast-growing end. The inherent polarity conferred by subunit asymmetry, and differing polymerization rates between filament ends, produces a polarized filament on which motor proteins can move in a unidirectional and defined manner. These motor proteins often carry cargo that must be efficiently delivered to specific locations, and therefore the unidirectional movement of the motor proteins is important. In general, the polarity of actin filaments and microtubules is critical for many of their functions. In contrast, intermediate filaments have anti-parallel subunits, are non-polar, do not facilitate motor protein movement, and assemble and are anchored in different ways.

9.8 Energetics of Polymerization and Depolymerization

As mentioned above, the polymerization requires energy and the ATP or GTP subunits hydrolyze the nucleotides after polymerization. This has several important implications; first, polymerization can produce force as is known for both actin filament and microtubule formation; second, the energy used to assemble the filaments can be recovered in depolymerization (seen for microtubules); and third, the filaments can treadmill by polymerizing and depolymerizing. One way to consider treadmilling is that high-energy subunits polymerize onto one end faster than they do on the other, so the subunits on the fast-growing end are more likely to contain the nucleotide triphosphate than the subunits at the slow end. Thus, the slowly polymerizing end will be more likely to have nucleotide diphosphate subunits that will depolymerize much more readily. Another consideration is that the nucleotides are hydrolyzed soon after the subunits polymerize and so the center of the polymers contain ADP/GDP-bound subunits that will more readily disassociate from the filament. When the slow-growing end has ADP/GDP subunits, depolymerization will proceed rapidly, yielding free subunits that will exchange the ADP or GDP for ATP or GTP. Those free subunits will now preferentially polymerize on the fast-growing end, driving the treadmilling of the filament. Thus, polymerization of cytoskeletal filaments involves energy consumption, and can be utilized by the cell to drive several functions. For example, extension of the cell's leading edge is often driven by force generated during actin polymerization. Similarly, the separation of chromosomes in anaphase is driven partly by force from microtubule depolymerization.

In several cases, actin filaments will undergo treadmilling where there is assembly typically at the periphery, and simultaneous disassembly toward the center of the cell. This can produce a net movement of the filaments inward from the cell membrane because of force from the membrane but primarily from myosin motors. Although we do not fully understand how hydrolysis is controlled, it is clear that it results in conformational changes within the filament and these conformational changes will weaken the interactions between the subunits, destabilizing the filaments. In terms of the energy available from polymerization, hydrolysis of a single ATP molecule releases about 80 pN-nm of energy, and for every actin monomer added, the filament will extend by 2.7 nm. Theoretically, the addition of an actin subunit could therefore produce about 30 pN of force. What limits the force is the detailed nature of the polymerization process. The force of actin polymerization has been estimated to be about 15 pN if the filament end is anchored to the membrane by a protein complex, like the protein VASP. This complex is postulated to form a porous cap over the filament, letting in ATP–actin subunits that elongate the filament and moving out when interior actin subunits hydrolyze the ATP because of a higher affinity for ATP actin (Dickson & Purich, 2002). Thus, the energy of polymerization of actin can be harnessed for the extension of the plasma membrane either in filopodia or lamellipodia.

9.9 Microtubule Dynamics (Nucleotide Hydrolysis, Polarity, Treadmilling, and Dynamic Instability)

In an analogous way to actin, the tubulin dimer contains two bound GTPs, one of which is hydrolyzed after polymer formation. The rate of hydrolysis is believed to follow a first-order decay process after the subunits assemble in the polymer. Following hydrolysis, the GDP subunits have a much higher K_d for binding (weaker binding) which results in filament instability and increases disassembly. Energy from the assembly of microtubules drives the movement of end-associated proteins toward the periphery of the cell along with some associated organelles. Thus, microtubule polymerization is an energy-requiring process and can do work.

Microtubules grow from the fast-growing end until the microtubule reaches the actin filament-rich cortex of the cell or until it is broken. The slow-growing or

Figure 9.4

Microtubule dynamics.

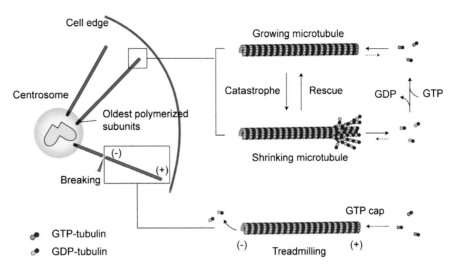

minus end is normally anchored by the gamma tubulin seed typically at the centrosome. At the cell edge, mechanical slowing of polymerization will increase the probability that the subunits at the growing end are bound to GDP, and this will stimulate depolymerization. Alternatively, the mechanical buckling of the tubule (often seen as tubules reach the edge of the cell) will encourage breaking of the tubule and a rapid depolymerization, because the subunits in the middle of the tubule will all be in the GDP-bound or lower-affinity state. When the slow-growing end is not capped with a seed, which may occur if the tubule has broken in the middle, GDP-bound subunits can disassociate from the slow end while GTP-bound subunits can be added onto the fast-growing end (also known as the 'plus end'). This gives rise to a treadmilling process that is similar to that observed in actin filaments. The addition of GTP-bound subunits to the plus end results in a GTP cap that is stable for a period. The stochastic rates of GTP-bound subunit addition and GTP hydrolysis can result in spontaneous depolymerization, a phenomenon called 'dynamic instability.' In cells both treadmilling and dynamic instability have been observed.

Description of Filament Systems

For tubulin: the critical concentration is 5 µM and the concentration in cells is on the order of 20 µM.

For actin: the critical concentration is 0.1 µM and the concentration in cells is on the order of 200 µM.

9.10   Microtubules as Cytoplasmic Organizers

Because microtubules need large seeds, and are relatively stiff, they can be used to organize the contents of the cell cytoplasm. In practice, microtubules originate from one or a few sites within cells, typically at the microtubule-organizing center (MTOC) or centrosome. With the help of motor proteins designed to move along microtubules, the microtubule network helps define the order in cytoplasm. As seen in Figure 9.5, there are several different patterns to the organization of microtubules in cells. Microtubules within fibroblasts originate from a central MTOC and radiate in many directions except through the nucleus. In neurons, there is a polar array of microtubules that extends into the axon and enables the transport of components to the axon tip and the repair of microtubules in the axons occurs through seeding complexes in the axons. Epithelial cells, on the other hand, have the microtubule minus ends at the apical surface often in a diffuse MTOC and the plus ends are in the basolateral area. Thus, material can be transported from the apical to the basolateral area simply with microtubule-associated motor proteins.

There is often an additional step in establishing the microtubule array, which occurs over time and involves covalent modification of microtubules pointing in critical directions. Covalently modified stable microtubules can provide an array of transport pathways that are used by motor proteins that interact specifically with modified tubulin subunits in the tubule, such as the kinesin motors. Stable microtubules often point in the direction that facilitates critical transport functions. For example, they may point toward the leading edge of a migrating cell, or extend the length of an axon. The major modifications that contribute to this organization are detyrosination, acetylation and glutamylation, although many others are possible (Wloga & Gaertig, 2010). In most cases the rationale for the modifications is not obvious, although they are believed to selectively modify the activity of motor proteins. Preventing these modifications *in vivo* by mutating tubulin or knocking out the modifying enzyme reveals significant phenotypic changes that are believed to occur through the alteration of cell organization. Thus, the stabilization of microtubule dynamics and the subsequent post-translational modification of the microtubules provide a simple means to regulate the transport of components within cells.

Figure 9.5

Organization of microtubules.

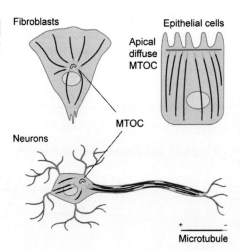

Microtubule dynamics are important for maintaining the required distribution of the microtubule array through random growth and stabilization. In growing fibroblasts, the microtubules will turn over in the range of 5–10 minutes. This was established by adding inhibitors of polymerization (e.g. nocodazole), which did not destabilize the microtubules but only prevented their regrowth. The majority of microtubules disappeared within 5–10 minutes of drug exposure, indicating that they depolymerized naturally in that period because regrowth alone was blocked. Further, studies of the growth and shrinkage of microtubules have confirmed that the dynamics occur as rapidly as predicted. These random dynamics allow active cells to determine where cellular components or material should be directed. For example, microtubules contacting cell-matrix adhesions will be stabilized, enabling modification of the tubulin that will preferentially support transport of components to the adhesion site. This dynamic method of determining transport delivery sites is robust and avoids the necessity of having a chemoattractant or similar system. In mitosis, the binding of microtubules to the chromosome kinetochores is a critical step for the delivery of the daughter chromosomes to the daughter cells. Microtubule dynamics ensure this step is carried out efficiently by enabling microtubules to find the kinetochores rapidly. For many cellular processes, microtubule dynamics is required and this means that long-term stability is rare, and only desired in a few cases, such as in cilia and flagella.

9.11 Cilia (Primary Cilia) and Flagella

In mammals, both cilia and flagella are relatively stable, microtubule-dependent structures that function in propelling materials out of the respiratory passages or propelling sperm to an egg. A human mutation in the motor protein dynein, which blocks motility of respiratory cilia and sperm flagella, is relatively benign in humans, but does cause the rearrangement of internal organs such that the heart is on the right as opposed to the left side, as well as causing infertility. The primary cilium (one per cell) is a related microtubule-dependent structure that senses the flow of fluid past cells. It extends 3–7 μm out from the apical surfaces of epithelia, or the luminal surface of endothelial cells, and is therefore ideally positioned to sense the flow of fluid past the cells. Mutant proteins associated with the base of the primary cilium have been linked to polycystic kidney disease, and the presumed alteration of sensory function. How the abnormal sensing of flow is linked to the necrosis of the kidney is not understood. Thus, cilia and flagella play important specialized roles in mammals, both in motile and sensory functions.

Figure 9.6

Primary cilia.

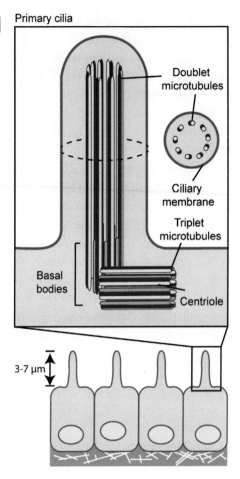

Primary cilia

Doublet microtubules

Ciliary membrane

Triplet microtubules

Basal bodies

Centriole

3-7 μm

9.12 Intermediate Filaments

The primary functions of intermediate filaments are to be mechanical stabilizers of tissue or cell integrity. *In vitro* studies using atomic force microscopy found that intermediate filaments can stretch dramatically at a constant force of about 200 pN (which is in the range expected at cell–cell and cell–matrix adhesions). Thus, intermediate filaments could restore the normal form of cells after tissue or matrix deformation. If the filaments were prestretched during their assembly, then they would be able to sustain forces on adhesion sites in the absence of continued myosin contraction. In most cells, however, the intermediate filaments appear relaxed and wavy.

9.13   Filament Mechanics

All of the different filaments in the cell behave mechanically in a similar manner that can be described by the stiffness of the filaments. An important aspect of the mechanics of a cell is the consideration of polymer or filament packing, which necessarily involves bending of the polymer or filament. A microtubule will have a considerably greater rigidity than an actin or intermediate filament.

The mathematical description of the bending resistance is given by the beam equation,

$$M = E^*I/R$$

where M is the bending moment (M = FX), and E*I is the flexural rigidity that is the product of the Young's modulus (E) and the second moment (I). The second moment is a geometrical factor that is related to the geometry of the beam. For a cylinder, $I = \pi r^4/4$. For a rectangular beam of width b and height a, $I = ab^3/12$. To estimate the bending of a beam, we need to adopt a different formulation and a common formulation is to describe the tangent angle at each point along the beam, $\theta(s)$. The angle of the bend is related to the parallel length displacement, dx, and perpendicular displacement, dy, for a given displacement along the beam, ds, by the relationships, $dx/ds = \cos\theta$, $dy/ds = \sin\theta$, and it follows that $d\theta/ds = 1/R$. The beam equation can then be rewritten as

$$d\theta/ds(s) = M(s)/EI$$

$$\text{or} \quad d^2y/dx^2 = M(x)/EI$$

and this formula can be used to derive the displacement of the end of a beam of length L, by a spring constant, K.

$$K = F/y(L) = 3EI/L^3$$

For a small glass rod of radius 0.25 μm and length of 100 μm, E = 70 Gpa, $I = (\pi/4)r^4 = 3 \times 10^{-27}$ which gives a spring constant K = 0.64 pN/nm. Note that increasing the length to 400 μm will decrease K to 0.01 pN/nm, whereas increasing the radius to 0.5 μm will increase K to 2.56 pN/nm.

9.14   Persistence Length

For a polymer it is useful to describe the rigidity in terms of the length over which the rod loses any correlation between the angles at the ends. The principle of

Equipartition of Energy can be used to derive the relationship that the persistence length, Lp, is proportional to the flexural rigidity

$$Lp = EI/kT$$

9.15 Freely Jointed Chain

Another way to treat the matter of the polymer length is to consider the polymer as a freely jointed chain with n links of a length b. In terms of the persistence length, the length of the links in the freely jointed chain can be described as

$$2Lp = b$$

Thus, some polymers in cells can be considered as a freely jointed chain [for example, DNA (persistence length of 53 ± 2 nm), RNA, and long fatty acid chains]. However, many of the filaments have a very long persistence length that exceeds the average dimensions of a cell (the persistence length of actin filaments is 10–20 μm and of microtubules is 1–6 mm).

9.16 Consequences for Cellular Mechanics

From many cell-deformation studies, it is clear that the actin cytoskeleton is the major factor that defines cellular mechanics. Because the cytoskeleton is dynamic and the cell is mechanoresponsive, the timescale of deformations of the cell are critical. Rapid deformations of cells often cause rapid responses in the cytoskeleton that typically increase the rigidity of the cell. Strain hardening is often found as well, both *in vivo* and *in vitro* with quasi-equilibrium networks. Hardening is due to alignment of filaments and greater forces on crosslinking proteins that can occur rapidly upon strain. However, longer-term increases in contractile forces are often stimulated by rapid strains. The cell type, matrix composition and the activity of neighboring cells will all have major effects on the cell mechanics. In epithelial cells with a cuboidal shape, for example, the apical actin network is critical for cell–cell adhesion-mediated shape changes and tissue mechanics. Furthermore, in fibroblasts, there is an extensive actin network that is responsible for developing matrix forces involved in matrix assembly. At the tissue level, intermediate filaments will strengthen skin and enable it to deform reversibly without damage by stretching and retracting. Macrophages and active immune cells are also very responsive to mechanical perturbations and can rapidly extend actin to engulf a bacterium upon stimulation.

9.17   **Molecular Mechanics**

In order to fully understand force transduction in the cytoskeleton, it is important to consider how forces affect individual molecules. Many recent studies have shown that cytoskeleton-associated proteins are stretched and unfolded in the course of normal cell motility. Most notable is titin, which is a muscle protein that links myosin filaments to z-lines in the sarcomere such that it is stretched with muscle stretch, and talin, which is in focal adhesions and binds vinculin after being stretched by actin flow. Because the cytoskeleton bears much of the tension in the cell, it is not surprising that many cytoskeleton-associated proteins are under tension and are unfolded as a result.

Protein unfolding can be described as the transition from a folded protein to an unfolded state. We will consider here a long protein such as talin, which has many similar domains that can unfold to produce about an eightfold increase in molecular length. There is normally an energy barrier that must be overcome to unfold a protein. A plot of the free energy of talin versus the overall length of talin will feature a peak that corresponds to the energy barrier that must be overcome to unfold the weakest domain. If we consider the fact that Brownian forces are working at unfolding the molecule, and domain motions are on the order of 10^9–10^{10}/s, even a high-energy barrier can be crossed occasionally. Pulling force on the molecule lowers the energy barrier for unfolding because the force times the displacement gives energy that lowers the barrier to unfolding (the force times the distance to the peak of the energy barrier gives the degree to which the free energy was decreased).

$$\Delta G_{force} = \Delta G_{norm} \text{-} F \cdot \Delta X$$

The rate constant for the unfolding will be increased by the Boltzmann relationship from the original rate constant

$$k_1 = k_1{}^0 \exp\left[F \cdot \Delta X / kT\right]$$

Applying force to talin using magnetic tweezers reveals that its domains will unfold reversibly with forces in the 6–8 pN range increasing the length of the protein to over 400 nm (Yao et al., 2016). Unfolding of talin domains will expose alpha helices that are normally bound with other alpha helices in the 4–5 helix domains of talin. Those helices have binding sites for another actin filament binding protein, vinculin, and perhaps binding sites for other proteins. After vinculin binds to exposed talin helices, the helices will remain in the unfolded form for a while and that is helped by vinculin binding to actin, which applies force to vinculin. Because different domains of a protein like talin unfold at different forces, a single protein can sense a range of forces, or perhaps more importantly, a range of displacements through the binding of different proteins to different helices. The force-induced effects on protein–protein binding interactions are difficult to study *in vitro* with whole proteins. However, new techniques enable analyses of binding interactions *in vivo* under force. There are other aspects of protein unfolding. For example, cytoskeleton proteins and matrix proteins can unfold at relatively low forces, but when the unfolded domains are fully stretched, the force rises dramatically with further stretch and that produces a behavior called strain hardening. This also means that the amount of unfolding is often limited, and is determined by the amount of strain, as opposed to the force level above a given threshold. Because of this, cells generally adjust the tension to a determined level that is related to myosin activation. Most confusing is the fact that the cytoskeleton filaments are constantly turning over and often moving, which will result in relaxation of stretched proteins after a relatively short period. Tension in the cytoskeleton causes protein unfolding and further unfolding-dependent protein interactions that can signal to the cell that the proper tension is established in the cytoskeleton. Thus, a dynamic tension is maintained in the cell by constant assembly of filaments and their contraction and that is critical for maintaining tissue shape and proper tone.

9.18   Organization of the Cytoskeleton

Because the cytoskeleton is involved in many cellular functions, it can take on many different forms. The modes of cytoskeleton organization, and the specific actin polymerizing proteins involved, will vary from cell to cell and they may also differ between cell phases. However, there are characteristic behaviors that are reproduced in similar circumstances. Those standard behaviors can produce emergent behaviors in response to stimuli from the microenvironment. There are several mechanisms associated with actin polymerization, or the anchoring of actin to other actin filaments, or to adhesions, that underlie some of the basic cell morphologies.

9.18.1   Formin-dependent Actin Polymerization

The polymerization of actin by formins first requires the localized binding of the formin dimer to a preferred site followed by activation of actin polymerization by the formin dimer. The formin dimer interacts with the barbed end, or growing end, of the actin filament, and it catalyzes the addition of actin monomers to that fast-growing end. Formin potentially interacts with small G-proteins, the plasma membrane, or membrane-bound proteins, and this enables it to anchor filaments to the membrane. Recent studies have shown that pulling forces on the filaments increase the rate of actin polymerization. Logically, if filaments are to originate from a specific site, then there should be a mechanism that activates formins at the same site. Alternatively, actin filaments should be crosslinked to that site. Because formins are found at the z-lines in muscle sarcomeres, they are logically responsible for assembling actin filaments at those sites. They are also localized along stress fibers at sites separated by 1.5–2 μm. As stress fibers turn over approximately every 1–2 minutes, it is possible that formins are responsible for assembling new stress fiber filaments. Several other proteins also contribute to formin-dependent actin polymerization. For example, in muscle z-lines, the actin filaments are capped at their barbed ends by the CapZ protein to prevent further growth of the filaments. Thus, formins must dissociate from the filament, and allow CapZ to cap the barbed end of the filament when it reaches the right length in the sarcomere. In addition, α-actinin is involved in anchoring the actin filaments at the z-line and needs to bind to the capped filaments after assembly. Local activation of the formins is the critical first step in the formation of many actin assemblies including sarcomeres, stress fibers, some lamellipodia, and filopodia.

9.18.2   ENA/VASP and WAVE

Consistent with the hypothesis that actin polymerization needs to be activated at important sites such as the sites of actin-dependent extension, the ENA/VASP proteins are found at the leading edges of lamellipodia and filopodia. In addition, they are found at adhesion sites. These actin-polymerizing proteins have multiple weak actin-binding domains located in long, finger-like regions that extend from their anchorage site. Weak actin-binding domains serve to concentrate actin monomers in a specific region and facilitate the polymerization process. As we discussed earlier, the presence of a multimeric protein complex that anchors the polymerizing end of the filament to the membrane can theoretically increase the force that actin polymerization generates on the membrane by about threefold compared to bare actin filament ends. This would mean that the lamellipodia could impose much higher forces on their environment, and hence move more rapidly. There is a logical question about how ENA/VASP is localized to sites at the membrane, where cytoskeleton dynamics is more pronounced. Recent studies indicate that ENA/VASP proteins are often anchored to the plasma membrane by BAR domain proteins, which help to bend the membrane at the lamellipodial tip. With ENA/VASP proteins polymerizing actin filaments against the membrane, protrusion will occur at that bent region. This is particularly relevant during fibroblastic migration, where several actin-polymerizing activities drive the critical step of membrane extension. Thus, the extension of membrane protrusions is often dependent upon the polymerization of actin by localized ENA/VASP proteins.

9.18.3   Arp2/3 Branching

In some cases, new filaments are produced from the side of an existing filament, rather than from a filament seed. This is termed 'branching,' and in lamellipodia, podosomes and bacterial comet tails, this is related to the activity of the Arp2/3 complex. Podosomes are actin-dependent protrusions that occur at adhesion sites in the absence of matrix rigidity or force and may help cells to invade through those weak matrix sites. Further, they are related to invadopodia that are produced by cancer cells and correlate with the metastasis of those cells. The Arp2/3 complex is recruited to filaments by N-WASP, and produces a branch that will then elongate. Initially, the branching of actin filaments was believed to be essential to push the membrane forward during motile functions; however, several recent studies have shown that the extension of cellular protrusions can occur with few branched filaments. However, some structures, like podosomes and bacterial actin comets, do not form when Arp2/3 activity is inhibited. Arp2/3 appears to bind more readily to bent filaments, and therefore, a mechanical factor may function in the recruitment of Arp2/3. In this case, higher forces cause increased actin bending, and hence, Arp2/3 binding. It follows then that the Arp2/3 system may not be needed in low force situations, but will be recruited as the force increases.

Figure 9.7

Formin-mediated nucleation of actin filaments and Arp2/3-mediated actin polymerization.

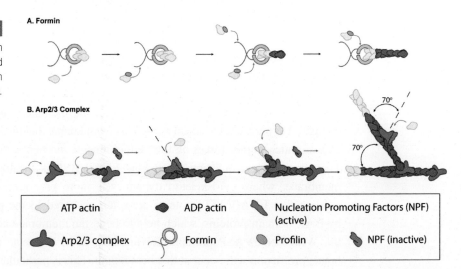

9.18.4 Actin Filament Disassembly

Actin cytoskeleton depolymerization is important because the actin subunits are continually needed at other sites to feed the polymerization of new filaments. Where filaments disassemble has an important effect on cytoskeleton organization and is critical for the dynamics. There are a variety of ways in which filament disassembly occurs. For example, the breaking of filaments, which is favored by a higher number of interior subunits being in the ADP-bound state, dramatically increases the depolymerization rate. In addition, there are factors like ADF/cofilin that will destabilize the actin in the filament by causing a rotation of the subunits relative to one another, i.e. changing the pitch of the filament from 75 to 57 nm for a complete turn. Fragmentation of the filaments can also be mechanical, as myosin can easily fragment actin filaments by bending the filaments during contractions. Although any off-axis stress tends to break the filaments and create ADP–actin ends that rapidly depolymerize, the effect of ADF/cofilin appears more reliable. When the activity of ADF/cofilin is increased by blocking its phosphorylation, the width of the lamellipodium is decreased. This indicates that the amount of active ADF/cofilin is inversely related to the final width of the lamellipodium.

Other factors also contribute to actin dynamics. For example, it remains unclear whether active mechanisms exist to move actin filaments to the leading edges of lamellipodia. Indeed, forward movement of actin has been observed; however, this may simply be the passive diffusion of actin monomers to the sites of polymerization where the free actin monomer concentration is low. Other proteins, like integrins, were also seen to move forward in lamellipodia and actin could be carried with those proteins. For active movement forward, there should be a myosin motor and some of the single-headed myosins move rapidly to the leading edges of active lamellipodia. Members of the myosin 1 family of proteins need to be crosslinked in order to have the multiple heads required for myosin-dependent movement. In this scenario, the fragmentation of actin filaments by off-axis forces from myosin contraction or other means would generate short filaments that could link to myosin 1 tails, thereby creating myosin 1 complexes that could move back to the leading

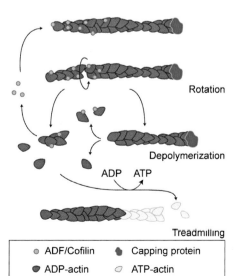

Rotation

Depolymerization

ADP   ATP

Treadmilling

| ⊙ ADF/Cofilin | 🌢 Capping protein |
| ⬤ ADP-actin | ⊘ ATP-actin |

edge. Additional elements would be needed to target the motors to fragmented filaments and to reverse the binding at the leading edge.

9.18.5   Actin Filament Crosslinking

Because actin filaments are relatively flexible and break easily when bent, the structural integrity of the cytoskeleton relies heavily upon actin crosslinking. Linkage of neighboring filaments by crosslinkers greatly increases the rigidity (strength) of the actin cytoskeleton and even enables the actin cytoskeleton to support significant compressive or bending forces. This occurs in filopodia where the actin-crosslinking protein, fascin, is correlated with filopodia growth from lamellipodia. Protrusion of actin filament bundles from sperm heads to reach the egg surface also requires bundling proteins to strengthen the projections.

Another type of actin crosslinking occurs in lamellipodia where the filamin A dimer forms a V-shape crosslink of two actin filaments. There is further evidence that mechanical force on the filamin A molecule can change the angle and cause the binding or release of other cytoskeleton proteins for mechano-sensing. In addition, filamin has a binding site for membrane proteins that are linked to the external matrix molecules. This can greatly strengthen the lamellipodium. Again, it is important that the crosslinking is dynamic and that it is localized to the correct regions of cytoplasm.

9.19   Microtubule Dynamics

Microtubules are also dynamic, but differ from the actin cytoskeleton in that they are, on average, much longer and much less abundant. Microtubules contribute to cellular organization by providing roadways on which motor proteins will transport specific components to the correct place in the cytoplasm. Because there is no way to control the direction of polymerization precisely, microtubules will elongate in various directions from the centrosome and those pointing in the correct direction will be stabilized. The growth of microtubules can be followed by monitoring the movements of end-binding complex proteins like EB1, which only binds to the ends of growing microtubules.

Figure 9.9

Actin filament crosslinking.

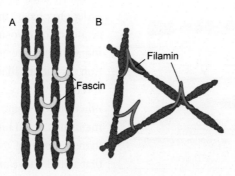

Because the end-binding complex at the tips of growing microtubules contains a dynein cofactor and peripherally bound dynein orients the microtubule array by pulling it toward the periphery (dynein moves toward the minus ends of the microtubules that are anchored at the centrosome), this complex may be an important component in keeping the microtubule array positioned properly within the cell. In neurons, there is an additional problem that different components need to be transported from the cell body down dendrites than down axons. Both types of processes rely upon microtubule-based transport and the question has been how they accomplish this sorting. Surprisingly, the polarity of microtubules in dendrites is mixed, whereas the polarity in axons is always with the plus ends out. Mixed polarity is observed in early axons and is postulated to be part of the early extension process similar to spindle elongation in mitosis. At a certain length, the axons switch to become polar arrays and this requires dynein. The current hypothesis is that dynein is needed for polarization and it may also aid in the modification of one orientation of the microtubules in dendrites to produce a preferred polarity. Thus, microtubules grow in multiple directions and those that reach the proper sites are stabilized and undergo post-translational modifications that then favor the transport of the needed components on those microtubules.

There are a number of circumstances where microtubules can be broken and when this occurs tubulin–GDP ends, which rapidly depolymerize, are exposed. Breakage can occur, for example, through the activity of the cutting protein, Katanin, and this is known to rapidly affect microtubule distribution. In addition, there can be mechanical bending and fracturing of microtubules in the periphery of cells, and this will result in the rapid depolymerization of the remaining fragments. Many of the microtubules that penetrate into the lamellipodial actin network will be pulled inward with the actin network, causing them to buckle or bend, and to sometimes break. Furthermore, individual microtubules are not stiff enough to stabilize the cytoplasm in the face of myosin motors, and bundling is usually not observed in cytoplasmic microtubule arrays. All of these factors contribute to the dynamics of microtubules and further reinforce the idea that cytoplasmic microtubules are randomly sampling cytoplasm and when they contact an important site, they will be stabilized causing subsequent modification of the tubulin.

An additional factor in microtubule dynamics is their depolymerization by certain kinesins. In this case, depolymerization can occur without requiring ATP as the motor protein translocates along the microtubule. Such a mechanism has been observed when the depolymerization of spindle microtubules by kinesins generates force to pull apart chromosomes without an input of additional energy. It should be noted that microtubule arrays can reform from fragments that arise during breakage or depolymerization in cells that lack a centrosome. In that case, specific complexes activate gamma tubulin and new microtubules are formed from the deploymerized tubulin at sites adjacent to existing microtubules. Thus, microtubule depolymerization is important for the dynamics of the structures and for providing dimeric subunits for the polymerization of new microtubules.

9.19.1   Mitosis and Cytokinesis as Cytoskeletal Functions

During mitosis and cytokinesis, division of the mother cell produces two daughter cells. These are complex processes that involve major reorganization of the cytoskeleton and the cessation of many interphase cell functions. This is also a great example of how cellular phases can change cell behavior totally with only a minor change in cell composition, mostly due to proteolysis of important cell-cycle kinases. Initially, the microtubules are reorganized by the duplication of the centriole and the movement of the two centrioles to the opposite ends of the cell to form the spindle driven by a multimeric kinesin complex that moves the centrosomes apart by walking toward the plus ends of the interpolar microtubules from each centriole. Microtubule dynamics are increased as the astral microtubules try to find the chromosome kinetochores so that all of the chromosomes can be aligned at the metaphase plate. When they find a kinetochore, forces on the microtubules will align the chromosomes at the metaphase plate. Once aligned, the bonds between the chromosome pairs are cleaved and one of each

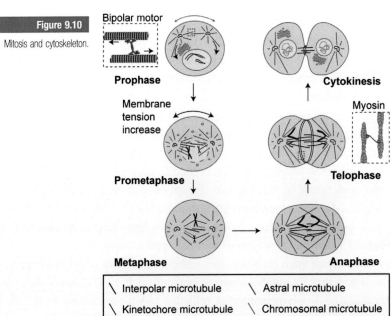

Figure 9.10

Mitosis and cytoskeleton.

chromosome pair is transported to each spindle pole along the chromosomal microtubules. To further separate the chromosomes the spindle is elongated by kinesin complexes pushing on the interpolar microtubules. Then the cell activates an actomyosin system that is positioned halfway between the poles to bisect the cell into the two roughly equal daughter cells. This whole process is very robust with many different sequential steps and checkpoints to assure that processes are completed before the next step is started.

From the analyses of mitosis, the process is clearly robust and several alternative pathways can complete nearly every step in the process. This means that although a single protein mutation in one of the pathways may change the process quantitatively, it may not change the overall outcome, because the process can proceed through alternative pathways. The presence of a number of redundant pathways to perform a function enables some cells to favor one or another depending on the conditions of the cell. In an evolutionary sense, the critical importance of mitosis means that efficiency of one pathway over another is less important than having an alternative pathway that can complete the process when conditions inhibit a more efficient pathway. Thus, robustness wins over efficiency.

9.20 SUMMARY The eukaryotic cell cytoskeleton is comprised of actin filaments, microtubules and intermediate filaments, and determines the shape of the cell based upon its response to environmental signals. Seeding of filaments is critical for defining where filaments form and how they are anchored. Because the cytoskeleton is dynamic on the timescale of seconds to minutes for mammalian cells, its organization at any given time is only a snapshot of a dynamic process. This means that external deformations of cells will result in adaptation of the cytoskeleton, which may be driven by forces applied to dynamic filaments. There are multiple mechanisms for actin polymerization that are tailored to particular cell functions. Furthermore, the specific system that is activated for a given function is critical for understanding that function. In other words, neither formins nor WASP nor any other actin-polymerizing protein is universally activated, and each function needs to be analyzed for the distinct set of actin polymerizing proteins that are important. Similarly, microtubules are seeded, but only from a single, or at most a few, sites within the cell. Microtubules are also dynamic, and their random extension samples the cellular space to find the proper sites where they will be stabilized and modified for the transport of materials to those sites. Because they transport the intermediate filament subunits, they also define where intermediate filaments will form. In the process of mitosis, the cell stops most interphase functions and reorganizes the cytoskeleton initially through the formation of the microtubule spindle to separate the daughter chromosomes and finally in the actomyosin ring that bisects the cell into two daughter cells. In all of these cytoskeletal processes, there appear to be many sequential steps that are controlled by the physical location of polymerization, the site of activation of motors and the region of filament depolymerization. Thus, the cytoskeleton provides the major components of the tool set that the cell needs to shape itself and aspects of its environment.

9.21 PROBLEMS

1. In the theoretical filaments that are linear assemblies of subunits where the binding constant for the dimer is the same as the binding constant for the n + 1 mer, the filament length is limited because of what? Describe how you would compute the average filament length from the dissociation constant and the concentration of the protein (assume that the on rate is diffusion limited, i.e. 2×10^7 M^{-1} s^{-1}).

Answer: The average length of the polymer would be simply the assembly rate divided by the dissociation rate because the polymer can be broken equally well at any point. The assembly rate will be the concentration times the on rate that is given and the off rate is calculated by multiplying the dissociation constant by the on rate.

2. If the K_d for the actin subunit–subunit interactions along a strand is 0.1 mM and the K_d for subunits at the ends of two-stranded filaments is 0.03 mM, then what is the K_d for a single inter-strand bond (assume that subunits that bind at the ends are bound by one intra- and one inter-strand bond)?

Answer: The free energy of the bonds should be additive, thus $\Delta G_{SS} + \Delta G_{IS} = \Delta G_E = -RT \ln K_{dE}$. Solving for $\Delta G_{IS} = \Delta G_E - \Delta G_{SS} = -RT (\ln K_{dE} - \ln K_{dSS}) = -RT \ln K_{dE}/K_{dSS} = -RT \ln 0.3$ mM or $K_{dIS} = 0.3$ mM.

3. A classic experiment in the microtubule field involved the dilution of the microtubule solution and the subsequent changes in microtubule number and length. The observation was that the dilution of an equilibrium solution of dimer and polymer resulted in the decrease in the number of microtubules but they had approximately the same length. How can you explain this result based upon the GTP hydrolysis after polymerization to destabilize the microtubule end?

Answer: This can be explained if the probability of starting growth depends heavily upon free tubulin concentration but the lifetime of the microtubule growth is dependent upon the probability of GTP hydrolysis at the growing end, which is assumed to be independent of total tubulin concentration. Further, once depolymerization of the microtubule starts, it depolymerizes completely. Under those circumstances the microtubules will disassemble until the free concentration is about the concentration needed for growth and then will grow to a constant length.

4. Consider a piece of spaghetti 1 mm in diameter. Young's modulus is ~10^8 J/m^3.
   a. What is its persistence length at room temperature (25°C)? Is the result consistent with your everyday observations?

Answer: The persistence length is described by the equation Lp = EI/kT where Lp is the persistence length and E is the Young's modulus and I is the second moment (I for cylinders is $\pi r^4/4$). At 25°C, kT is 4.11 × 10^{-21} J which then gives (10^8 J/m^2 × 3.14 × 1.56 × 10^{-14} m^4)/4.11 × 10^{-21} J = 1.2 × 10^{15} m which is surprisingly long. (Check this with DNA, which has a Young's modulus similar to spaghetti but has a radius of about 1.6 nm and you should get about 50 nm.)

   b. Please calculate spring constant for a 1 cm spaghetti piece. If you consider spaghetti as a linear spring, how many molecules of ATP have to be hydrolyzed in order to displace the end of a spaghetti piece by 3 mm?

Answer: The spring constant K = 3EI/L^3 where E is Young's modulus and I is second moment (I for cylinders is $\pi r^4/4$). At 25°C, kT is 4.11 × 10^{-21} J which then gives (3 × 10^8 J/m^3 × 3.14 × 1.56 × 10^{-14} m^4)/1 m^3 = 1.4 × 10^{-5} J/m^2 or 1.4 × 10^{-2} pN/nm. To calculate the energy for a displacement of 3 mm, we do the integral of Kxdx from 0 to 3 mm, which is 1/2Kx^2 or 0.7 × 10^{-2} pN/nm × 3 × 10^{12} nm^2 = 2.1 × 10^{10} pNnm. Because a single ATP has 80 pNnm of energy, this will require 2.62 × 10^8 ATPs.

5. Suppose that the ratio of substrate to product in a mixture is 10 times greater than the ratio at equilibrium. How much mechanical work could be obtained by converting one molecule of substrate to one molecule of product? Suppose that you have a total of 10^7 substrate plus product molecules in a cell and you can convert 10^6 molecules of substrate to product. What is the approximate amount of mechanical work that could be done by that cell?

Answer: ΔG = kT ln 10 = 2.3 kT = 2.3 × 4.11 pNnm = 9.45 pNnm. For the 10^6 molecules, you will get less than but on the order of 9 × 10^6 pNnm or about 10^5 ATP molecules.

6. Actin filaments and microtubules are typically formed from seeds because (explain why you consider the answer false or true):
   a. they have multiple strands in the filaments that interact;
   b. the site of polymerization can be controlled by the localized activation of seeds;
   c. multiple bonds between subunits stabilize internal subunits over those at the ends of filament; and
   d. motors need multiple strands for directed motility.

Answer:
(a) True because the energy of forming a dimer is less than the energy of adding a subunit to a filament;
(b) True, because this enables the cell to control where filaments form and even the direction of the filaments;

(c) True, because the free energy of removing a subunit from the filament is the sum of the free energies of the bonds; and

(d) False, because motors like kinesin follow a single protofilament in a microtubule.

7. A common diagram that describes the behavior of actin filaments shows binding and dissociation from both the barbed and pointed ends of the filament. I have copied the appropriate constants from a review article by Pollard (1986). If the concentration of globular actin is 0.5 mM and it is all ATP actin, then what is the rate of addition of subunits to the barbed end ($C \times k_+ - k_-$)? Compare that with the rate of addition to the pointed end. If the rate of addition is negative (the filament is shortening), then ADP actin subunits in the filament will often be at the end. How will that affect the rate of shortening?

Actin filament elongation rate constants in 50 mM KCl, 1 mM $MgCl_2$, 1 mM EGTA, pH 7.0 (from Pollard, 1986).

	ATP-actin Barbed end	ATP-actin Pointed end	ADP-actin Barbed end	ADP-actin Pointed end
K_+ ($\mu M^{-1} s^{-1}$)	11.6	1.3	3.8	0.16
K_- (s^{-1})	1.4	0.8	7.2	0.27
K_d (μM)	0.12	0.62	1.9	1.7

Answer: Using the numbers given, $0.5 \text{ mM} \times 11.6 \times 10^3 \text{ mM}^{-1} \text{s}^{-1} - 1.4 \text{ s}^{-1} = 5800 \text{ s}^{-1}$. Thus, the rate of elongation is not affected by the loss of subunits from the barbed end. For the pointed end, $0.5 \text{ mM} \times 1.3 \times 10^3 \text{ mM}^{-1} \text{ s}^{-1} - 0.8 \text{ s}^{-1} = 650 \text{ s}^{-1}$, the rate of addition will be nearly 10-fold lower. Because the rates of ADP-actin dissociation are greater than those of ATP-actin for both the barbed and pointed ends, the exposure of ADP-actin at either end will increase the rate of depolymerization.

8. If we think about the microtubule polymer, each tubulin dimer binds at its ends to the next subunit along the protofilament and along its sides to adjacent protofilaments (a total of 15 protofilaments is common in cytoplasmic microtubules). If the binding energy between neighboring dimers along the protofilament is −3 kcal/mole and between adjacent dimers in different protofilaments is −4 kcal/mole, then what is the energy needed to bring a dimer out of the wall of a microtubule? At the ends of the microtubule, the dimers only have one protofilament and one adjacent dimer bond; therefore, what is the energy of removing a dimer from the end? Using the diffusion limited on rate constant for binding (given in problem 1), what is the off-rate for subunits coming out of the wall of the microtubule or out of the end?

Answer: For each dimer there are two adjacent protofilaments (−8 kcal/mole) and two adjacent dimers in the same protofilament (−6 kcal/mole) giving a total of −14 kcal/mole

$\Delta G = RT \ln K_D$ and $2.3 \log X = \ln X$ or $2.3 \log K_D = \Delta G/RT$

$\log K_D = -14/0.002 \times 310/2.3 = -9.82$ or 1.6×10^{-10} M for the wall.

For the ends $\log K_D = -7/0.62/2.3 = 1.2 \times 10^{-5}$ M

$K_{off} = K_D \times 2 \times 10^7$ M^{-1} s^{-1} = 3.2×10^{-3} s^{-1} for end & 2.4×10^2 s^{-1} for end.

9. If we assume that the energy from actin hydryolysis of ATP in the filament can power filament elongation, then what is the force that can be developed by actin filament polymerization? Why might the actual force generated by actin filaments be greater if the filaments are restrained to polymerize perpendicular to the membrane?

Answer: Energy of ATP is 80 pN nm and the elongation of the filament per actin monomer addition is 2.5 nm, which gives a maximum force of 32 pN if the filament is perpendicular to the membrane. In the second part, the bending of the filament when it contacts the membrane at an angle would lower the maximal force that is generated because of the low bending stiffness of the actin filament.

10. Bacteria use the cytoskeleton to evade host defenses in general and lysteria specifically uses the actin cytoskeleton by:
    a. assembling a myosin on the bacterial surface when it enters the cytoplasm;
    b. assembling an actin-polymerizing complex on the bacterial surface when it enters the cytoplasm;
    c. secreting an actin-depolymerizing protein to fluidize the cytoskeleton; and
    d. crosslinking cadherins on the surface to stimulate actin-dependent endocytosis.

Answer: Only (b) is correct because it uses actin polymerization to propel itself into the neighboring cells without contacting serum proteins that would excite the immune surveillance systems.

9.22 REFERENCES

Alberts, B., Johnson, A., Lewis, J., Raff, M. & Roberts, K. 2014. *Molecular Biology of the Cell*. Philadelphia, PA: Saunders.

Dickinson, R.B. & Purich, D.L. 2002. Clamped-filament elongation model for actin-based motors. *Biophys J* 82(2): 605–617.

Pollard, T.D. 1986. Rate constants for the reactions of ATP- and ADP-actin with the ends of actin filaments. *J Cell Biol* 103: 2747–2754.

Pollard, T.D., Earnshaw, W., Lippincott-Schwartz, J. & Johnson, G. 2016. *Cell Biology*. Philadelphia, PA: Elsevier.

Wloga, D. & Gaertig, J. 2010. Post-translational modifications of microtubules. *J Cell Sci* 123(Pt 20): 3447–3455.

Yao, M., Goult, B.T., Klapholz, B., et al. 2016. The mechanical response of talin. *Nat Commun* 7: 11966.

9.23 FURTHER READING

Actin filament polymerization: www.mechanobio.info/topics/cytoskeleton-dynamics/
actin-filament-polymerization/

Actin nucleation: www.mechanobio.info/topics/cytoskeleton-dynamics/actin-nucleation/

Cytoskeleton: www.mechanobio.info/topics/cytoskeleton-dynamics/cytoskeleton/

Cytoskeleton dynamics: www.mechanobio.info/topics/cytoskeleton-dynamics/

Tubulin complex assembly: www.mechanobio.info/topics/cytoskeleton-dynamics/
tubulin-complex-assembly/

10 Moving and Maintaining Functional Assemblies with Motors

The generation of force and the active transport of material over long distances are processes that are important in a number of complex functions. Both force generation and transport are driven by molecular motors that utilize the energy derived from ATP hydrolysis to move along polar filaments of the cytoskeleton. This movement generates the forces needed for synthesis, packaging, and transport of cellular components, and enables the cell to do work and maintain tension in tissues. The best-understood motor proteins are myosin and kinesin, which move on actin filaments and microtubules, respectively. Other motor proteins that move on DNA, on RNA, and on the protein backbones have roles in synthesis, packaging, and degradation processes. In a broad view, molecular motors couple the Brownian movements of proteins (vibrational, rotational, or translational) to the dissipation of energy (from ATP or GTP hydrolysis, or from coupling to a proton gradient). The three major types of motors, myosin, kinesin, and AAA superfamily of ATPases, are significantly different in their designs and functions. Although single-molecule studies have provided many insights into the mechanisms of motor protein function, the rules that control their motility are not well understood. The mechanisms that regulate motor protein activity are complex yet highly important because the level of force and/or displacement that is generated must be tightly controlled in order for the complex functions to be carried out properly (for more background on motors, see Howard, 2000).

Functions Where Motors Are Essential

1. Motility (contractile force generation) (flagella in bacteria to muscles in humans).
2. Shaping of cells and tissues (moving cell–cell boundaries, traction forces).
3. Transport in cells (DNA to daughter cells, mitochondria in axons, large protein complexes, vesicular movements).
4. Packaging (DNA into phage capsids, chromosome condensation).
5. Synthesis of molecules (DNA, RNA, proteins by polymerases and ribosomes, respectively) ATP by F_0/F_1 ATPase.

Motor protein complexes constitute a major class of proteins that are needed for many functions, as outlined in the textbox above. Analyses of the genetics of motor proteins have been somewhat frustrating because knocking out individual motor proteins often produces 'unremarkable' phenotypes. For example, individual myosin I motor proteins, which represent a diverse group of single-headed motor proteins, can be knocked out in the amoeboid cells of *Dictyostelium discoideum*, with little effect. Only when several myosin I proteins are knocked out do cells show significant changes in function. In the early days of protein knockouts, this result was viewed by many as an indication that the myosin I proteins were unimportant and young faculty were discouraged from working on them because it would be doubly hard to get tenure. However, as described in the earlier chapters, we favor the explanation that the myosin I proteins are so important for the survival of the organism that they have several backup systems in place with overlapping functional capabilities. The lab environment does not stress the organism in the same ways it would be stressed in its natural environment and having many myosin I's could confer robustness. Further, specialized wild-type functions can involve slightly modified motor proteins, and overlap of activities may be sufficient to confer a similar, albeit suboptimal, performance for mutant cells grown in the lab. Thus, because rapid motility, synthesis and long-range transport are essential for survival in many cases, the motors are often redundant, with backup systems in place.

Large cells such as neurons and muscle cells can only survive if there is a system of rapid internal transport of components. As discussed in earlier chapters, such systems involve three primary components; specifically, lipid bilayer vesicles or large complexes of proteins that constitute the cargo, a network of filaments 'roadways,' and the motor proteins that carry the cargo along the filaments. In contrast to plant cells, which use myosin movement on actin cables for intracellular transport, animal cells use kinesin/dynein movement on microtubules. The opposing roles for these motor proteins in plants versus animals is further extended with the extracellular matrix/cell wall being organized by actomyosin contraction in animal cells but by kinesin/microtubules in plant cells (the cellulose synthetase system moves on microtubules as it deposits cellulose). For a long time, the micrometer-level movement of cells and organelles that could be seen using light microscopy constituted proof of life. Today, despite the presence of nanomanufactured devices that can produce movements with inanimate materials, such motor protein-driven movement is still considered a major hallmark of cellular life. Furthermore, the mechanisms of integration and control provide important insights into how stochastic events can be regulated at the cell and tissue levels to give the desired movements and forces.

10.1 Defining Linear Motor Systems

There are several classes of motor proteins that move on linear filaments. These include the polymerases, protein processing enzymes, DNA–RNA packaging enzymes, as well as myosins and kinesins, which move along actin filaments and microtubules, respectively. Common to almost every linear motor system is the requirement for a polar filament, and at least two ATPases in the active complex. ATP hydrolysis is usually linked to a cyclic process of motor function, which will differ for each protein. For example, skeletal myosin functions through cycles of ATP binding–release from actin → activation → binding → force generation and ADP release.

Linear Motor Systems

Common types (motor – polar filament)

Myosin – actin filament
Kinesin or dynein – microtubule
DNA polymerase – DNA
RNA polymerase – DNA
Helicases – DNA
Ribosome – RNA
Protein folding – translocation complexes in ER or the proteasome

Basic elements of linear motor systems

Polar filament with repeating asymmetric subunits
Motor molecule that moves in one direction on the filament
Cargo or object to be moved

Steps coupling ATP hydrolysis to movement (shown for myosin, but order differs for other motors)

1. ATP binding (typically the step that uses most of ATP's energy)
2. Release of motor unit from and diffusion along the filament
3. Transition to an activated state (often coupled with ATP hydrolysis)
4. Binding and movement
5. Release of ADP and Pi

In each case, two or more motor protein monomers are linked and form active complexes that maintain contact with the filament and sequentially generate force. Without continual contact, diffusive movements would rapidly separate the cargo

Figure 10.1

Linear motor system.

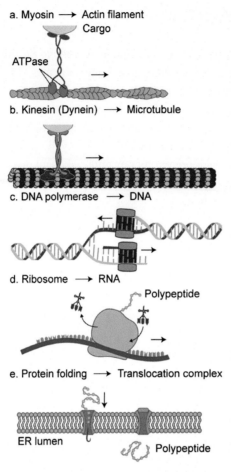

a. Myosin ⟶ Actin filament
Cargo
ATPase

b. Kinesin (Dynein) ⟶ Microtubule

c. DNA polymerase ⟶ DNA

d. Ribosome ⟶ RNA
Polypeptide

e. Protein folding ⟶ Translocation complex
ER lumen
Polypeptide

and motor proteins from the filament. Regulating motor activity is critical not only because force must be applied at specific sites in the cell, but also because there is usually a great excess of motor proteins present in the cytosol. If even a fraction of these 'free' motors were active, the cell would be rapidly dysfunctional and ATP consumption would increase dramatically. For this reason motor proteins are usually maintained in an 'off state,' and are only activated when they are integrated into the functional complex. Furthermore, the time in which they are active is usually limited, and motile events are generally intermittent as with term limits in robust devices.

The majority of cellular movements, and active cargo transport, are facilitated by three major classes of motor proteins, namely the myosins, kinesins, and AAA type of ATPases. For this reason, we will focus only on these classes in this chapter. In each case these motor proteins operate in two important regimes: the non-processive regime, which relies on highly organized arrays of motors and filaments, and the processive motor regime, in which a typically dimeric motor must remain attached to the filament as the motor moves in basically a hand-over-hand fashion. Although synthesis and processing systems involved in DNA–RNA–protein dynamics also utilize motor proteins, they are discussed in Chapters 13, 14 and 15, respectively, of this book. When examining the basic features of linear motor systems, it becomes clear that they all rely upon a consistent set of steps to fulfill their function (see Additional Reading). Despite the general motor protein activity occurring through roughly the same steps, the steps may be altered where a specialized motor is involved, or a specific function is to be performed. For example, muscle myosin is non-processive because it releases from the actin filament upon ATP binding, whereas the processive myosin V stays bound to actin in the ADP form until it is pulled off by the forward head. Control of motor protein function is a critical issue that has been studied much less than the mechanism by which motor proteins move. However, this control must involve sophisticated systems that activate/inactivate the motors in the right place and for a defined period of time. In addition, cells must integrate the pulling effect of millions of motors to develop the tension in a region of a tissue. Thus, the major area of future interest is not the details of motor movements at the nanometer level, but rather the control and integration of motors needed for proper cellular and tissue function.

10.2 How do Motors Move?

The details of motor movements have been studied extensively with subnanometer resolution and there is an extensive literature detailing the cycles and the degree of motion involved in each step in the cycles (see Figures 10.2 and 10.4). We will discuss myosin and kinesin here to highlight the differences between processive and non-processive motors. There are three major differences between myosin and kinesin motility; specifically, the mode of force generation, the degree of mechanical coupling between the motor heads, and the nucleotide form in the tightly bound state. However, both kinesin and myosin rely upon repetitive binding, movement and release from a polar filament substrate, in a process that is coupled to ATP hydrolysis.

The ATPase cycle begins with the binding of ATP. In the case of muscle myosin without ATP bound, the head is strongly bound to actin in a rigor (nucleotide-free) state. This term was derived from rigor mortis, where after death the muscles lose ATP and become rigid because the myosin heads are all strongly bound to actin filaments. When ATP binds to the myosin head, it releases from the actin and starts the longest part of the myosin cycle, while in the low-affinity state for actin, myosin will undergo a major change in conformation that will result in the priming of the head for movement. This change is often thought to involve ATP hydrolysis, and ends when myosin reaches a state where it can bind strongly to actin and move. Binding and movement are hard to uncouple because movement without binding represents a futile cycle. In the tightly bound, force-generating state, myosin has ADP in the nucleotide pocket and possibly a phosphate that is relatively rapidly released. If there is tension on the head from the neck region, then ADP will be retained and the head will remain bound. Only when the tension on the head decreases can ADP be released, and a fresh ATP molecule bound so that the cycle will start again. As mentioned, the increase in the lifetime of the ADP bound state with increased force will slow the ATPase cycle and will increase the fraction of the motors in the bound, force-producing state. Thus, the unloaded myosin cycle is more rapid than the loaded myosin ATPase

Figure 10.2

Kinesin movement cycle.

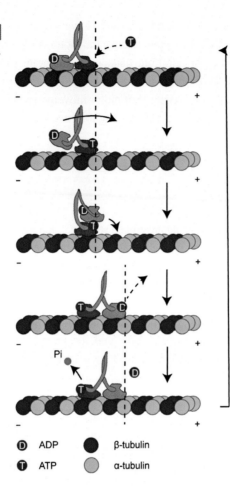

Ⓓ	ADP	●	β-tubulin
Ⓣ	ATP	●	α-tubulin

cycle and the difference is due to increased duration of binding in the ADP state with force.

In the case of kinesin, the ADP is released from one head in the tightly bound state allowing ATP to bind, here designated the first head. This will cause a change in conformation without dissociation from the microtubule. The change in conformation moves the free (second) head to a position where it can bind to the next tubulin dimer. This will be in the direction of the microtubule plus end. At this point, the second head, which is in the ADP-bound state, is able to bind to the microtubule and develop the tension required to promote ATP hydrolysis in the following head. The release of the following (first) head occurs after ATP is hydrolyzed and the leading (second) head will release ADP once the tension is relaxed. The cycle will then repeat.

Communication between the heads will ensure that a single dimeric molecule is able to hold onto the filament as it moves in a unidirectional fashion. ATP binding leaves the head in a high-affinity state. Historically, kinesin was initially purified by using the non-hydrolyzable analog of ATP, AMP-PNP, to lock the motor to the microtubules so that they would co-sediment. Then the motor was released from the microtubules with salt and high ATP concentrations. In normal motility, the free head, which exists in an ADP-bound state, communicates with the head that is interacting with tubulin, and in an ATP-bound state. As the ADP-bound free head binds and moves forward, a forward mechanical force pulls on the lagging ATP-bound head, and this facilitates the hydrolysis of ATP. The forward head will release ADP; however, if there is a high drag force, the release of ADP in the forward head will be inhibited and the hydrolysis of ATP in the lagging head will be slowed. Thus, under very high loads both kinesin heads should be bound to the microtubule, creating a stronger link to the microtubule.

In the case of cytoplasmic dynein, the molecular details of the movement are not known. Force measurements indicate that it is a weaker motor but moves over longer distances in each stroke. The recent crystal structure of cytoplasmic dynein dimer has spawned a number of models of its mechanism of motility. Many models focus on movements of the strut that supports the long microtubule-binding arm as the element that drives the movement on the microtubule. What remains a mystery is the coordination of the binding, movement and release steps between the two motor molecules in the dimer. The long length of the microtubule-binding arm (~20 nm) and its structure as an alpha helical coiled coil raise questions about how the mechanical cycle will drive the motor forward and generate significant force. Cofactors of cytoplasmic dynein, such as dynactin, can play a significant role in controlling the movements and aid in linking cytoplasmic dynein to cytoplasmic vesicles in motility assays. Thus, the unusual architecture of cytoplasmic dynein indicates that it moves by a significantly different process than either myosin or kinesin.

10.3 AAA Family of ATPases

The AAA superfamily of proteins is the largest class of motor proteins and includes the F_0/F_1 ATPase, helicases, proteasome complex, membrane fusion complexes, microtubule severing, and the pilus retraction motors in bacteria. Members of this class typically have six subunits, and all six can have ATPase activity. Dynein is a special adaptation of the AAAs, which has one hydrolytic and three additional non-hydrolytic ATP binding sites. The six subunits form a ring that facilitates the respective function of the protein. For example, to carry out their function in unwinding DNA, helicases have to assemble as a collar around one strand of the double helix. Similarly, the packaging of DNA into bacteriophage involves an ATPase that pushes the DNA into the hollow phage capsid with considerable force (measured at about 50 pN). The largest measured motor force is that of the pilus retraction motor in bacteria, which can develop up to 100 pN when pulling the pilus through the central pore. Even polypeptide chains are transported by AAAs, the 26S proteasome grabs a ubiquitinated protein and pushes it into the core of the proteasome through mechanical denaturation that enables it to be rapidly proteolysed.

Because AAAs drive a very diverse set of functions, this class represents a basic motor that can be adapted to many different functions by different coupling mechanisms. Why this superfamily is used in so many functions, as opposed to other motor proteins, is not clear. The sixfold symmetry of the ring complex provides for flexibility as it can be modified easily through mutations to have a two- or threefold symmetry that can match the functions as needed. Again, the critical elements for motor function are the cyclic deformations of the motor that must be coupled to the repeating pattern of the polymer or filament upon which it moves.

In the case of motor proteins the hydrolysis of ATP (or other high-energy compounds) is coupled to the movement of the protein along a filament. Kinesin movement on microtubules is a good example in which the details are known. For kinesin, there appears to be one ATP hydrolyzed per 8 nm of movement. The equilibrium energetics of the process brings out some important general aspects of reversible force-dependent processes.

If we assume that the Substrate (S) is hydrolyzed to the Product (P) by the motor enzyme (E), then a simple description of the process is

$$S + E = ES = EP = E + P$$

We have described this previously in terms of the free energy of the reaction; however, it is obvious that the relative concentrations of substrate, [S], and product, [P], affect the free energy of the hydrolysis reaction. The standard free energy of hydrolysis of ATP is defined as the free energy of the conversion of ATP to ADP where both are at 1 M concentration. The normal cellular values are 2–4 mM and 10–30 μM for ATP and ADP, respectively. Because ATP is much greater than ADP in concentration, there is more energy available (mass action favors the hydrolysis of ATP to ADP). Thus, the free energy of conversion of S to P can be described as noted below.

$$\Delta G_{S-P} = \Delta G^0 + kT \ln [P]/[S] = kT \ln K/Keq = kT \ln [P][Seq]/[Peq][S]$$

As a motor, the protein generates force through a cycle of ATP hydrolysis and ADP release. When the force on the motor stalls the forward movement of the motor, the energy of hydrolysis is theoretically equal to zero. The additional term of the force times the distance traveled per ATP can theoretically give the stall force from the ATP and ADP concentrations.

$$\Delta G_{S-P} = 0 = \Delta G^0 + kT \ln [P]/[S] + F\Delta X \text{ or by rearranging}$$

$$F \Delta X = -\Delta G^0 - kT \ln [P]/[S]$$

The energy from a single molecule of ATP is estimated to be 80 pN·nm, which would correspond to a force of 10 pN on a kinesin molecule that moves 8 nm. In fact, the measured stall force for kinesin is 5–7 pN and the estimated distance of movement is 8 nm. The motor can then be said to be 50–70% efficient.

One AAA protein is a rotary motor rather than a linear transporter and it has the surprising mechanochemical role of converting rotary movements driven by an electrochemical gradient into ATP. The F_0/F_1 ATPase synthesizes ATP by coupling the rotation of a central subunit to the transport of protons through the membrane-bound 'F$_0$' domain. In this case, the protein possesses an F$_1$ ring (AAA structure) with threefold symmetry that sits above the membrane. The rotary motion of the central subunit has been analyzed extensively and laser tweezers have been used to artificially rotate the central subunit to synthesize ATP. Thus, the AAA structure supports rotary as well as linear motor activity that can be coupled with important enzymatic reactions such as ATP synthesis.

10.4   Energetics of Motor Movement

Both myosin and kinesin are approximately 50% efficient in converting ATP energy into mechanical work. Isometric contractions of myosin generate about 4 pN of force, and the lever arm moves 10 nm forward. In the case of kinesin, the motor generates about 5 pN of force and moves 8 nm to the next tubulin dimer. It follows then that both of these motors can generate 40 pN-nm of mechanical energy. ATP has about 80 pN-nm of energy in the high-energy bond. Others have often compared the efficiency of biological motors with internal combustion or electrical motors. Biological motors of the kinesin and dynein families compare favorably with internal combustion engines that are less than 20% efficient in converting chemical into mechanical energy. The most efficient biological motor is the F$_0$/F$_1$ ATP synthase complex, which is over 95% efficient and matches electrical motors that approach 100% efficiency. However, the motor systems are probably not selected for maximum efficiency, because the critical issue at the cellular level is controlling the amount of movement and the force generated, not the amount of ATP required to drive the process. Skeletal muscle is perhaps the major exception wherein efficiency can benefit survival of the organism.

10.5   Motor Types

Because motility is a critical biological function, there are many different myosins and kinesins. With the myosins, there are two fundamentally different types: non-processive, and processive. Processive motor proteins can move as single molecules along a filament and most kinesins and cytoplasmic dyneins fall into this category. Skeletal muscle myosin is a non-processive motor in that it spends most of the time off the actin filament and relies upon the organization of the muscle sarcomere to keep it in close proximity of actin. Of the about 20 different types of myosin that have been identified, the majority are processive motors, because non-processive motility requires that the motile complex be highly organized and that is found primarily in striated muscles.

Each myosin contains a characteristic ATP-binding pocket and possesses a lever arm that develops force. The ATP-binding domain and the actin-binding site are both located within the myosin head. Despite major differences in amino acid sequence, the polypeptide folding patterns for the ATP-binding domains of both myosins and kinesins reveal that they are structurally similar. It is therefore possible that the domain movements needed to couple ATP hydrolysis to changes in protein conformation are optimal in such a geometry. This does not, however, reflect similar motile functions at the single molecule level. Indeed, the mechanisms of movement of multiple myosin isoforms are significantly different from the mechanisms of movement of the kinesin motors that have been analyzed.

Most myosins move toward the barbed ends of actin filaments, except for myosin VI, which moves toward the pointed ends. Because most actin filaments are anchored at their barbed ends, motors moving toward the pointed ends will potentially compress the filaments, and single actin filaments are much weaker than microtubules in resisting compressive forces. Thus, myosin VI is found in the periphery and is thought to move primarily upon bundled actin filaments. In the case of kinesin, the majority of isoforms in mammalian cells move toward the plus or fast-growing ends of the microtubules. However, there are a significant number of isoforms that can move toward the minus ends, as does cytoplasmic dynein. In higher plants, there is no cytoplasmic dynein and, consequently, the minus-end-directed kinesins support microtubule-dependent transport toward the minus ends of the microtubules. Thus, although the majority of myosin and kinesin isoforms move toward the fast-growing ends of actin filaments or microtubules, there are notable exceptions, particularly in plants.

Myosin II is the myosin found in muscle, and a variant is responsible for the large forces generated by fibroblasts for matrix and tissue reorganization. Unlike the other myosin isoforms, myosin II assembles into bipolar filaments, which enables them to pull apposing actin filaments together, and they create the major contractile forces of cells. Although myosin II is a non-processive motor protein in non-muscle cells, the cytoskeletal filaments are not well organized. To be able to move for long distances on disordered actin filaments, the non-muscle myosins have a slower velocity and a longer attached period in the motility cycle. This means that non-muscle myosin II may be somewhat processive. In addition, there is evidence of sustained tension in the actomyosin cytoskeleton that involves myosin II holding onto actin for long periods under tension in what is called a catch-bond state. When tension is relaxed, myosin can then contract to restore the tension that will again lock it into the catch bond. Thus, the bipolar

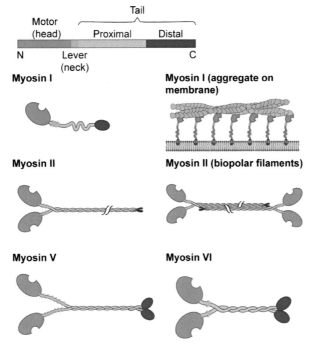

Figure 10.3

Myosin.

myosin filaments appear to have a major role in the developing and sustaining tension in tissues.

Like kinesin, myosins can carry vesicles and other cargo in cells. For example, myosin V is a processive myosin that is involved in vesicle motility. Other two-headed myosins have a variety of cargos that are not all known. In the case of the single-headed myosins (myosin I's), they appear to form aggregates with their cargo, which will produce a complex that has multiple heads for ATP hydrolysis and, subsequently, for movement. Thus, myosin motor domains are linked to many different tails and this enables the controlled movement of many different components in cells through differential binding and different mechanisms of controlling motor activity.

10.6   Processive vs. Non-processive Myosin

To better understand the differences between processive and non-processive motors we will consider the force-generating cycles of muscle myosin II, which is non-processive, and myosin V, which is processive. The critical force-generating step in the myosin movement cycle (see Figure 10.4) is a conformational change in a region called the lever arm, or neck region, relative to the

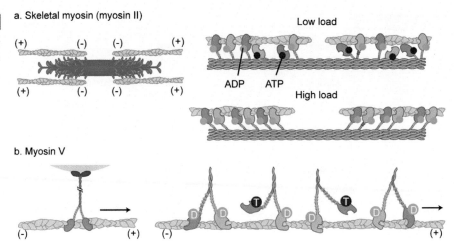

Figure 10.4

Processive and non-processive myosin.

myosin head, which is bound to the actin filament. Single-molecule measurements show that skeletal myosin, which contains a lever arm with two light chains bound to two IQ motifs, will be displaced by about 10 nm per ATP cycle along an actin filament. In contrast, myosin V moves by 35 nm with each ATP hydrolyzed, and has a lever arm containing six light chains bound to six IQ motifs. In skeletal muscle, the unloaded velocity of actin filament movement past myosin is 10,000 nm/s, but the individual myosin heads only hydrolyze 20 ATPs/s. Thus, if one of the two myosin heads in a molecule remained in contact with the filament at all times, it should only move about 400 nm/s. However, as a processive motor, myosin heads do not maintain contact with the actin filament for much more than 2% of the cycle, or 1 ms, and are free for the remaining 49 ms. If that is correct, then 50 myosin heads acting in sequence could move the filament at 10,000 nm/s. In unloaded muscle, the force-generating heads are balanced by lagging heads that have not released. The release rate of the lagging heads is accelerated by the force-generating heads and must approach the 1 ms lifetime (the force versus negative displacement curve is not well described for myosin). For myosin V the ATPase rate of 15/s per head is multiplied by the displacement per ATP of 35 nm, and further multiplied by two heads to give the observed velocity of movement of about 1000 nm/s. Thus, an advantage of the non-processive movement is that contraction velocities can be much higher than for a processive motor.

Another advantage of the non-processive motor systems is that more heads can be recruited to generate higher forces as the load on a muscle increases. This is known as the high-force regime. In unloaded contracting muscle, most of the heads are not attached. However, when force increases, the release of attached myosin heads is slowed because the ADP release step is slowed. In a muscle sarcomere, there is a high concentration of free heads that can attach to support the load. When the loading force increases, more heads bind until a threshold is crossed where the myosin heads can generate a force equal to the load. At this point there are enough force-generating heads bound to overcome the resistive force, and the heads can begin to release from the filament as movement resumes. Because the heads are attached for longer periods under load, a larger fraction of their ATPase cycle is spent in the bound force-generating state. If we consider a situation where the load equals the force generated by 10 myosin heads, then 10 myosin heads will be bound at all times. If there are 50 heads, then 20% of the heads will be bound at any one time, which translates to an average time 10 ms in which the heads are bound in the attached state. Assuming that the portion of the ATPase cycle when the head is not attached to the filament remains constant at about 49 ms, the overall cycle time is 59 ms. The velocity will be reduced to 2000 nm/s, and the number of ATP molecules hydrolyzed per micrometer of movement will increase to 1000. This highlights that the non-processive motor system is better at bearing large loads and generating large forces within tissues than the processive motor system. This is because processive motor systems only utilize a single molecule, with only one force-generating head bound to the filament at any one time. The only way to increase the motive force is to recruit additional motors to the complex, which may take time.

10.6.1 Non-muscle Contraction Systems

In non-muscle cells, the contractile actomyosin systems that generate traction forces are more complicated. There are two major contractile systems that have been described: a contractile actomyosin network, and stress fibers (analogs of sarcomeres). The contractile networks are dynamic and have nodes with actin filaments radiating out that are connected by bipolar myosin filaments. Actin polymerization and myosin contraction are important for maintaining the networks, as is actin crosslinking. On the other hand, stress fibers often extend from one side of the cell to the other and have alternating bands of α-actinin and myosin. Although stress fibers are analogous to muscle sarcomeres, the majority of actin in fibroblasts is in less-ordered actomyosin networks where those filaments are also moving in myosin-driven processes. Furthermore, stress fibers are not needed for fibroblasts to generate myosin II-dependent forces on substrates. Myosin II always moves toward the barbed ends of actin filaments and actin filaments typically assemble at the cell edge with their barbed end pointed outward. Thus, bipolar myosin II filaments will draw the filaments inward. Unlike the situation in the cell, the actomyosin networks assembled randomly in the test tube to form a 3D array of filaments can contract to condense the network into a small knot. This phenomenon is called superprecipitation. However, superprecipitation is not normally observed in cells, because filaments are continually assembling actin filaments from the periphery and those filaments that become condensed are disassembled. Thus, both stress fibers and the actomyosin networks contribute to the dynamic tension generated by cells.

It is remarkable how robust the actomyosin systems are within non-muscle cells. Stress fiber release, which can follow photoablation of adhesions or spontaneous breakage in the middle of the fiber, will cause the stress fibers to contract a small distance, creating a gap. However, new filaments subsequently assemble in the gap, and tension is restored. Even before cells assemble stress fibers, they are able to generate high forces on matrix adhesions located at opposite sides of the cell through the cohesive actomyosin network. Inhibition of myosin contraction blocks formation of the network and blocks the healing of breaks in stress fibers. Furthermore, addition of the g-actin sequestering compound, latrunculin A, causes dramatic superprecipitation of stress fibers. Thus, the cell is able to block major contractions that would disrupt the network by a healing process involving actin filament assembly that maintains cytoplasmic cohesion. Several lines of evidence indicate that this healing is facilitated by the formin actin-polymerizing proteins, which generate a branched filament network that is then bridged by myosin II bipolar filaments to form the cohesive structure. Thus, because cohesion of the tissue depends upon cohesion of the cell, there are very robust mechanisms to maintain a cohesive network of actin filaments throughout the cell despite the rapid filament turnover.

10.7 Kinesin

There are about 30 different classes of kinesins defined on the basis of their sequences similarities. Surprisingly in plants, kinesins that move toward the filament minus end appear to be the major motor in the cytoplasm moving in such a direction. As in the case of myosin, there are classes of single-headed kinesins (e.g. Kif1a) and they appear to have two microtubule-binding sites. This property would enable them to move as single molecules, much like an inchworm. However, recent studies indicate that when transporting vesicles, they will dimerize. Thus, there is no accepted exception to the rule that processive motors move as dimers.

Because of its small size, kinesin has been a favorite motor for biophysical studies and is perhaps the best understood of all motor proteins at this point. Although we understand a lot about the molecular events involved in coupling ATP hydrolysis with movement, much less is understood about how the motors

Figure 10.5

Kinesin superfamily.

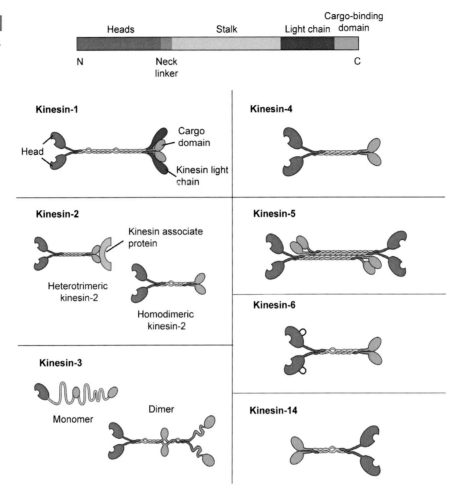

are controlled to organize the cytoplasm. Vesicle movements are typically inter-mittent and transport events often only last for 2–4 μm before the vesicles fall off the microtubule and often diffuse. Proteins to be secreted need to move from the Golgi to the periphery and need to preferentially activate kinesin for that transport.

Regulation of the kinesin motors is critical for several reasons. First, their cellular concentration is sufficiently high to deplete ATP within the cell if all were active. Second, as a kinesin motor can move from the cell center to the distal (plus) end of a microtubule in 5–10 s, wholescale rearrangement of the cytoplasm would occur very rapidly should their activity go unregulated. Third, both kinesin and cytoplasmic dynein are simultaneously bound to vesicles in the cytoplasm, such as mitochondria; however, the positioning of those vesicles is determined by the relative activation of kinesin versus dynein. Finally, even in the most active cells, the number of moving vesicles is over 1000-fold lower than the number of kinesin motors in the cell. Thus, there is an excess of motor activity in cytoplasm and the critical issues become how to attach to the right cargo, move that cargo, and leave the cargo at the right location.

It has been difficult to reconstitute the motility activation process *in vitro* because motility will be switched off or on in cells in a position-dependent way, and is tightly regulated in cytoplasmic extracts. Saltatory vesicle movements (periods of rapid movement followed by diffusion) are commonly observed, even in axoplasm, and the dilution of components *in vitro* makes the probability of rebinding and moving very low. This could be explained in part by the extent of processivity of kinesin, because kinesin-coated beads only move an average of about 1.2 μm before they fall off a microtubule. Alternatively, the motors may return to a default OFF state before they can be activated for another round of movement. This second mechanism fits well with the engineering principle of term limits and would help to explain how the localization of the organelles could be controlled. In the axoplasm and some cytoplasmic regions, the density of filaments is very high and occasionally the moving organelles will slow down, stop, and then recoil backwards. This is interpreted as elastic recoil of the filament, which would occur after stretching elastic proteins that will carry the vesicle backwards. It is important *in vivo* for the vesicular cargo to carry the signal for transport and to signal where the cargo should stop moving. Thus, until physiologically active positional information can be produced *in vitro* and the material maintained at high concentrations, it will be very difficult to reconstitute physiologically relevant vesicle transport *in vitro* that will function properly.

In the process of testing for protein aggregation, movements of motor proteins can provide a mechanism for clearing aggregated proteins on the basis of the aggregate size, either through attachment to aggregates on vesicles or within tubular membranes. As aggregates, they will not be able to pass into transport vesicles and may be concentrated with other components that would signal their degradation. Transport of these aggregates to autophagic vacuoles or the lysosomes could help to rapidly clear this cellular waste.

During mitosis, kinesins also play important roles, where a variety of motors are present in the spindle to help to organize the microtubules as well as to move chromosomes. In alignment of chromosomes at the metaphase plate, the chromosomes may initially move toward a single pole when attached to only one kinetochore microtubule through dynein activity. However, they are prevented from going to the pole by an outward ejection process that involves the binding of kinesin motors to the chromosome arms and outward movement on adjacent microtubules. When a microtubule from the other spindle pole binds to the other kinetochore, the chromosome will be aligned at the metaphase plate by dynein

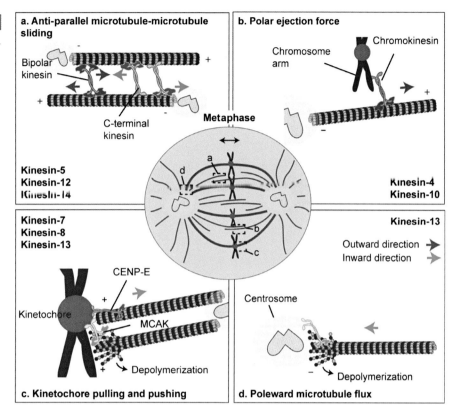

Figure 10.6

Mitotic kinesins.

a. Anti-parallel microtubule-microtubule sliding

Bipolar kinesin

C-terminal kinesin

Kinesin-5
Kinesin-12
Kinesin-14

b. Polar ejection force

Chromosome arm

Chromokinesin

Kinesin-4
Kinesin-10

Metaphase

Kinesin-7
Kinesin-8
Kinesin-13

CENP-E

Kinetochore

MCAK

Depolymerization

c. Kinetochore pulling and pushing

Kinesin-13

Outward direction
Inward direction

Centrosome

Depolymerization

d. Poleward microtubule flux

activity or a minus ended kinesin. It is still a mystery how the cell knows when all of the chromosomes are aligned and can then initiate chromosome separation. As the chromosomes move to the poles, kinesin molecules in a region where the anti-parallel microtubules overlap drive spindle elongation. Surprisingly, a kinesin motor protein in the kinetochore can follow depolymerizing microtubules as they move to the poles without hydrolyzing ATP. Energy is still required for this process; however, it is derived from the energy that was put into the microtubule through GTP during its polymerization. The motor simply follows the depolymerizing tubule in a hand-over-hand fashion. Thus, the process of mitosis provides many examples of the different but important motor functions of the kinesins that even provide for redundancy in some cases.

Axonal transport of mitochondria is particularly important for the maintenance of axons, because they provide the energy needed for motor transport and the local repair of the microtubules. An interesting feature of the early movement of mitochondria in developing axons serves to illustrate the sensitivity of the system. Mitochondria can be tracked for long periods with the fluorescent dye, mito-tracker. As they move down the axon from the cell body, the mitochondria will often stop and become anchored in the axon. When the positions of anchorage are plotted relative to the neighboring anchored mitochondria, a pattern emerges whereby the moving mitochondria are most likely to stop halfway between the anchored mitochondria. This makes sense in terms of providing for the most even distribution of mitochondria in the axon, but raises a lot of questions about how the decision to stop is made. Logically there could be a diffusive signal produced by the anchored mitochondria that would have a limited lifespan such as ATP. The lowest ATP concentration would be halfway between the anchored mitochondria. This diffusive signal could indirectly modify the matrix to decrease its affinity for mitochondria. Several alternative models are possible, but this serves to highlight how sensitive the control of motor function can be. Many of the important neuropathies have been linked to alterations in axonal mitochondrial distribution, which provides an important selection criterion for positioning the mitochondria correctly.

10.8 Dynein

The dynein motors, whether cytoplasmic or axonemal, differ dramatically from myosin and kinesin. Although the recent crystal structure of cytoplasmic dynein does give a much clearer view of its possible mechanisms, it also reveals the many unusual structural features that dynein possesses. For example, dynein contains a long microtubule-binding arm, and a basic hexagonal organization of the motor body, which is unusual for motor proteins. Furthermore, there are four ATP binding sites in each of the two heavy chain subunits, yet it appears that only one of the four ATPs is hydrolyzed during motion. As might be expected from the long microtubule-binding arm, the stall force of cytoplasmic dynein is quite low (1–2 pN), yet the velocity appears similar to kinesin.

In axonemes, dynein can function as a non-processive motor. However, there are aspects of its mechanical activation that indicate that specific elements retain contact with the neighboring microtubule during the motility cycle. Here, movement in axonemes is attributed to the mechanical coupling of the motors to each other as a result of the relative displacement of neighboring tubules. Different patterns of flagella beating are driven by different mechanical signaling patterns, which are derived as a result of motor protein organization, and the activity of associated proteins. In contrast to axonemal dynein, cytoplasmic dynein is a processive motor, and can carry latex beads on microtubules, in a similar manner to kinesin, for at least 1 μm before falling off.

In vitro motility assays for kinesin and cytoplasmic dynein rely upon the ability of the motors to bind strongly to microtubules. However, the strong binding was

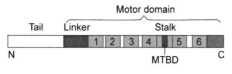

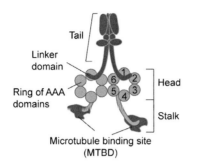

Figure 10.7

Dynein.

observed initially in low ionic strength media. As such, most *in vitro* motility assays include less than 20 mM of chloride ions, which is similar to the cytoplasmic concentration. If cellular concentrations of sodium and potassium are used, then very little motility is observed because the motor proteins are shielded from binding to the microtubule surface by the counterions. However, even though the number of transport events decreases dramatically with increasing salt concentration, the average run length of the motors on the microtubules does not change dramatically. Thus, the stepping of the motors on the filaments is not highly dependent upon salt concentration, which implies that the interaction between the motor and the filament may be largely hydrophobic after binding, but there is an initial charge–charge interaction that helps the motor bind to the microtubule.

In vitro, microtubule arrays will form with a similar organization to spindles when pure microtubules are mixed with pure motor proteins and ATP. This implies that the motor proteins interact with microtubules at multiple sites or will aggregate. When cytoplasmic dynein is mixed with microtubules, the microtubule plus ends face outward as in spindles. In contrast, when kinesin is mixed with microtubules, the microtubule minus ends face outward. Thus, it appears that the motor aggregates move to the appropriate ends of all filaments to subsequently create a radial array. This is illustrated nicely in cytoplasmic extracts from frog eggs where the plus ends face out, as in the case of spindles, and chromatin-coated beads move to those plus ends.

10.9   Why Is There So Much Microtubule Motor Traffic in Cells?

The trafficking of intracellular organelles can help us to better understand how microtubule motors organize the cytoplasm and what emerges is a sloppy sorting system. Firstly, the number of movements per minute appears very high, but is 100- to 1000-fold lower than the number of motors in the cytoplasm. This shows that activation of motor transport is a rare event and there is a large excess capacity for additional transport. Secondly, there is randomness to the movements, i.e. they are salutatory, intermittent movements of 2–6 μm followed by confined diffusion, stasis, or even reversal. This indicates that the motors are not continually active, but rather are programmed to turn off rapidly with a first-order decay rate. This is even observed in axonal transport. As noted previously, this can be considered as a term limit on motility and indicates that the default state for the motor is 'off.' In order for organelles being transported by motor proteins to reach their proper destination, the motor protein must continually be activated until the cargo reaches the proper destination. Thirdly, in actively growing cells, there is a countercurrent of vesicle traffic that appears balanced as almost equal numbers of organelles rush past each other in opposite directions. This means that the traffic is largely bidirectional and that the outward movements of secreted material represent just a small fraction of the total traffic. In fact, endocytic and recycling vesicles may account for most of the observed traffic (>95% of endocytosed membrane is recycled or ~2% of plasma membrane area per minute). Cells appear to use the motor protein facilitated movement of material as a way to sort components. This includes sorting secreted proteins from retained proteins, damaged molecules from native molecules, material for repair from material to be degraded. Basically, the directed transport of materials in the cell does not occur in a simple linear process. Rather, many successive movements occur in any one direction with low discrimination for the material to be moved and outward movements balance inward movements that return missorted components to the proper compartments. The process is not economical, but can produce a high fidelity of transport through a reiterative sorting process. Thus, getting it right involves repeated activity, but it is important to get components sorted properly in cytoplasm.

10.10 SUMMARY We understand a lot about the molecular and even atomic details of the motor systems in cells. However, for many efficient and robust functions, motor proteins must be controlled in a dynamic feedback fashion and that places the primary importance on the control of motor activity and the location of the active motors. In actin systems, the activation of bipolar myosin filament assembly is strongly linked to the filament dynamics. The type of motor, processive or non-processive, determines the types of functions that it can perform and non-processive motors require a highly organized cytoskeleton as in skeletal muscle. How the decisions are made to activate one motor protein, and inactivate another, are critical for membrane trafficking, and transport of components to the proper location in the cell. In the larger context, each motor system needs a different set of controls to work properly, with some being mechanical and others being chemical. Furthermore, the consequences of motor activity need to feed back on the motors themselves most often through a physical or physical chemical process. For a robust and active system, the motors provide essential elements to speed up processes by providing a way to couple energy to force generation for DNA unwinding, whole organism motility, or for vesicle transport to get components to the desired location.

10.11 PROBLEMS

1.
$$\text{MtK} + \text{ATP} \underset{}{\overset{k_1}{\rightleftharpoons}} \text{MtK} \cdot \text{ATP} \underset{}{\overset{k_2}{\rightleftharpoons}} \text{MtK} \cdot \text{ADP} \cdot \text{P} \overset{k_{3(-P)}}{\rightleftharpoons}$$

$$\text{MtK} \cdot \text{ADP} \underset{}{\overset{k_4}{\rightleftharpoons}} \text{MtK} + \text{ADP}$$

$$\Updownarrow k_{dis}$$

$$\text{K} \cdot \text{ADP}$$

SCHEME 1

We discussed processive motors in terms of the kinetic scheme for kinesin (Scheme 1 with MtK above). If we assume that the concentration of ATP is high, then ATP binding will not be the rate-limiting step (K_1). What is the rate-limiting step, if $K_2 = 1000\ s^{-1}$, $K_3 = 100\ s^{-1}$, and $K_4 = 400\ s^{-1}$? If the step length is 8 nm/ATP, then how fast can the motor move if it has one head (assume that between steps 3 and 4, the kinesin head will come off the microtubule)? (Same question for two heads like the normal molecule?)

Answer: The lowest rate step is K_3 and that will take on average 10 ms. If it has one head, it can only move 8 nm. After it makes one movement, it will not be able to make the second movement without falling off the microtubule and diffusing away. In the case of the two-headed kinesin, the maximum rate should be 80 s^{-1} for each head giving the maximum velocity of 1280 nm/s, which is similar to what is observed experimentally.

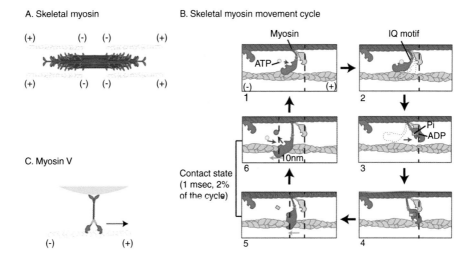

A. Skeletal myosin

B. Skeletal myosin movement cycle

C. Myosin V

Contact state (1 msec, 2% of the cycle)

2. For muscle myosin, which is non-processive (see Scheme 2 with AM above), the rate-limiting step is the hydrolysis of M.ATP to M.ADP.Pi (this is the forward rate constant of transition from 2 to 3, which is typically 20 s^{-1}). M.ADP.Pi then binds to actin to produce a force by the swing of the crossbridge (4 to 5), Pi and then ADP are released (5 to 6). Myosin bound to actin then waits for another ATP to release it and start the cycle again. The series of steps from force production to release takes about 2 ms without load in a maximally activated muscle. Assuming 0.002 s is the time that the active heads are bound to the actin filament and that the forward swing of the crossbridge is 10 nm, then what is the maximal velocity of muscle contraction (you should also assume that without an external load, the myosins are pulling against themselves; i.e. half of the time is spent producing force to pull other myosins forward and half of the time is spent being pulled forward by other myosins)?

Answer: If the muscle is contracting at a constant velocity and the heads are binding randomly, then each head will pull forward for 1 ms and will cause a 10 nm displacement. This will be a velocity of 10,000 nm/s or about the maximum velocity of sarcomeric contractions.

3. What happens to the time of force production in the muscle when a load is applied that slows contraction? Assuming that the velocity is slowed to 1/10 of the maximal velocity determined in Problem 2, what is the average time that a head is bound and producing force before it is released? Compute the fraction of the myosin heads (on average) that are bound to actin under these conditions.

Answer: In this case the muscle is contracting at a rate of 1000 nm/s and will take 10 ms to travel 10 nm. Because myosin heads cannot release from actin until they complete the 10 nm power stroke, they will remain bound for 10 ms. The rest of the ATP hydrolysis cycle takes 49 ms and so the overall rate of ATP hydrolysis will drop by 50/59 = 0.84 of the original rate. Because the heads remain bound for 10 ms, other heads will also bind and the fraction of overall time in the bound state will be 10/59 = 0.168 or 16.8% of the cycle corresponding to 16.8% of the heads bound.

4. Force will slow the myosin contraction. If the force per actin filament increases to 60 pN and each myosin head can generate 4 pN before the motor velocity drops to zero, then about how many myosin heads will be bound to each actin filament (explain your reasoning)? What will be the approximate velocity of contraction, assuming that there are 30 possible myosin heads that could attach to the filament?

Answer: Because the myosin head forces should add linearly, there should by about 15 myosin heads bound to produce a force of 60 pN. If half of the 30 possible myosins are bound, then the bound portion of the cycle should correspond to 50% of the overall cycle or 49 ms (gives equal time unbound and bound) and this corresponds to a displacement of 10 nm or a velocity of 204 nm/s.

5. Kinesin and cytoplasmic dynein are often both bound to single vesicles, creating a potential tug-of-war on the vesicle surface. If kinesin has a stall force of 5 pN per head and cytoplasmic dynein has a stall force of 2 pN per head, then which direction should the vesicle move in if it has four kinesin and eight cytoplasmic dynein molecules bound to its surface? In the cell, many bound motors are inactive and often cycle between active and inactive states. Further, vesicles often fall off microtubules and rebind at a different site on the vesicle (a process called saltatory motion). If only 20% of the motors on the vesicle surface can contact the microtubule when it diffuses to the microtubule and there are two kinesin and three cytoplasmic dynein molecules per vesicle, then what is the approximate fraction of movements that go to the plus end of the microtubules?

Answer: It should move in the direction of the kinesin motors initially. On average, one motor will be active in the 20% binding case and as there are two kinesins, the fraction of movements to the plus end should be 40%.

6. Intracellular motor proteins move by:
   a. changing conformation due to thermal energy;
   b. binding high-energy compounds and releasing low-energy compounds;
   c. cyclic interactions with polar filaments; and
   d. nanometer-level steps.

Answer: (a) True, all movements are thermally driven and ATP hydrolysis biases the thermal motions; (b) True, and this provides energy for movement; (c) True, the cyclic steps to successive filament subunits drive the movements; and (d) True, the order in the filaments are repeated at a nanometer level (about 5 nm for actin filaments and 8 nm for microtubules).

7. Studies of mitotic motors indicate that they will move with microtubule ends as the microtubules depolymerize. Please describe the steps in the motor process in an analogous way to the diagram in problem 1.

Answer: The scheme would be simplified because there is no ATP hydrolysis and only the kinesin binding to microtubules is relevant. Such a scheme would then propose that the last tubulin dimer would release from the microtubule and unbind from kinesin. Then the unbound kinesin head would move forward past the bound kinesin head and rebind to the microtubule. A necessary hypothesis to make this work is that the release of the terminal tubulin dimer will wait until the free head binds to the second tubulin dimer.

8. If the energy from one ATP molecule in the cellular environment is 80 pNnm (piconewton-nanometers), then what is the theoretical limit of the force that a single motor head can generate for kinesin that moves along single microtubule protofilaments? If we suggest that myosin filaments can move on both sides of an actin filament making the effective step length the distance from one actin on one

side to the next actin on the other side of the filament, then what is the maximum force that myosin can generate?

Answer: Because the tubulin dimer occupies 8 nm along the protofilament, the minimum step length for the kinesin is 8 nm/ATP, which then gives a maximum force of 10 pN. In the case of myosin, the repeat distance for actin along a protofilament is 5.4 nm and the two protofilaments are staggered, giving a possible step length of 2.7 nm. This means that the maximum force theoretically should be about 30 pN, which has not been reported.

9. In general, the rate of muscle contraction is proportional to the maximal ATPase activity of the isolated myosin (slow myosins have low ATPase activities). Which explanation is most logical and why? (1) The motors are all processive; (2) the fraction of the ATPase cycle when the motor is not bound to actin is constant; (3) the time that the motor is not bound to actin is constant irrespective of the time for the overall cycle.

Answer: Explanation (2) is the most logical because the slowing of all steps in the cycle would result in a slower velocity as well as lower ATPase activity. Explanation (1) is in theory possible, but the observations include skeletal muscle myosin types and they are not processive.

10.12   REFERENCE

Howard, J. 2001. *Mechanics of Motor Proteins and the Cytoskeleton.* Sunderland, MA: Sinauer Associates.

10.13   FURTHER READING

DNA replication: www.mechanobio.info/topics/genome-regulation/dna-replication/
Kinesin and cytoskeleton-dependent intracellular transport of cargo: www.mechanobio .info/topics/cytoskeleton-dynamics/cytoskeleton-dependent-intracellular-transport/
Motor protein activity: www.mechanobio.info/topics/cytoskeleton-dynamics/motor-activity/

11 Microenvironment Controls Life, Death, and Regeneration

In the adult organism, the cellular microenvironment is critical in controlling cell growth, death, or regeneration. Although the fluid microenvironment includes several factors that influence cells, such as nutrients, salts, and hormones, we will focus here on the extracellular matrix and neighboring cells. The cellular microenvironment must support the tissue functions in the organism and should enable cell turnover and repair. Repair necessitates controlled growth to the proper cell density relative to matrix; however, safeguards are then needed to avoid uncontrolled growth, which can result in cancer. In this chapter, we will discuss the organizing principles in (1) cell–matrix interactions, (2) cell–cell interactions, and (3) the interplay between the two. In the case of the extracellular matrix (ECM), the composition of the ECM and its architecture sets the program for the adjacent cells through both the chemical nature of the ECM molecules and the shape of the crosslinked matrices in the adult. The detergent-resistant matrix of a tissue contains a lot of information that can direct the regeneration of stem cells when added to the matrix. In addition, the ECM will direct the cellular repair process following the loss of individual or small numbers of cells and the mechanical characteristics are critical. Matrix provides mechanical support through basement membranes, ECM fibrous networks, or linear fibers. The number of cells in the given matrix will be set by the physical aspects of the cell–cell contacts and the amount of ECM available. In the case of cell–cell junctions, there is a more limited set of components in adhesions, but some of the same cytoskeletal elements are important. Again, mechanical signals derived from the reciprocal communication between neighboring cells as well as the matrix are critical for directing the growth of the organism and controlling repair processes. Most importantly, the cell must only grow in an environment that is defined by a combination of cell neighbors, the matrix as well as nutrients and hormones. If the cell doesn't respond properly to the local cell and matrix cues, then it is likely that dysfunction will result. Major trauma and damage in the adult is usually bandaided with collagen fibers that cover the damage site (scar tissue) but do little to restore normal function in that region. Normally there is a delicate balance between growth and apoptosis that depends heavily upon the density of cells relative to the matrix (too much matrix will stimulate growth and too many cells will stimulate extrusion – apoptosis). The challenge is to understand how the balance is achieved for a given cell type in a particular organism.

11.1   Microenvironment in Regeneration

After the organism is formed during development, there are many different circumstances where repair is needed, such as disease, trauma or cycles of starvation and replenishment. Cell growth during repair processes is primarily directed by the surrounding microenvironment, but hormones can signal that the system has been stressed and should respond. In a particular microenvironment, the extracellular matrix proteins are the major components that will provide the cues for the type of cells that will grow. Both the density of the cells relative to the amount of matrix and matrix shape will influence the amount of growth to give the proper density. ECM interacts with cells primarily through cell integrins that produce the proper chemical signal. Cell–cell interactions depend initially upon the cadherins, but there are many additional contacts that form over time. In the case of humans, we have the ability to repair a few damaged tissues (bone, muscle, skin, and liver are relatively easily repairable tissues) and this is unlike salamanders that can regrow entire limbs. For tissues like heart, kidney, and lung, there is an extensive ECM that is covalently crosslinked and it will withstand protein-solubilizing detergents *in vitro*. Thus, organs from adults can be treated with detergents to remove cellular material to prepare decellularized matrices. When stem cells are grown in heart-decellularized matrix, they will differentiate into heart cells that will produce a beating structure, whereas most chemically synthesized matrices have a poor success rate. Because similar success in regenerating the tissue-specific properties was obtained after putting stem cells back into other decellularized matrices, we will restrict our considerations here to *in vitro* studies of ECM cell interactions or regeneration using decellularized matrix models. Thus, the individual cellular responses to ECM are particularly relevant during migration, development, and localized repair.

During development, cell–cell responses to shape and force are important for creating proper tissue shape in multicellular complexes. The shape is then reinforced through further matrix deposition. Matrices are generally dynamic in juveniles to allow for further growth and only after reaching adulthood are the matrices sufficiently crosslinked to stabilize the tissue. In this chapter, we will consider the molecular bases of the cell–matrix and cell–cell interactions and describe how they relate to the cytoskeleton to help shape the organ. Because much more is known about cell–matrix interactions, we will focus initially on them and then later consider the cell–cell interactions. In some cases specific proteins, and their functions, are shared between the two types of adhesion.

Extracellular Matrix

Types of matrix:

Basement membrane; contains laminin, type IV collagen, and many minor components;
Fibrous; primarily type I collagen; and
Embedding; primarily type I collagen with crosslinks of elastin.

Matrix receptors: integrins (24 different forms), CD44 (hyaluronan-netrin).

11.2 Cell–ECM Receptors and Mechanics

If the cells in a tissue are to function properly, they must be able to respond to matrix molecules, as well as neighboring cells, as both provide information about the cellular microenvironment. The cellular ability to interact with the ECM is conferred by matrix receptors that span the plasma membrane and form the basis of cell–matrix adhesion complexes. The most common matrix receptors belong to a family of transmembrane proteins called integrins. These dimeric proteins have large external domains that contain matrix-binding sites and small internal domains that bind to a number of signaling components as well as cytoskeleton-binding proteins. Deletion of some integrins is lethal at an embryonic stage because they play a central role in the organization of the tissues. Integrins also play important roles in the growth of cancers and the activity of associated tyrosine kinases links matrix biochemistry and/or mechanics to cell growth. Integrins are sensitive to both matrix and cytoskeletal signals and are inherently dynamic in cells during wound repair, migration, or other active processes. Recent studies show that cellular sensory machines use the integrins as mechanical links between the cytoplasmic complexes that sense matrix mechanical properties and the matrix itself (Wolfenson et al., 2016). In all, there are commonly 24 different integrins, each of which will bind to specific matrix proteins such as fibronectin, collagen, laminin, vitronectin, and fibrinogen, although several integrins can bind to the same matrix molecules. In addition, several integrins bind to growth factors such as TGFβ, and VEGF or to cell-adhesion molecules such as ICAM or VCAM. For the ECM to control cellular differentiation, the response of a cell to matrix properties culminates from the integration of multiple integrin-mediated signals, both chemical and physical. In addition, the cell will naturally experience

Figure 11.1

Twnety-four integrins and
the major extracellular
ligands.

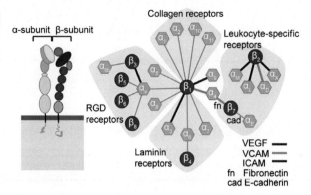

mechanical signals from neighboring cells, and the cell can further probe the mechanical properties of the matrix by generating tension on the substrate, and sensing the shape of the matrix.

Although much has been written about the formation of integrin-dependent adhesions on planar glass or plastic surfaces, there is controversy about whether similar adhesions actually form *in vivo*. It seems that a similar formation is likely, but the size of the adhesion structures appears to be considerably smaller. This is because matrix fibers (e.g. collagen fibrils are 0.1–1 µm in diameter) are generally smaller than the adhesions on glass; however, the basic integrin adhesion clusters are only 100 nm in diameter, making it possible for them to form on even very small matrix fibers (Changede et al., 2015). The mechanical issues of matrix clustering, geometry, and rigidity will be discussed below, and they clearly play major roles in the final response of a cell to the matrix. If we are to understand how tissues function, we will need to understand the roles of both the mechanics and the biochemistry of different matrix molecules in cell behavior. Furthermore, soluble hormones often activate cell motility and appear to cause cells to recheck their matrix microenvironment. For each tissue, it will be important to define the matrix microenvironment and the significant hormone signals that are part of regeneration.

11.3   Clustering and Geometry

Although matrix proteins such as fibronectin and vitronectin are present in significant amounts in serum, and can bind to integrins from solution, the soluble forms have little or no effect on cells. Instead, matrix proteins must be immobilized in order for an integrin to properly recognize and respond to them. Originally immobilization was believed to induce chemical alterations to the molecules; however, it is now clear that oligomerization and force (rigidity) sensing are critical physical aspects of matrices that cells recognize. In the case of fibronectin, the circulating form is dimeric and does not activate cells; however, it readily aggregates if mechanically sheared and sticks to surfaces. Trimers or tetramers of fibronectin will activate cells and they can support the forces generated during cell spreading. Thus, the formation of multiple integrin–ECM bonds are important for cell–ECM adherence and force serves to recruit additional components that stabilize the adherence sites.

At a molecular level, the minimal binding complex that will support cell spreading on glass has been worked out for fibronectin. The integrin β3 binds to the RGD tripeptide (RGD, arginine–glycine–aspartate) in fibronectin (in the fibronectin type III domain 10) and the cyclic tripeptide binds with almost equal affinity as full-length fibronectin. With new nanofabrication technologies, we can make 3–5 nm dots (size of a single integrin) that contain a single RGD molecule and array them in virtually any pattern that we wish. From previous studies, when the RGD sites on surface are separated by more than 60 nm, cells will no longer stay spread on that surface. Further, it was found that there needed to be about four fibronectin RGD domains spaced less than 60 nm apart in clusters to support cell spreading and motility (Schvartzman et al., 2011). The major integrin-binding protein in the cytoplasm, talin, is also about 60 nm in length and talin is needed for cells to adhere to fibronectin and RGD sites. Thus, it seems that multiple integrin bonds are needed in close proximity for adhesion.

The rationale for clustering is that multimeric integrin units will interact with multimeric binding proteins on the cytoplasmic surface, and the stability of the multimeric bond will depend upon the summation of the free energies of each individual bond. When there is an excess of RGD, clusters of about 50 integrins form within a region of approximately 100 nm on both RGD-coated glass surfaces and fluid bilayers containing RGD. This indicates that the clusters are a basic adhesion unit (Changede & Sheetz, 2017). From recent studies, it appears that clusters recruit formins to polymerize actin filaments from the cluster sites. These can then be pulled on by myosin and if a force is generated on the matrix, then the cell will proceed to develop further adhesion complexes. However, this will only occur if the immobilized matrix is within about 3 μm of another matrix cluster with actin filaments. Such multistep processes in complex functions are believed to be common because they provide a means to test if the matrix is properly organized.

Another important physical aspect of matrix organization is the 3D spatial arrangement of the sites. For example, collagen fibers range in diameter from 20 nm to micrometers, and smaller fibers can cause the associated plasma membrane to bend. This can promote the binding of membrane curvature-sensing proteins, such as the BAR domain proteins, which can then signal to the cell to

Figure 11.2

Fibronectin and clustering.

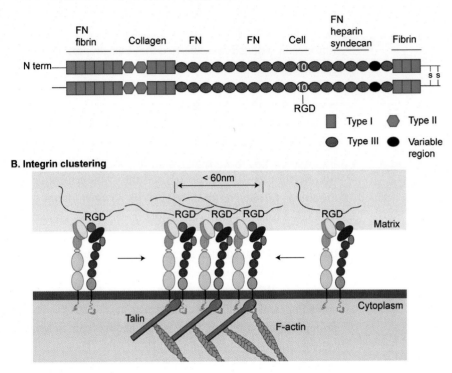

A. Fibronectin structure (dimer) and binding domains

B. Integrin clustering

activate a fiber motility mechanism. For example, myosin IIb can be activated in order to move collagen fibers. Indeed, the deletion of myosin IIb blocks collagen fiber transport but does not significantly alter the movement of fibroblasts on 2D collagen-coated glass surfaces (Meshel et al., 2005). Another factor that may affect matrix architecture is the roughness of the surface on which the matrix is immobilized. Here, dramatic effects on immobilization patterns are observed at the nanometer level. In this area, as with clustering, we expect that the normal matrix architecture of the tissue will be preserved through crosslinking. As cells die or divide, the organization of the matrix can activate the required cell–matrix response, which will preserve the proper tissue order.

11.4 Matrix Adhesion Signaling

The matrix adhesome, which can be defined as the collective group of proteins that bind to integrins, or other focal adhesion components, has been shown to include at least 150 different proteins (Zaidel-Bar et al. 2007). Because the matrix biochemistry and mechanics control many aspects of cell behavior, including differentiation and growth patterns, it is logical that proteins included in the matrix adhesome influence a range of different signaling pathways. For example, a variety of adhesion proteins can move from the cell periphery to the nucleus as part of signaling pathways that ultimately modify gene expression and they are integral parts of the sensory machines that test matrix mechanical properties.

Cell–matrix adhesions are reasonably dynamic in that on matrix-coated glass surfaces they will gradually disassemble over tens of minutes. In contrast, photobleaching recovery experiments show that most of the proteins in the adhesions will turn over with a half time of tens of seconds. The adhesion complex can therefore be viewed as a processing center, where components come and go to the rest of the cell with signals they receive from tests of matrix properties. The steps involved in the assembly of adhesion complexes will be discussed later, but it is beneficial to appreciate here that each step is complicated, and involves many components. This is logical because there are so many cell functions that depend on inputs from the cell microenvironment. Many of these signals are contextual, and specific adhesions for one matrix environment will differ greatly from another. For example, we find that mouse fibroblasts spread to a smaller area on soft collagen and soft fibronection surfaces than hard, but an equal mixture of collagen:fibronectin on a soft surface will result in greater spreading like a rigid surface. Thus, the matrix adhesions involve many components to produce integrated signals that are dynamic on short time scales but evolve over longer times, with hormonal or other changes in the environment.

11.5 Rigidity

Matrix rigidity is another factor of the environment that can be measured by cells and influences growth and differentiation. For cells to measure rigidity, they need to measure force per unit of matrix displacement or vice versa. From studies that assessed the displacement of polydimethylsiloxane (PDMS) pillars that were 0.5 μm in diameter, it was found that cells produce a constant matrix displacement over a wide range of matrix rigidities and the force generated is a measure of rigidity. This appears to be coupled to local sarcomere-like contractile units in non-muscle cells that contract to a fixed length of about 50 nm (described in more detail in Chapter 16). If a force threshold is exceeded, then the rigid matrix signal is activated for growth or appropriate differentiation. The rigidity-sensing machinery is regulated by tyrosine kinases and provides an important link from mechanosensing to cell function.

11.6 Matrix Configurations in Tissues

There are several basic configurations of matrix, each of which is customized for relevant tissues. The most abundant protein in the ECM, as well as in the human body, is collagen, of which there are several isoforms. Collagen I, for example, can be found in highly aligned ligaments or three-dimensional matrices in skin, whereas collagen IV is a major component of two-dimensional basement membranes. Laminin is found in brain axonal pathways and basement membranes. Fibronectin is circulating in blood and associates with wounded regions to help healing. Hyaluronan is abundant in cartilage as it surrounds the cells and provides a compressible structure. Thus, the different types of matrix molecules are designed for the different roles that they play in tissues.

During development, matrix fibers are deposited by cells to stabilize tissue structure after the basic movements have occurred. Because those fibers are assembled at a relatively early stage in the life of the organism and do not exchange readily, they can be considered as the scaffolding, or framework, on which the tissue is organized. When cells in the tissue die for any reason, the matrix will remain in place, thereby assisting the remaining cells to repair the tissue and maintain its shape. At early times the matrix is flexible and weakly crosslinked, but over time crosslinks form that stabilize the structure and inhibit further remodeling. Crosslinking of matrix proteins occurs through post-translational modifications such as oxidative crosslinking or through the activity of transglutaminase, which can catalyze the exchange of the amine group of lysine from one protein with the amide group of glutamine on another. Because the ammonia formed will quickly dissipate, the bond is essentially irreversible and it requires no energy input. Lysyl–glutamyl and oxidative linkages will ensure that the shape of the matrix is kept in place until it is degraded or damaged by trauma. Thus, the matrix crosslinking occurs primarily in adulthood to stabilize the mechanical and biochemical properties of the matrix.

The very property that enables the ECM to provide long-term structural support to tissues also raises questions about its repair in adult organisms. In particular, how can aberrant matrices be replaced with good ones? Because of the limited turnover of normal matrices, they are not easy to repair. Furthermore, matrix proteins like elastin are only made during infancy and get altered or degraded with time. Similarly, in the case of tendons like the Achilles tendon, collagen deposited in development remains for the life of the organism. Once the matrix pattern is established, it is hard to rebuild or repair properly. Despite this, matrix receptor-based mechanisms exist to help cells interpret the matrix organization at least intermittantly, and tell the cell how to respond to any anomaly or mechanical cue within the matrix.

11.7 Fibrous Matrices

The major component of extracellular matrix is collagen, which forms fibers that can be arrayed in many different configurations, from tendons to the three-dimensional arrays in the dermis. In the heart, the collagen fibers provide most of the structural integrity by forming layered arrays of fibers at different angles. Collagen fibers are assembled outside of the cell from procollagen molecules that are cleaved during the assembly process. The growth and deposition of collagen matrices is dependent upon mechanical factors. Furthermore, fibroblasts are programmed to assemble fibers in a tissue-specific manner, such as along existing tendons or expanding the depth of the collagen layer in skin. In sports medicine studies, the repair of tendons is aided by periodic stretching of the tendons with a light stress. The process of repair is mechanical, but we don't understand how repair can be facilitated by periodic forces from the matrix. Several explanations for the effect of periodic stress could be: (1) growth of the fibroblasts could be stimulated by the periodic contractions; (2) there could be a cinching process where the cell develops high transient forces during stretch before the relaxation allows the cell to draw in any slack; (3) high stress could order the fibers periodically, which would allow them to be reinforced in the right position; or (4) stress-dependent signals are beneficial for the healing process and activate the healing cells.

Long-lived fibrous matrices are also present in tissues such as skin and lung where the elastin that is deposited during development gradually degrades over time, and is only poorly replaced over decades, by newly synthesized elastin. Thus, when there is major trauma to the lungs or skin, the repair process will not yield tissue with the same elasticity as the original. This means that there is a major problem with the body's ability to repair tissues containing an elastic matrix, particularly in aged adults. The cosmetic industry has spent a lot of time and effort to develop ways to stimulate elastin production in skin, but most remedies produce modest results at best.

Because in major tendons such as the Achilles tendon the majority of collagen does not turn over in the life of the individual, there will be problems if injury occurs. Natural repair mechanisms will be slow, and in reality, will fail to restore the original functionality of the tendon.

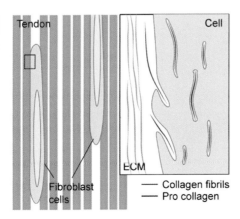

Figure 11.3

Collagen fibrils.

Tendons that are under a very significant tension, like the Achilles tendon, appear to remodel very slowly because the core is permanent. Although the body can add more fibers to the surface of the tendon, it is unable to repair the inner core, particularly when the system frequently experiences very high loads that can reinjure the tendon. The process of collagen fiber addition has been recently analyzed using the electron microscopic. From this work it appears that cells migrate on a fiber bundle, and deposit new fibers as they move. Deposition will preferentially occur on the surface of the tendon and remodeling over time can consolidate the fibers; however, it is unlikely that the repaired tendon will have the same structural strength as the original.

The mechanical aspect of fiber assembly is particularly interesting as it can reveal how fibers can be assembled so as to constrain tissues. In the heart, for example, long collagen fibers provide much of the force for relaxation, i.e. they are stretched during contraction, and can power the relaxation of the cardiac muscle. During development, and other growth periods like age-related cardiac hypertrophy, the volume of the heart expands and the fibers need to reorganize appropriately. This raises the question, then, how is cardiac function maintained, when the collagen fibers are lengthened or shortened? A system must therefore sense the tension in the fiber, and elongate or shorten it according to whether the heart is growing or shrinking. Little is known about how tension is generated on collagen fibers; however, there must be a resting tension even in fibers that are part of tendons.

A major problem that arises during the process of wound healing is fibrosis, which is the deposition of excess collagen that serves to strengthen traumatized tissue. However, the excess collagen often remains once healing is complete, i.e. a scar. By restricting tissue elasticity, and potentially occluding certain regions, the buildup of collagen can have major effects in heart, liver, and lung function. This is of particular concern in the aged population. Although fibronectin, fibrinogen, and other serum-based matrix proteins are involved in transient wound-healing processes (clotting and initial wound healing) only collagen is deposited for long periods. This assembly of collagen fibers in later stages of healing commonly results in scar formation and scar tissue.

A related matter is the body's response to the presence of foreign bodies. For example, when bullets are left in tissues or tumors form that are not vascularized. In these cases, fibroblasts can be recruited to the sites where foreign materials are embedded, and will assemble a wall of collagen around the material to isolate it from the rest of the tissue. This can prevent further damage from the foreign material, but leaves excess collagen that can stiffen tissues or cause fibrosis.

11.8   Basement Membranes

Another form of matrix is the basement membrane that is planar and aids in structuring many epithelia and endothelia. The thickness ranges up to several hundred nanometers and the surface has gaps and roughness in the range of less than a hundred nanometers. In the adult tissues, the basement membrane and the neighboring cells serve as orientation markers for cellular level repairs after local cell death. Contraction to cover the gaps created by cell death cause expansive tension from the neighboring cells that signals growth and growth also requires that the cells are in contact with the matrix. Thus, basement membranes that are assembled in adult tissues aid in maintaining the structure of the tissue as cells come and go.

Very little is known about how basement membranes form, and in particular, how they come to include such a vast array of different components, which include laminins, type IV collagens, nidogens, perlecan, agrin, and other macro-molecules. In early development, epithelial sheets form naturally, and recent studies indicate that laminin deposition serves to stabilize the structure. As the epithelium matures, other components are added, and the structure is stabilized further through covalent cross-linking of different matrix components. These links will be produced by transglutaminase or lysyl oxidase. Thus, the basement membrane is initially formed on the expanding epithelial basal surface and grows

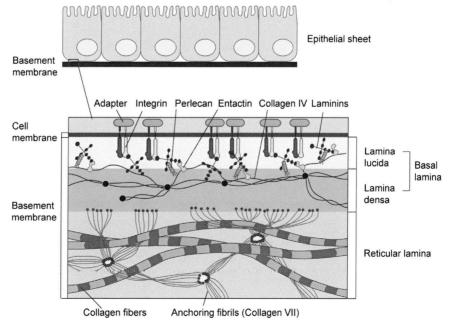

Figure 11.4

Basement membrane.

with the tissue during early stages. Once adulthood is reached, the basement membrane is stabilized to aid in the mature functions and in the natural turnover of individual cells.

In severe trauma where significant portions of a tissue are damaged or degraded, the basement membrane template may also be lost. This makes the process of regeneration much more difficult. Unfortunately, the molecular order of basement membranes is hard to replicate using artificial materials, and this makes treatment of damaged tissues challenging. However, understanding the critical elements in the assembly process may change this. Embryonic cells may one day serve to recapitulate the normal development process or simply stimulate laminin assembly, thereby improving treatment options. In both cases, an immature basement membrane would be formed that could then be developed and stabilized over time.

11.9   Cell Encompassing Matrices

11.9.1   3D versus 2D

There are tissues in the body where cells are largely embedded in the surrounding matrix. For example, in bone, osteocytes are embedded in channels in the bone. Similarly, chondrocytes are surrounded by proteoglycan and elastin fibers in cartilage while fibroblasts are surrounded by a 3D matrix of collagen in the dermis. These matrices provide important mechanical functions, allowing bone to support the body and limbs, or, in the case of cartilage, providing a cushion for bone–bone joints. The cells are needed for the turnover and repair of these tissues; however, controlling the tissue regeneration is difficult. Certainly in the case of bone, the turnover and repair are important functions.

The loss of bone density that is associated with osteoporosis is a significant clinical problem. In osteoporosis, excessive osteoclast activity, which weakens bone, is exacerbated by a lack of osteoblast activity to maintain bone density. Although bone formation must be controlled locally, a global balance between the activities of the two cell types is critical. As in osteoporosis, loss of bone mass is experienced by astronauts in space flight under zero gravity. Here, a 1–2% loss of bone mass per month is experienced. Nevertheless, the repair of bones is critical for organism survival and analyses of wild monkeys have shown that repair is possible in the wild without the benefit of proper splinting or other aids. Because the factors that stimulate osteoblasts vs. osteoclasts are known, there are antibodies that will inhibit bone loss and now can stop the degradation of bone with aging. However, the major issue is how to preserve local repair processes while keeping the proper bone density for the whole body.

An important implication in any repair mechanism is that there are many mechanical factors that must be sensed to enable proper maintenance or repair. For example, there are small channels in bone that contain long processes from osteocytes and those processes potentially sense bone compression or strains by the flow of fluid past them. If the strains are large, then the osteocytes signal to the osteoblasts to build more bone in that area and osteoclasts to reduce the rate of disassembly of bone. In studies of rats, the bending of a bone need only occur once a week and still there will be a significant response in bone growth. If there are insufficient strains, then the osteocytes will activate the osteoclasts and inhibit the osteoblasts. Thus, the bone has a built in mechanosensory system to regulate bone mass locally. Why the overall level of bone mass decreases with age is unclear.

In contrast to bone repair, there has been only limited success in the repair of cartilage. In fact, it is not clear if the cartilaginous matrix can be repaired in a useful way. These tissues are poorly vascularized and the rates of metabolism are consequently low, making natural repair particularly difficult.

The role of matrix in cartilage highlights a critical role of matrix in lubrication. When force is applied to cartilage, the matrix can be compressed, driving out some water. This is similar to matrix glycoproteins on the cell surface which provide a protective coating to the plasma membrane and inhibit interactions with potentially damaging surfaces in the region. They can compress and reversibly expand with concomitant changes in the local extracellular volume. This can increase the rate of fluid exchange in cartilage and will aid cellular metabolism. However, large-scale repair is much more difficult. In general matrices, surrounding cells provide cushioning and lubrication to enable better movement between the surfaces. Thus, for the repair of cartilage, a way to repopulate cells in dense matrices is needed and that will require both the expansion of cells *in vitro* and the preparation of suitable matrices.

Figure 11.5

Cartilage matrix.

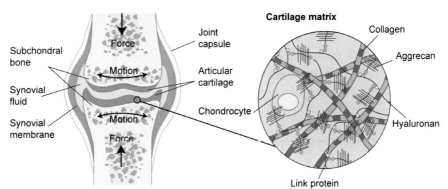

11.10   Tissue Formation through Cell–Cell Adhesions

11.10.1   Conveying Mechanical Information between Cells

Specialized cell–cell junctions are needed in many tissues to enable tissue functions. For example, cells can form regulated barriers that facilitate filtration in the kidney or serve as a permeability barrier in the skin. To achieve this they need to form cell–cell adhesions with their neighbors, called tight junctions. In cardiac muscle, on the other hand, the cells need to communicate electrically with their neighbors in order to depolarize. This process is mediated by a type of cell–cell adhesion called a gap junction. Many specialized types of adhesions are needed for the variety of tissue forms present in the body.

As cells in the tissues die or divide, the junctions must turn over or be adjusted to maintain cell–cell contacts. This must occur without impeding the normal tissue dynamics or tissue repair mechanisms. Several obvious activities are needed for this: first, cell–cell junctional complexes must transmit force from one cell to another to maintain proper tissue tension. Second, the death of a neighboring cell should be sensed so that barriers can be restored. Third, growth of cells is needed to replace dead or diseased cells; and finally, the shape of the epithelium (curvature mostly) must be delicately controlled. Thus, the junctions and the mechanical aspects of the system must be controlled dynamically.

Perhaps one of the most important physical parameters conveyed between cells is the tension of the cell monolayer. This is because individual cells need to sense tension in order to determine whether to grow. As one thinks through the problem from a physical point of view, measuring tension in a monolayer appears relatively difficult. Epithelial monolayers, for example, need to stretch to accommodate transient tension pulses, but growth should not be activated. If the rise in stretching tension is sustained, then cells should grow. Cell stretching occurs primarily through elastic elements in the cytoskeleton that passively resist further stretching (intermediate filaments) and try to contract the tissue laterally, creating tissue turgor. Also, actomyosin networks can create a dynamic tension when myosin pulls on actin to a constant level of force. Clearly, the relative contributions of intermediate filaments and the actomyosin network can change with different circumstances. Cell–cell junctions will transmit the tension, but will peal apart should the cells within the monolayer begin to separate. For example, removal of calcium will disrupt cadherin complexes and cells will separate as they endocytose the cadherins. However, secretion of new cadherin molecules will rapidly restore the epithelium. Because it is such an important parameter, the stretching tension on the junctional complexes needs to be integrated over time and then transduced into growth or apoptosis signals.

Asymmetric tension-dependent morphological changes in epithelia have been studied recently and they show interesting behaviors. In developmental processes, the tissue often invaginates through apical constriction and the degree of invagination is critical for determining subsequent morphological changes. Contraction comes from an actomyosin network in the apical region that is connected to junctional complexes. Cycles of contraction–consolidation occur until the proper morphology is reached. Thus, it appears that the cells have a feedback mechanism whereby they establish what the desired morphology should be, and then activate rounds of contraction until the epithelia reaches the proper morphology, and in particular, the correct degree of curvature. Similar contractile processes are involved in adapting the morphology of other tissues. It is not known whether the cells are measuring the relative area of the apical versus the basolateral surfaces, or some other parameters. Furthermore, it is also unclear which biochemical system is involved. Many genetic screens for effects on tissue morphology have been performed, but it is hard to know exactly where in the process of morphological change the proteins are involved. However, several mutations have been linked to changes in the contraction or consolidation processes. Thus, the tensions at the apical surfaces of epithelia are sensed differently from the basolateral and cells have delicate ways to measure the differences in apical and basolateral area that are not known.

11.11 Cell–Cell Adhesions

Although there are a variety of cell–cell junctions in epithelia, the primary components involved in initial and dynamic adhesions are the cadherins. These proteins can form homodimeric bridges between cells (see Figure 11.6) that are calcium-dependent. However, isolated dimer bridges are unable to support forces transduced between cells. For this reason, cadherins form clusters laterally, similar to integrin clustering, and these can support much larger forces. Also similar to focal adhesions, the cadherins will serve as the integral components of adhesion complexes known as the cell–cell adhesion complexes. These complexes can include over 100 different proteins that have either signaling (transient) or structural roles. Very prominent in the cadherin adhesions is the protein beta-catenin, which moves to the nucleus to alter cell growth or expression patterns in response to mechanical and chemical signals. Beta-catenin highlights the importance of the cell–cell adhesions as signaling centers. Because the forces derived from the pulling of neighboring cells converge on these adhesion complexes, it is logical that they transduce the force into a biochemical signal. Again, the major question is do they integrate mechanical force over time, and how is the resting tension in the tissue determined? In mature epithelia, there are other cell–cell adhesion

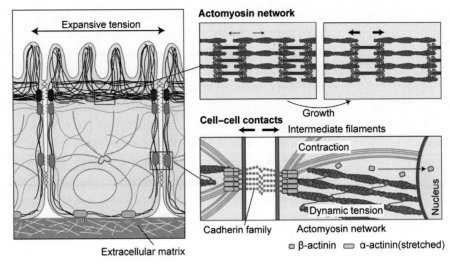

Figure 11.6

Problems of network tension and cell–cell contacts.

complexes that could also act as signaling centers. Again, dynamics is critical and it is likely that the generation of sustained tension in the epithelium will stimulate remodeling as well as growth.

Recently, a mystery surrounding cadherin-based junctions was solved and the solution provides interesting insights into the importance of mechanics in understanding cell–cell contacts. Previous studies had rigorously shown that alpha-catenin (a binding partner of beta-catenin) would not bind to actin as a monomer, which is the form in the junctions. This raised the question of how actin filaments could be linked to the junctions. Understanding this was essential, because the transmission of forces from neighboring cells depended upon a physical link existing between the adhesion complex and the cytoskeleton. The solution, however, was revealed using laser tweezers to move the filaments past alpha-catenin. This experiment showed that the generation of force on the alpha-catenin bond with actin was sufficient to increase the lifetime of the bond (Buckley et al. 2014). Because force strengthens the connection between alpha-catenin and actin filaments, it follows that a loss of force will weaken the binding and increase dynamics. Thus, not only is mechanosignaling occurring at cadherin-based cell–cell adhesions, but the stability of the connection to the cytoskeleton is dependent upon the forces converging on the junction.

11.12   Processing by Adhesions

Cell–cell adhesion complexes contain many components that contribute to the functions of the epithelium, from mechanosensing to barrier formation (see the cadherin adhesome paper of Murray & Zaidel-Bar, 2014). Often the accessory components have much shorter half-lives in the junctions than the major junctional cadherins. One interpretation is that like the cell–matrix adhesions the junctional complexes are processing centers where forces and other external signals are converted into biochemical signals, primarily through conformational changes of components in the complex. Coordination of the various functions that must be activated by a single signal is much easier if the components are in a large complex. However, the presence of so many components (potentially >100) in these complexes makes it very difficult to determine how processing and coordination could actually occur. As in the case of clathrin-mediated endocytosis, the temporal order in which different components associate and dissociate with the complex can tell us a lot about their potential roles. Still, much more effort is needed to understand such processes.

11.13   The Role of Cell–Cell Adhesions in Cell Death

When a cell in an epithelium dies, the neighboring cells rapidly fill in the gap. Studies of this process in developing *Drosophila*, where there is natural apoptosis, or in mammalian epithelia after laser killing both show similar processes. The neighboring cells employ two mechanisms to close the gap. In one mechanism, known as the 'purse-string' mechanism, an actomyosin ring forms near the apical surface of cells surrounding the dead cell. This ring can then contract, thereby closing the gap. In the second mechanism, lamellipodia extend from the basal surface of the cells surrounding the gap. These protrusions will then cover the gap.

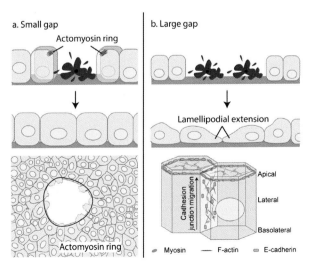

Figure 11.7

Actomyosin ring contraction or lamellipodial extension in wound closure.

If many cells die and the gap is very large, the first mechanism is of little use and therefore the cells use lamellipodial extensions to migrate into the wound and cover it. Importantly, the cells maintain their adhesions with the cells further back from the wound. It is still not known how neighboring cells know to move into the gap formed by the dying cell. It is possible that a loss of the normal mechanical response from the dying cell, or the movement of the lipid, phosphatidyl serine, to the external surface (a hallmark of apoptosis) of the dying cell could act as signals. The mechanical communication through cell–cell adhesions must be continual, for example, to maintain tension on the α-catenin–actin bonds and that is a favored mechanism of sensing changes in the neighboring cell. Thus, there is a very robust process to repair gaps in epithelia that involves at least two distinct motility systems.

11.14   Junction Migration

Cadherin-based junctions appear to form at the basolateral region of the cell–cell contacts, and then migrate apically over time. In the process of apical movement, the initial contacts aggregate such that the density of cadherins increases towards the apical region. The junctions in the apical region then either turn over, or in some cases, translocate back towards the basolateral region. These movements are dependent upon actomyosin contractility, which is in turn dependent upon signaling between the cells (Wu et al. 2014). Junction migration is supported by a relatively regular array of actomyosin aligned along cell–cell junctions. This array maintains a very close correspondence between the sarcomeric units in neighboring cells through the cell–cell junctions. The epithelial form of myosin IIC is preferentially found in these apical actomyosin bands and presumably the movements of junctions are driven by myosin IIC.

11.15   Epithelial Wound-healing

Skin is continually turning over and in many cases is capable of repair. Should it be damaged by injuries such as a cut or burn, a clot will form and the cells that surround the wound will begin to migrate. A complex process of skin restoration then occurs and this will hopefully replenish the different skin layers (see Figure 11.8). Without intervention to close the edges of the wound, there is often imperfect healing and scar tissue, which will last for the life of the organism. The repair and growth processes within skin are highly dependent on the mechanical properties of the underlying tissue. This dependency is exploited, for example,

Figure 11.8

Healing process.

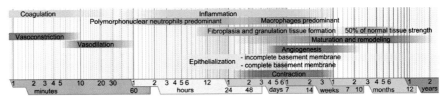

Approximate time periods for each wound-healing phase. Due to different wound sizes and healing conditions, substantial variations are represented in faded intervals.

when growing extra skin for cosmetic surgery. In this case, a balloon may be inflated underneath the skin to cause it to expand. Over the course of a month or two, the skin will grow to a much larger area due to the increased tension caused by the balloon. The step-wise process of skin regeneration includes several steps that involve the formation of new collagen matrices by fibroblasts in the underlying dermis and the basement membrane for the keratinocytes to follow and re-establish the epithelial barrier. Our increased understanding of this process can aid in reducing scarring and facilitating the rate of healing.

11.16 SUMMARY   Extracellular matrix as well as neighboring cells in a mature organism both serve to preserve the tissue organization, even during the normal turnover of cells. Cell–matrix and cell–cell adhesions are the primary sites through which signals from the microenvironment are transduced. From cellular studies with fibrous matrices, we know that the matrix shape, mechanics, and composition all contribute to the phenotype of cells. At the tissue level, the regeneration of fibrous matrices, basement membranes or cells embedded in matrices is difficult in adults. The major successes in regeneration have been in seeding stem cells into tissue-derived matrices that will then direct the proper differentiation of the cell. Thus, the extracellular matrix can direct stem cell differentiation, but it is difficult to have cells restore the normal matrix. In epithelia or endothelia, the cell–cell adhesions will control a number of parameters such as morphology and growth. Morphology of epithelia is tightly regulated in embryos through control of apical versus basolateral cell areas by actomyosin contractions of the apical surfaces. Loss of cells in an epithelium increases the tension in the epithelium as the remaining cells try to close the gaps and higher tension on cell–cell adhesions stimulates growth. Skin and gut epithelia are continually being shed and renewed. However, even in those cases, wound-healing over large areas is problematic, particularly because some matrix components like elastin are not renewed. In general, the ECM of adults directs cell phenotype and the cell–cell junctions direct proper growth and other aspects of tissue morphology.

11.17 PROBLEMS

1. It is important for integrins to associate in clusters of 3–4 integrins within a distance of 60 nm. We want to know how much more tightly a protein like talin will bind to an integrin cluster of three than to a cluster of two. Talin is a dimer and has four binding sites for the integrin beta tail (one in each head domain and one in the middle of each rod domain). If we suggest that two head domains bind to two integrin beta tails with an affinity constant (K_A) of 3000 (3×10^3 M^{-1}) for each head, then what will be the off rate for the complex? With the addition of one more bond from the rod domain that binds to an integrin tail with an affinity constant of 2000, what will be the new off rate? Now we want to add a fourth bond with the same affinity as the third. What will be the new overall affinity constant and new off-rate? (We assume that the on rate of binding is 2×10^7 M^{-1} s^{-1}.) If we assume that an adhesion complex requires 30 s to form and talin must remain bound to integrins for that whole period, then how many integrin–talin bonds are needed for the formation of an adhesion?

   Answer: $K_A = k_{on}/k_{off}$ and $k_{off} = k_{on}/K_A$
   $k_{off} = 2 \times 10^7$ M^{-1} s^{-1}/K_A
   With 2 bonds and $K_A = 9 \times 10^6$ M^{-1} the $k_{off} = 2.2$ s^{-1}
   With 3 bonds and $K_A = 1.8 \times 10^{10}$ M^{-1} the $k_{off} = 0.0012$ s^{-1}
   With 4 bonds and $K_A = 3.6 \times 10^{13}$ M^{-1} the $k_{off} = 5.5 \times 10^{-7}$ s^{-1}
   To maintain a bond for 30s, three integrins must be bound.

2. When cell–cell contacts are broken because of calcium depletion or mechanical overstretching, the adhesion complexes are typically endocytosed and processed before new cadherins are released to the cell surface. Develop a hypothesis about why the endocytosis might be the most efficient way to prepare for the formation of new cell–cell contacts.

   Answer: If the cadherin complexes are formed through a basically irreversible step, then they cannot simply dissociate and reform. There are a number of irreversible steps in the assembly of other adhesions emerging that provide a precedence for this hypothesis.

3. Cell matrix adhesions produce signals to the cell. A logical hypothesis is that the adhesion proteins are modified in the adhesion and then leave the adhesion to go to sites in the cytoplasm or nucleus to convey the signal. If there are 20 μm^2 of adhesions with 2000 molecules of paxillin/μm^2 and the lifetime of paxillin is on average 20 s, then how long will it take for the cell to generate 2 μmolar of modified paxillin in the cytoplasm (assume cell volume is 3000 μm^3)?

   Answer: First, we will compute the number of protein molecules in a cell with a 2 micromolar concentration of that protein. This gives 3.6×10^6 molecules needed and there will be on average 2000 modified paxillin molecules released per second. Thus, it will take 1800 s or 30 minutes to produce that concentration.

4. Aggregation of integrin receptors with bound ligand directly leads to:
    a. assembly of a stable link from the integrins to the cytoskeleton;
    b. turning off signaling pathways necessary for growth;
    c. cytoplasmic adhesion protein binding to integrins; and
    d. a positive change in the free energy of binding to multivalent cytoplasmic proteins.

Answer: (a) and (c) are true but (b) and (d) are false. In (a) and (c) the multimeric nature of the integrin clusters favors lower free energy of binding through multiple weak interactions. In (b) it is false because binding to matrix generally turns on growth. In (d) the free energy is negative for binding.

11.18 REFERENCES

Buckley, C.D., Tan, J., Anderson, K.L., et al. 2014. Cell adhesion. The minimal cadherin–catenin complex binds to actin filaments under force. *Science* 346(6209): 1254211.

Changede, R. & Sheetz, M. 2017. Integrin and cadherin clusters: a robust way to organize adhesions for cell mechanics. *Bioessays* 39(1): 1–12.

Changede, R., Xu, X., Margadant, F. & Sheetz, M.P. 2015. Nascent integrin adhesions form on all matrix rigidities after integrin activation. *Dev Cell* 35: 614–621.

Meshel, A.S., Wei, Q., Adelstein, R.S. & Sheetz, M. P. 2005. Basic mechanism of three-dimensional collagen fibre transport by fibroblasts. *Nature Cell Biology* 7: 157–164.

Murray, P.S. & Zaidel-Bar, R. 2014. Pre-metazoan origins and evolution of the cadherin adhesome. *Biol Open* 3(12): 1183–1195.

Schvartzman, M., Palma, M., Sable, J., et al. 2011. Nanolithographic control of the spatial organization of cellular adhesion receptors at the single-molecule level. *Nano Lett* 11: 1306–1312.

Wolfenson, H., Meacci, G., Liu, S., et al. 2016. Tropomyosin controls sarcomere-like contractions for rigidity sensing and suppressing growth on soft matrices. *Nat Cell Biol* 18: 33–42.

Wu, S.K., Gomez, G.A., Michael, M., et al. 2014. Cortical F-actin stabilization generates apical–lateral patterns of junctional contractility that integrate cells into epithelia. *Nat Cell Biol* 16(2): 167–178.

Zaidel-Bar, R., Itzkovitz, S., Ma'ayan, S., et al. 2007. Functional atlas of the integrin adhesome. *Nat Cell Biol* 9(8): 858–867.

11.19   FURTHER READING

Cadherins: www.mechanobio.info/topics/mechanosignaling/components-of-cell-adhesion/#cadherin

Cell–cell adhesion: www.mechanobio.info/topics/mechanosignaling/cell–cell-adhesion/

Cell–matrix adhesion: www.mechanobio.info/topics/mechanosignaling/cell–matrix-adhesion/

Collective cell migration and wound-healing: www.mechanobio.info/topics/development/collective-cell-migration/

Extracellular matrix: www.mechanobio.info/topics/mechanosignaling/cell–matrix-adhesion/

Integrin clustering: www.mechanobio.info/topics/mechanosignaling/cell–matrix-adhesion/integrin-mediated-signalling-pathway/#clustering

Substrate rigidity: www.mechanobio.info/what-is-mechanobiology/substrate-rigidity/

Talin: www.mechanobio.info/topics/mechanosignaling/components-of-cell-adhesion/#talin

Tight junctions: www.mechanobio.info/topics/mechanosignaling/cell–cell-adhesion/tight-junctions/

12 Adjusting Cell Shape and Forces with Dynamic Filament Networks

The shape of a eukaryotic cell or tissue is determined by the actions of the dynamic cytoskeleton pulling or pushing on the ECM or neighboring cells through a set of complex functions. In the previous chapters, we have considered the important elements of cytoskeletal filament dynamics and myosin activation as well as how cell–matrix and cell–cell interactions might affect the cytoskeleton. However, these motile systems must be orchestrated spatially and temporally to produce the correct types of motility for the needed functions of individual cells or of cells in tissues. Because of the dramatic differences in the behavior of individual cells and cells in tissues, we will discuss each type of behavior separately, making this a two-part chapter. In the first part, we will describe individual cell migration and motility, whereas in the second part, we will consider the cooperative processes that occur in tissue formation and remodeling. In both cases, the actin cytoskeleton is dynamic and the major active forces are generated by myosin II. However, there is a different organization of motile functions through the activation of different sites of actin polymerization and myosin activation in the different motile processes. At the individual cell level, we will consider the mechanisms of the few characterized types of motility and how the different elements of the cytoskeleton might be coordinated to produce migration or matrix remodeling. In tissues, the types of motility are dramatically different and the subcellular forces are difficult to measure. Thus, there is a lot of speculation, but some important principles such as the cohesion of the cells in the tissue have emerged. Modeling of the motility processes has started and such models will be very useful in prioritizing experiments to focus on the critical elements that will control each type of motility selectively. The goal in this area is to define the steps in motile functions, how they are coordinated as well as altered by forces or mechanics, and the roles of specific molecules involved.

Cell and tissue shape is critical for the survival of the organism. Because eukaryotic cells form without an exoskeleton, organism structure and shape must be developed using the tools that we have discussed in previous chapters; namely, the cytoskeleton, motor proteins, and extracellular matrices. There are approximately 300 different cell types in the human body and each has a number of different mechanical properties that are needed to form the appropriate tissues. We believe that those different properties can be derived from a smaller number of tools that are controlled to produce the final form. For example, the tension of skin can be set differently from the surface tension of the liver, when the corresponding cells use a different control mechanism for force transduction. Although the basic components of the actomyosin networks may be similar, each mechanism will produce different degrees of myosin activation and/or concentration in the actomyosin network, which generates the tension. Importantly, there are some basic motility processes that are commonly found in many different types of cells (see text box). These processes are discussed in more detail on the MBInfo site (www.mechanobio.info) and I refer you to that website if you have further interest in individual processes. Our discussion here, however, will be of the fundamental aspects of the processes, as there is not sufficient space to treat them in detail.

From the viewpoint of robust devices, the types of motility have been developed through evolution and a few integrated motility processes have emerged as the most reliable for mammals. The stereotypical behavior in each integrated motility process can be analyzed for a specific cell type and the modifications that will occur in another cell type can be readily worked out by analogy. In practical terms, this means that each motility process can be broken down into the individual 'tools' or components that the cell uses to control the cytoskeleton. There are a finite number of different types of motility (tens rather than hundreds), which makes the problem tractable, and each is typically activated in local cellular regions of just 1–3 μm. These motility types will be dependent upon one or more of the approximately 20 different actin polymerization proteins found in mammalian genomes. With the different polymerization factors, motor proteins, and actin crosslinkers, it is therefore possible to produce a wide variety of different single cell shapes and motility behaviors. Many of these motility programs are activated by small G-proteins, but when the 135 different small G-proteins are expressed individually in fibroblasts, only 10 different types of cell shapes are observed. Thus, there are only a few basic motile behaviors that are observed commonly in fibroblasts.

The critical elements in any actin-based motility process are the sites of actin polymerization, the anchorage or crosslinking of the filaments, and myosin positioning plus activation. Starting with actin filament assembly, polymerization can cause membrane extension through lamellipodia or filopodia that will enable the cell to reach new matrix sites. Once the cell becomes anchored to adhesion complexes, actin filaments from those adhesions can transmit myosin-generated forces to those sites from all around the cell. For the tissue or the cell to continually generate tension on the adhesion sites, there must be a dynamic feedback between the motor proteins, actin assembly and the actin filament crosslinking proteins. This has been described as 'cohesion of the cytoplasm' and enables mechanical connectivity across cells and continuously through tissues. The major elements needed for continuous cohesion are bipolar myosin filaments, actin filament arrays from formins or other actin-polymerizing proteins and crosslinking proteins such as filamin. In addition, links between the dynamic actin cytoskeleton and intermediate filaments and microtubules are required to prevent overstretching of the cell. These also provide transport of components that are critical in maintaining tissue integrity, as well as polarized migration. Ordering of the cytoskeleton is through the localized assembly of filaments. Where these filaments form and how the cytoplasm is organized is, in part, determined by microtubules, which often transport components of the assembly complexes to their required location. Because there are relatively few polymerization processes, the ways in which a cell can change shape is also limited. Thus, the question shifts to how each process is localized and controlled, both in terms of activation of assembly and disassembly.

Basic Motility Tools

1. Filopodia extension/contraction.
2. Lamellipodia extension/contraction.
3. Bleb extension/contraction.
4. Cytoplasmic network contraction.
5. Sarcomeric contractions rigidity-sensing, stress fibers, muscle sarcomeres.
6. Contractile rings in epithelia.
7. Apical constriction in epithelia.
8. Convergent rxtension in epithelia.
9. Tube formation in endothelia.
10. Matrix contraction by fibroblasts.

12.1   Motility Processes in Isolated Cells

In thinking about how to robustly perform motile functions, it makes sense that the cell uses a set of tools that works reliably under a variety of different conditions. Those tools need to be under local feedback controls, to ensure they perform reproducible and controllable tasks that produce the desired force or displacement. Although these processes need to be very dynamic during development, in the adult organism they primarily serve to respond to the local microenvironment and trauma situations such as injury or disease. A surprising result from recent observations is that cells can activate many different motility mechanisms even when the cell type is not normally using that type of motility. Thus, it is useful to know how to activate and modify many different types of motility because each type may be activated in a wide variety of cells.

12.2   Matrix Shape can Control Cell Behavior

Recent studies using microfabricated surfaces show that the shape and behavior of cells can be controlled by the shape of the matrix environment. In contrast, cells plated on continuous fibronectin-coated glass will adopt a variety of shapes because there are no constraints on the matrix environment. When fibroblasts spread on defined patterns, they will adapt to those patterns and will express thousands of different proteins on a fibronectin-coated circular disc than on a fibronectin triangle of the same area. The nature of the matrix adhesions and the stress fibers are totally different on those two shapes. Thus, it is perhaps not surprising that totally different cellular programs will result. In considering cell shape it is important to keep in mind that the type of cell, the composition of the matrix, and the shape of the matrix all interplay to activate the type of motility that will then create the cell shape.

12.3   Cell Migration Mechanisms

There are three major types of cell migration that have been characterized: specifically, mesenchymal migration, process extension in neurons, and amoeboid migration. Fibroblast migration primarily involves the extension of lamellipodia – neuronal growth cones extend filopodial whereas amoeboid cells extend membranous blebs that are rapidly stabilized by the assembly of actin filaments. These are in order of decreasing forces generated on the substrate. Fibroblast motility is designed for high-force applications such as matrix remodeling or cell sheet movements (also known as collective cell migration). Neuronal growth cones typically migrate through neural tissue that is relatively soft, and they do not generate particularly high forces in this process. Macrophages and other immune cells (as well as some cancer cells) that move by amoeboid migration are primarily designed for invading tissues and penetrating porous matrices. They typically generate only very small forces on the substrate unless a region of the cell is stuck and then the cell tries to pull away from that site.

The key in any form of migration is cell polarization and movement in only one direction. In the case of fibroblast motility, the polarization is often dependent upon

Figure 12.1

Three types of motility.

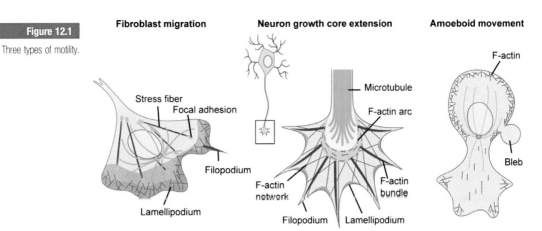

Fibroblast migration **Neuron growth core extension** **Amoeboid movement**

hormone activation, rigidity-sensing, or the polarized activation of cell extension. Polarization of the cell generates leading and trailing edges as the cell migrates. Extensions grow preferentially at the leading edge and newer contacts are made within this region. For the cell to change direction, it often turns slowly by sending out extensions on the side of the leading edge toward the direction it intends to go. Growth cone movements normally follow guidance cues that are often bound to matrix proteins, and thus follow a path that is detected primarily by filopodial extensions. In the case of amoeboid movements, the cell blebs in the direction of movement and polarization appears to depend upon membrane tension that serves to restrict movements in other regions but allows blebbing in the correct direction. Each different type of migration requires a different set of proteins and a different mechanism for polarization and guidance. Again, defining the major steps in the process is critical for developing an understanding of the overall mechanism.

12.3.1 Mesenchymal Cell Migration

Although many different types of motility have been observed in fibroblasts, the most studied has been fibroblast migration. The basic steps in fibroblast migration are shown in Figure 12.2 and involve extension, attachment, contraction, release, and recycling. This is a bit oversimplified because cells will often move toward rigid surfaces, which means that they are sensing the matrix rigidity as they are moving on it (see Chapter 16 for a detailed treatment of rigidity-sensing). Taking these steps as a starting point, it is important to understand how the steps are coordinated in time and space.

For example, cell edge analysis programs have shown that 2–5 μm portions of lamellipodial edges extend at a relatively constant rate, but only for about 1–2 min

Figure 12.2

Basic steps in fibroblast
migration.

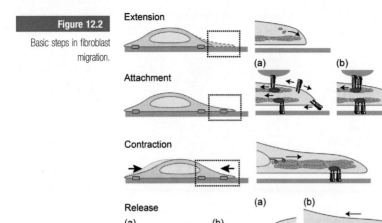

before they stop or retract. Thus, it appears that lamellipodial extension is a
discrete event in time and space. Once initiated, the motile process continues
through a choreographed set of steps that move the edge forward for a period and
then concentrate on adhesion formation or retraction to the previous adhesion.
Control of the extension is linked to activation of Rac1-GTP formation through a
GTP-exchange factor (GEF) localized to the leading edge. Cessation of extension
is linked to the hydrolysis of GTP by a Rac1 GTPase-activating protein (GAP)
that is itself activated by Rac1-GTP, perhaps indirectly. Other small GTP-binding
proteins like Rho and Cdc42 are also activated in that region. A hypothesized
order of activation is that Rac1 is activated at the start of extension and the RhoA
is activated to start the retraction phase, and this subsequently involves an
activation of Cdc42 for consolidation of the adhesion. Because this is a relatively
common type of cell motility, it is logical to presume that this process involves an
automatic sequence of events. In other words, activation of Rac1 and the initiation
of extension will cause activation of RhoA automatically to induce contraction.
This will automatically trigger the activation of Cdc42 to help in consolidation.
We suggest that this is one type of motility that is preferentially activated in
fibroblasts in response to hormonal, mechanical, or matrix-dependent signals.
Once the right signal is received, the process will initiate, and will proceed

automatically. For such processes to proceed automatically, but in a step-wise fashion, a stringent method of coordination must be built into the pathway. This will most likely exist in the nature of the complexes that carry out the individual steps. In the most simple scenario, this coordination would be a natural series of activations, where A activates B, B activates C, and so on. Another signal could of course alter the progression of this pathway or change either its rate or its outcome. For example, if the new surface reached by extension is found to be soft in the contraction step, then consolidation may be compromised and the cell will possibly change direction. Thus, many motility functions involve a set of linked steps that will normally proceed to completion but can also produce signals that will affect future motility.

For migration, the next step is the overall cell contraction to bring the cell body forward. This should not be confused with the local contractions in the lamellipodia that sense rigidity, although those contractions will create the adhesions needed to support migratory contractions. There is a general flow of actin from the leading edge to the nuclear region that is almost universally seen in fibroblasts during migration. That contraction can create large forces on the substrate, but those forces are isometric, i.e. the force at the back balances the force at the front. Because the velocities of cell movement are only 1 to at most 40 μm per minute, the fluid drag on the cells creates less than 1 pN of force on the cell, whereas the contractile forces are typically 100,000 pN or more. Thus, the mesenchymal cells could move much faster if they could release the adhesions at the rear of the cell that slow them down. Indeed, amoeboid migration involves much less adhesion to the substrate and much lower forces. The large mesenchymal cell forces can move matrix fibers and larger structures.

The release step in fibroblast migration involves focal adhesion disassembly and the physical disruption of the remaining adhesive contacts. Focal adhesion disassembly is linked to the attachment of microtubules at the adhesion sites and presumably involves the active transport on microtubules of components needed for disassembly. When adhesions in migratory cells are observed, there are many adhesions just behind the leading edge of the cell that disappear about 4–5 μm back from the leading edge. Some of the adhesion components remain in adhesions at the back of the cell, but those are much smaller and under very high tensions because the forward-pulling forces from all of the adhesions in the front are balanced by force from the few adhesions (often just one) at the back. Sometimes cells leave behind the rear adhesion components as they move forward, indicating that mechanical forces are driving final release. Before that time, however, most adhesion complexes are disassembled by biochemical processes. Thus, fibroblastic migration depends heavily on adhesion disassembly and most are disassembled biochemically in the front half of the cell.

Recycling of components is necessary because migration does not require the proteolysis of many of the proteins involved. For example, actin monomers are believed to diffuse from sites of disassembly of filaments at the back of lamellipodia to the front. There are active movements of integrins toward the front of lamellipodia where they can rebind to matrix and support further migration. The process of active transport to the front that carries integrins to the front can also carry other components. Myosin 1e will move to the leading edge and is known to transport actin capping protein and several other proteins in a complex. Actin crosslinking proteins are also likely to be transported to the leading edge. There are many cyclic processes in migration, and it is important to continue the cycle, whether it be actin assembly–disassembly–reassembly or adhesion assembly–disassembly–reassembly. Thus, recycling is an important element in migration.

12.4   Lamellipodial Extensions

Because lamellipodia are critical elements in mesenchymal cell migration and other motile processes, it is important to understand them in detail, e.g. they are thin (~200 nm) but broad (2–4 μm wide) extensions of the plasma membrane supported by crosslinked actin filament networks. For a fibroblast to extend lamellipodia, it typically must be activated as described in Figure 12.3. Basically, round cells must bind to the matrix, assemble actin and then pull on the matrix. If the matrix is attached to a substrate, then the generated force should be sufficient to activate the extension of lamellipodia. In contrast, round, rat basophilic leukemia cells can bind a crosslinked activating ligand at many sites on their surface and that can stimulate lamellipodial extensions in 3D. Thus, lamellipodia can extend above the surface or on the surface. The proteins involved in the extension normally include VASP, N-WASP, and the ARP2/3 complex, which will produce a branched actin network. In addition, formin activity is also required for lamellipodial extension. This suggests that formin-dependent actin filament assembly is also involved. The relative importance of the two different actin polymerization systems is not yet fully understood. A more important issue to understand is how of actin polymerization can be activated over a 2–4 μm region of the cell edge. The spread of the extension wave is very rapid and must involve a specific lipid, or a rapidly diffusing small G-protein, to trigger an almost instantaneous activation of extension. Importantly this will occur only over a 2–4 μm region of the lamellipodia, and not over the whole cell.

Because lamellipodia will only extend for a period of 40–200 s before stopping, the cell must actively choose to continue the process by reactivating actin polymerization. The process of reactivation can come from the surface or from a hormone signal such as EGF. The next step in lamellipodial extension is an active contraction phase (Figure 12.3). During this step, local actomyosin-based contractions at the cell edge test the rigidity of the matrix. If the surface is rigid, then the fibroblast will form adhesions and the lamellipodia will extend further; however, if it is soft, then further extension will cease and adhesions will not assemble. The consequence of this is that fibroblasts will be able to move easily on hard surfaces because the adhesions are formed and traction can be gained. On soft matrices, however, where the adhesions are lacking, the lamellipodium will retract, and movement will be impeded. These steps have been studied extensively for fibroblasts because they grow well on solid substrates. Although other cell types can also extend lamellipodia, they will not necessarily employ an identical set of steps. It is to be expected that there can be several different variations on the theme that will give similar behavior.

Also characteristic of lamellipodia formation is the movement of actin filaments toward the cell center. This phenomenon is quite universal and is associated with a cofilin-mediated turnover of those filaments. Here, cofilin promotes actin severing and disassembly of the actin monomers from the pointed end. Because of this, the width of the lamellipodium is inversely related to cofilin activity. Phosphorylation of cofilin by the Lim kinase causes its inhibition, which in turn leads to an increase in the width of the lamellipodium. This is observed, for example, following the activation of Rac1, which activates Lim kinase. For growth of actin at the leading edge, G-actin must be transported back to the leading edge after the filaments disassemble towards the cell center; however, calculations indicate that the rate of depolymerization after filament breakage may not be sufficient to account for the required levels of G-actin. Consequently, short filaments may be transported back to the leading edge with crosslinking proteins, which are needed to stabilize the actin network within lamellipodia. In addition, some of the filaments are condensed by myosin filaments at the back of the lamellipodium and myosin can cause filaments to break, thereby speeding depolymerization. The remaining actomyosin structures can form stress fibers or circumferential arcs of actomyosin. Thus, the filaments that originate at the leading edge are remodeled to support interior actin-based structures such as stress fibers.

12.5 Filopodial Extensions

In neuronal growth cone migration, filopodia are the primary form of extension and act both as sensors and part of the motile machinery. Filopodial extensions are characterized by small fingers of actin-supported membranes, which extend for 2–4 µm (limit is 20–30 µm). Filopodia are used extensively in the migration of neural growth cones. In this case, the filopodia probe the surrounding region for chemoattractive and chemorepulsive signals, which will guide the direction in which the growth cone moves. When a chemoattractant is found attached to a matrix, a bond is formed with the attractant and force is generated on that bond. This will usually cause the growth cone to turn, and encourage more filopodia to extend in the direction it is now facing. However, when a chemorepulsant is found, there will be a major collapse of the filopodia, resulting in the growth cone moving away from the repulsant. In some cases, the growth cone will collapse altogether. Filopodia make ideal chemical sensors as they can probe a much larger area with fewer resources, compared to lamellipodia.

The formation and function of filopodia takes place over a number of steps, and involves a series of components that drive cytoskeleton dynamics. Many of these components are localized either to the tip or base of the filopodia. At the base, for example, are actin polymerization and crosslinking systems that push the tip outward against membrane tension, but are also supported by polymerization at the tip. Myosin motors are also localized to the base, as they carry components to the tips, following the parallel actin filament arrays toward an outward-facing barbed end. Myosin motors at the base of a filopodia also serve to continually pull the actin inward. When a repulsive signal is encountered, a block of actin polymerization at the tip will result in a myosin-mediated retraction of the filopodia. Binding to an attractant-coated site, however, can cause focal adhesion kinase and other adhesion proteins to strengthen the bond to the actin cytoskeleton and the movement of actin inward by myosin will pull the attractant site inward and the growth cone outward. Again, the cell is using the extension process to probe the environment, and chemical and mechanical signals will produce the proper morphological changes to the cells.

In defining filopodial processes, it is important to be able to reproducibly observe the same motility. This will allow quantitative changes in the behavior of cells to be linked to molecular changes within the filopodia. The major components involved in actin polymerization in filopodia are the formins and the VASP proteins. In cases where filopodia form from lamellipodia, there are bundling proteins like fascin that crosslink filaments near the leading edge of the lamellipodia, to produce a tight bundle that can support the filopodium. Recent characterizations of the process of filopodial extension showed that it was enhanced dramatically by the overexpression of myosin X, and that long filopodia normally formed with full-length myosin X. However, overexpression of a form of myosin X lacking the membrane-binding FERM domain caused shorter filopodia to form. The difference in filopodia length was attributed to the number of extension events that took place during filopodia formation. While multiple extension events occurred with full-length myosin X, only one extension event occurred when myosin X lacking the FERM domain was expressed (Watanabe

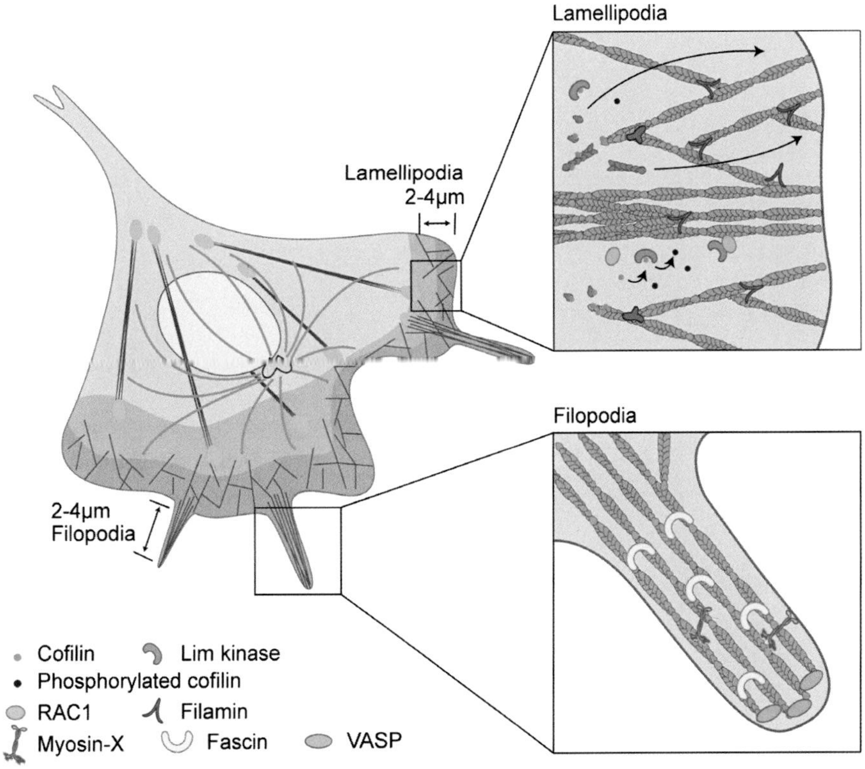

Figure 12.3

Lamellipodia and filopodia.

Figure 12.3

Lamellipodia and filopodia.

et al., 2010). Myosin X moves to the tips of filopodia and it was therefore thought to be primarily involved in component transport. However, the fact that it stimulates filopodia formation, and is required for the multiple elongation steps, indicates that it has multiple roles in different steps in filopodia elongation. The ways in which filopodia actually function may be far from our initial assumptions about mechanism. As for other types of motility, further quantitative analyses of filopodial extension will provide a more complete picture of how steps are performed, the order of activation of components and the signals that can be generated by each step.

12.6 Amoeboid Extensions

Amoeboid movement is characterized by the formation of blebs at the leading edge. These protrusions commonly occur where matrix adhesions are weak and cells are responding to soluble chemoattractants or non-matrix cues. A bleb forms when high turgor pressure in the cell causes the membrane to separate from the actin cytoskeleton and balloon outward. Because the back of the cell is contracting

Figure 12.4

Bleb-based migration.

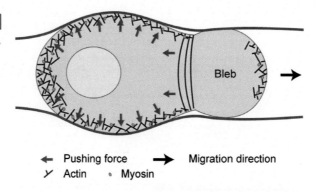

to create the pressure for blebbing, and chemoattractant binding to the front causes weakness in the membrane–cytoskeleton attachment, the cell will bleb in the desired direction. Again, the contraction of the cytoskeleton that raises the turgor pressure in the cell is driven by myosin, and the inhibitor of myosin contraction, blebbistatin, is so named because it inhibits blebbing at the cellular level. Initially, the bleb membrane contains no actin and membrane proteins are freely mobile in the membrane. However, actin rapidly polymerizes on the bleb membrane in an ezrin–redixin–moesin (ERM)-dependent process. Once polymerized, the actin network in the bleb connects with the rest of the actin cytoskeleton and is drawn rearward. Repeated blebbing and rearward contractions can drive the cell forward with very little force on the substrate. This mode of migration is used exclusively in cases where migration occurs without apparent adhesion molecules such as integrin. Further, amoeboid cells can move through small pores in the matrix as blebs extend through the pores, but cell movement is limited by the ability of the more rigid nucleus to pass through the pore. Although bleb extensions don't normally form matrix or other contacts, they do constitute a sensory region that is responding to the binding of chemoattractants by disassembling the cytoskeleton–membrane attachments. Further, there is often a persistence of movement in a given direction for several extension retraction cycles until a signal for movement in another direction finally causes the cell to change direction.

12.7 Sarcomere Contraction Units

The major structure for force generation in muscle is the sarcomere. Because muscles are common in most organisms, the assembly of sarcomeres is a process that is well developed. It is therefore not surprising that sarcomere-like structures are also found in non-muscle cells. Unlike those in muscle sarcomeres, however, non-muscle contractile structures exist as local contractile units that test matrix rigidity. Alternatively, they will exist as stress fibers, which generate large forces on adhesion contacts. The local contraction units are the most transient of the sarcomeric structures and, in fibroblasts, undergo a cycle of assembly, contraction, relaxation, and disassembly in 60–100 s. Stress fibers are long, actin-rich fibers that extend typically across the cell from an adhesion site on one side of the cell to another adhesion on the other side of the cell. Like muscle sarcomeres, stress fibers contain a 1.5–2 μm repeat pattern of α-actinin and myosin. They last much longer than the local contraction units (tens of minutes to hours) but are also dynamic, as revealed by photobleaching recovery experiments where fluorescent actin in the stress fibers recovers in 60–150 s after photobleaching. Smooth muscle sarcomeres are similar to stress fibers in that the myosin filaments are not well ordered in linear arrays but are offset in adjacent sarcomeric units. In all of these cases, anti-parallel actin filaments are connected by bipolar myosin filaments. The actin filaments are also at least partially anchored by α-actinin. Other common elements in the sarcomeric units include tropomyosin, which coats the actin filaments, tropomodulin, which caps the pointed ends, and CapZ, which caps the barbed ends. Thus, non-muscle cells use analogs of striated muscle sarcomeres to generate force in a variety of situations.

When large forces are needed for cells to move matrix fibers in the heart, in tendons, and in skin, the fibroblasts involved utilize sarcomeric structures that can produce large forces. Because large forces are needed for many important functions in the organism, it is critical to assemble these sarcomeres in the right places for high-force applications. This raises a lot of questions about how the cells respond to matrix forces or receive signals to generate large matrix forces through localized assembly of myosin filaments.

12.8 Podosomes and Invadopodia

When adhesion complexes form under weak forces and are therefore not activated for rigidity-sensing, they can produce podosomes. These structures form in osteoclasts and are arrayed in circles to help create local regions on the bone surface where resorption of the bone can occur. They are formed by actin polymerization in a core bundle that extends 1–2 μm perpendicular to the attachment surface or basement membrane, largely dependent upon Arp2/3 activity. Similarly, in cancer cells, when adhesions form under weak forces, invadopodia will form. These structures will push into new areas such as weak basement membranes and enable cancer cells to metastasize from the tissue. Thus, the podosomes and invadopodia have specialized functions that depend upon adhesion formation in the absence of contractile force.

12.9 Cohesive Network Assembly and Dynamics

Early in the fibroblast spreading process, inward forces at the cell edge are balanced by forces from the opposite side of the cell. In this case, forces are transmitted across the cell through a cohesive actomyosin network that is maintained by cytoskeletal structures termed 'nodes.' These nodes contain formin and the actin crosslinker, filamin (Luo et al., 2013). Actin filaments will polymerize from nodes in multiple directions; thus, enabling bipolar myosin filaments to form a cohesive network that is both contractile and dynamic. New nodes and new bipolar myosin filaments must continually form in order to keep the network stable. The actin filaments must also be anchored to external matrix proteins near the cell edge otherwise it will condense as illustrated in Figure 12.4. Such networks are critical for the cell to be able to transmit forces over long distances and to maintain structural coherence in the cytoplasm.

12.10 Adhesion Site Assembly and Dynamics

The formation of adhesion sites involves the recruitment of many different proteins into a large reversible complex. This formation involves a gradual maturation of the adhesion site, with nascent adhesions developing into focal adhesions as a result of force. Focal adhesions may then mature into fibrillar adhesions. During this progression the components change, and this means the signals coming from the adhesions will also change. An important aspect of adhesions is that the structure as seen with fluorescent adhesion proteins lasts for tens of minutes, whereas most of the components in the adhesions have exchange rates in the tens of seconds. An explanation for this discrepancy is that the adhesion is a processing center that sends components to other regions of the cell once they have been modified in the adhesion. This would, in effect, allow the adhesions to control wider cell behaviors.

The major purposes of the adhesions, however, are to transmit cytoskeletal forces to the matrix, and to generate signals that inform the rest of the cell about

Figure 12.5

Actin network.

Actin network

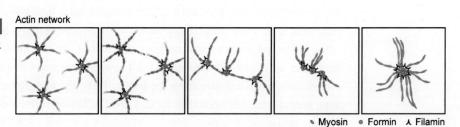

Myosin   Formin   Filamin

the nature of the matrix. When myosin contraction is inhibited, then adhesions will often disassemble and no further signals will be generated. This means that the adhesions depend upon force for their persistence. Formins are found at adhesion sites and they typically polymerize actin from the adhesions that is incorporated into stress fibers. Thus, stress fibers move from adhesions toward the center of the cell. Interestingly, adhesions are discreet complexes that only form every 1–3 μm along the cell periphery. This means that major gaps exist in between them, even though the matrix proteins exist as continuous layer on the glass. Thus, the adhesions must inhibit the formation of other adhesions in their immediate vicinity.

12.11 Endoplasmic Cages Confer Coherence at the Cell Periphery

Another type of actin-based structure seen in many cells is a boundary region that separates the plasma membrane-associated actin cortex from the microtubule-rich endoplasm. This is often associated with the cell cortex, and is organized by the crosslinking of actin filaments. The density of crosslinks in the cortex increases toward the center of the cell and can assume a geodesic dome appearance at the boundary between the endoplasm and the ectoplasm in some cases. In this area, we find proteins that bind to microtubules, actin, and intermediate filaments that are under tension developed by myosin. What results is effectively a cage around the endoplasm. Signals that alter the activity of actin-polymerizing enzymes, such as cAMP, also alter the thickness of the ectoplasm, and the position of the boundary. When cAMP levels rise, microtubules extend out to almost the plasma membrane. This makes it easier for secretory vesicles to fuse with the plasma membrane. In fish keratocytes, the boundary is particularly pronounced producing an elliptical endoplasm. As the cells migrate, the elliptical endoplasm can rotate with the nucleus, microtubules and vesicular organelles inside. Thus, the components that organize an actin-based network at the endoplasm–ectoplasm boundary are effectively creating compartments in the cytoplasm that are not membrane-bounded but formed by cytoskeletal cages.

12.11.1 Coupling of Growth Cone Motility with Microtubule Movements

Another example where coherence is important for major functions is in the actomyosin-dependent movement of growth cones that are supplied with materials from the cell body by microtubule transport through neurites. As growth cones migrate to form neuronal connections, they leave behind an axon that is organized around microtubules. Extension of the microtubules during growth cone migration requires them to be mechanically connected as a single coherent network. Actin polymerization is required to maintain these connections, and the inhibition of actin polymerization causes microtubule retraction from the growth cone. What is left behind after retraction is only a very small membrane tube that connects the growth cone with the rest of the cell. This example serves to highlight the importance of actin filament–microtubule connections as well as the dynamics of those connections.

12.12   Modeling of Actin-based Motility

Because a system cannot be fully understood until we can model it and predict the outcome of the loss of important components, it is useful to model motile functions. In the types of motility that have been characterized, there appear to be discrete activation events that start the process in a specific region of the cell. Thus, it is important to model both the localization and the timing of the activation event. Once started, and depending on the cellular conditions, the process will proceed until it automatically stops after a period (typically as short as 30–60 s). Thus, we need to understand how the propagation is limited in space and how the duration is limited. Treating each event as a discrete motile element then enables the modeling of the total process. Quantitative measurements of the process can be made under different conditions, and these can be used to test the model. Further, it is possible to correlate discrete events with the overall outcome of a completed motile process.

For most of the motile processes described above, the activation of actin polymerization is a critical step. Therefore, it is logical to suggest that the first step may be the activation of a GEF (see Chapter 9) to produce an active Rac1-GTP or other G-protein that would bind to and activate an actin-polymerizing factor. The GEF is probably immobilized near the site where activation occurs, and thus the greatest concentration of Rac1-GTP will be in the vicinity of the GEF. As in the protrusion–retraction motility process (see Figure 12.6), the initial active Rac1 may activate events that over time activate a second GEF, such as a RhoA GEF. This may activate a third, and so on. Although the forward cascade is relatively easy to imagine because subsequent steps are activated by preceding steps, the mechanism of inhibition that limits the propagation of the signal in time and space is poorly defined. Further, experimentally testing hypotheses *in vitro* is often difficult. Although the process should run upon activation when conditions are ideal, there can be hidden steps such as mechanical feedback that are difficult to reproduce *in vitro* or even can change *in vivo* for cells in different environments.

In the case of actin filament polymerization, the important conditions for propagation include the concentrations and activities of GAPs (GTPase-activating proteins that cause the inactivation of small G-proteins by facilitating GTP hydrolysis to GDP) (Welch et al., 2011). Furthermore, the downstream activation of other GEFs needs to be in the proximity of the previous GEFs such that the concentration of the active G-protein is sufficient to trigger it. Another condition will be the cell geometry. For example, in a thin lamellipodium, diffusion of the active G-proteins is limited by the restricted volume, which means that their concentration will remain much higher for longer distances.

In addition to these biochemical criteria, some types of motile processes require that mechanical criteria also be satisfied. For example, periodic retractions at the cell edge will stop when the cell encounters soft matrix surfaces. It is believed that an initial retraction determines that the matrix is soft, and a signal is generated to stop further extension in that region. Thus, both biochemical and mechanical steps can be involved in the performance of a given motile process. When modeling a motile process, it is therefore important to include the diffusivity of the components, mechanical parameters that the cell senses, as well as the activation and inactivation rate constants.

When experimentally subjecting fibroblasts to unusual mechanical environments, such as supported lipid bilayers that do not support lateral forces, unexpected types of motility can also be elicited. For example, a change in cell state, or a change in the biochemical signaling that activates other motile processes, may occur. This raises the question of whether two similar motility processes can be fit by the same model. This can be better understood by considering the formation of podosomes. Recent studies show that podosomes can form in fibroblasts when they bind to lipid-linked matrix molecules, i.e. in the absence of matrix forces (Yu et al., 2011). Furthermore, a single cell can form both podosomes on supported bilayers, and regular adhesions at sites of attached matrix when the areas are separated by more than 2 μm. Mechanical signals appear to be local and will influence the processing of matrix signals. Thus, rather than changing overall cell state, the cell can respond by modifying the type of cell motility locally. This implies that the signals do not disperse over large distances, but remain at the site where motility is altered; however, the integration of the different signals from the different microenvironments will occur at the cellular level.

To best construct a model of podosome formation, it is necessary to look at some specific signals that could modify the process. For example, the phosphorylated inositol lipids have been implicated in podosome assembly, and their limited diffusion in the membrane could determine where podosomes are localized. Thus, the phosphoinositol lipids (particularly PIP2 or PIP3) are of interest as signaling

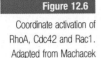

Figure 12.6

Coordinate activation of RhoA, Cdc42 and Rac1. Adapted from Machacek et al. (2009).

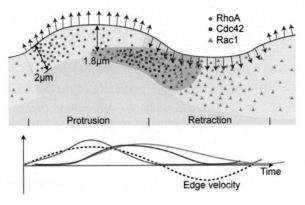

components that diffuse relatively rapidly but will be diluted rapidly and would restrict the response to the one micrometer area that is normally occupied by a podosome. Other signaling systems such as the small GTP-binding proteins (Rho, Rac, Cdc42, etc.) can diffuse more rapidly in cytoplasm and should activate larger areas of the cell than is normally the case with podosomes. Thus, the crosslinking of integrins and local mechanical factors appears to cause downstream activation of PIP2 or PIP3 synthesis that will rapidly alter the type of motility within a small region of the cell. At least two different types of motility are observed in the presence and absence of local forces. It is not clear how such switches occur and what the mechanotransduction mechanisms are. However, for the podosome it is logical to focus on membrane signals that could produce local activation processes.

In different cell states, the sequence may be different because of the presence of another active component, but such changes will be relatively easy to work out once a description of the basic process is developed. Because many motility events are accomplished through the sequential completion of individual but ordered steps, a similar approach can be taken to understanding other types of motility.

12.13 Multicellular Movements as Coordinated Motile Processes in Tissue Formation

In the previous section, we discussed the motile mechanisms common in single cells. In the organism, however, most cells are normally in contact with neighbors as part of an epithelium or endothelium. Although there are some aspects of isolated cell–matrix contacts and cell–cell contacts that are very similar, many types of movements in the tissue context are different. We will discuss here how the cell–cell junctions are involved in migration, in endothelial growth and in epithelial shape changes such as convergent extension or bending. Again, the basic paradigms of motility appear to apply to these motile processes such as having multiple steps, a limited on period, and multiple initiation events. In the case of epithelial cells, the shape is defined by interactions with neighboring cells. These cells are polarized by a basement membrane, and exhibit primarily a cuboidal morphology. For endothelial cells that form capillaries, the shape of each cell helps to shape the blood vessels, particularly at the level of the capillaries.

What must be understood for these self-forming cellular complexes is how basic differences in the cellular morphology are sensed and coordinated between cells to produce epithelia and endothelia of desired morphology. In the case of curved epithelia such as in neural tube or gut formation, the cells are adherent through E-cadherins and then must adopt an asymmetric area at the apical versus baso-lateral surfaces to produce the desired curvature. The curvature will vary from cell to cell and it is difficult to imagine how the cells can sense the asymmetry to produce the curvature and know their place in the tissue. Endothelia, on the other hand, can produce a tube from a single cell, potentially by stiffening the actin on one surface and contracting it on the other, thereby causing the flattened cell to bend, forming a capillary tubule. At present, we understand relatively little about the molecular mechanisms that produce the observed shapes, but are starting to understand the processes in a few cases.

12.14 Maintaining Shape through Adhesions and Volume

As in the case of single cells, the major determinants of cell shape change are the distribution of adhesion sites (cell–cell as well as cell–matrix), the forces on the sites, and the cell volume. As we have described in other chapters, the development of adhesions is a complex process involving feedback between the cell forming the adhesion and the extracellular contacts, either matrix or other cells. There are a multitude of different types of motility in tissues, but we will consider only three major ones here. The migration of cell sheets in wound healing and development is relatively well understood and we will discuss that process first. The second process is the migration of endothelial cells into a tissue to form a vascular

Figure 12.7

Adhesions and volume define shape.

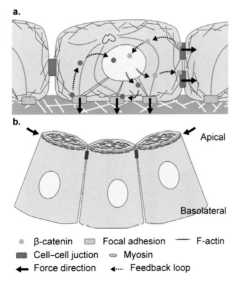

bed, which involves not only migration but also proliferation and ultimately the formation of a continuous flow system from the arterial to the venous circulation. The third process of interest is the bending of epithelia. During development, the contraction of the apical surface will cause the expansion of the basolateral surface because of the constant cell volume. It is still a mystery how the degree of contraction or expansion of the apical versus basolateral cell surfaces is sensed by the cell in the tissue. There are many proteins linked to shape control that have been defined genetically because the mutation of those proteins alters the shape of the tissue; however, the steps are still unknown.

12.15   Cohesive Cytoplasmic and Cellular Networks

Long-range forces that span many cells will develop as multicellular tissues assemble. These forces are supported by cohesive actomyosin networks as described in the discussion of single-cell motility above or more extensive networks that also contain microtubules and intermediate filaments. This can be trivialized to say that cells are continuous; however, there are many circumstances where the continuity of the cell or the tissue is broken due to lack of continuity in the mechanical links in the cell or tissue. Cohesion is critical in tissues, with many of the proteins that maintain this cohesion having important roles in the overall function of cells and tissues. Because there is continual disassembly of filaments within the cell and contacts between cells in young cells and tissues, there must be a continual replenishment of the links. As cells and tissues mature, the rate of turnover decreases and more stable linkages through intermediate filament networks can provide an elastic tension that will hold the tissue together without continual myosin contraction needed to maintain tension in cohesive actomyosin networks. In epithelia that surround an embryo, the cohesive tensions often are asymmetric and this helps to organize the appropriate shape changes. Thus, we know that tissues are cohesive and maintain a reasonable resting tension that may have contributions from myosin contraction as well as intermediate filament elasticity. That cohesion is a prerequisite in the following motile processes.

12.16   Migration of Cell Monolayers

In a number of situations such as the lateral line cell movements, tumor migration, and the healing of wounds in epithelia, groups of cells in monolayers migrate together maintaining cell–cell contacts as they move. From *in vitro* studies of the wound-healing in epithelial monolayers, there are important aspects of the migratory behavior. Leading cells generate most of the tension for migration, but following cells also contribute to the force of migration. When healing a gap in a cell monolayer, the cells that are concave move more rapidly than the cells that are convex or extending further into the gap. This is explained as a result of the formation of contractile actin bundles in the concave regions of cells that will help the concave cells to move into the gap. As a result of this effect, the cells that are initially moving rapidly into the gap will slow down so that a relatively smooth line of cells will cover the gap. In finger-like extensions of epithelial cells, the lead cells behave differently from the following cells and laser oblation of the lead cell enables the following cell to change behavior and take on the characteristics of the lead cell. Further, when a following cell is oblated and creates a gap in the line of cells, then the leading group slows down until the following cells catch up. Thus, the continuity of the cell–cell contacts is important and when cells lack contacts on all sides and the surface is coated with matrix, they will move on the matrix until they form new cell–cell contacts, often circularizing around gaps.

12.17 Endothelial Cells and Tensegrity

Endothelial cells have very fascinating motility properties that enable them to form capillaries from a single cell and to create transcellular holes through the cytoplasm. The task of the endothelial cell system is to create the vasculature of the organism. This system is composed of a network continuous tubes that go from the heart to the tissues (typically reaching to within 200 μm of all active cells) and back to the heart. When tissues grow, or metabolism is increased, the lack of oxygen triggers an angiogenic response. This will cause fingers of endothelial cells to sprout from existing blood vessels to create capillaries that pass through the anoxic tissue. These endothelial cells are generally very thin, yet they maintain strong cell–cell contacts behind them as the lead cell moves forward. Furthermore, numerous *in vitro* studies of capillary growth show that individual endothelial cells, which are initially flat, are large enough to curve to form a tube that becomes the capillary. Because the tube is relatively rigid, the mechanism of tube formation is postulated to be a contraction of the actin cytoskeleton in the half of the cytoskeleton facing the lumen of the capillary, while the cytoskeleton on the outer half is stiff and will therefore be bent to form a tube. This is one of the few examples of cells using a rigid cytoskeleton to produce a morphological change. Almost all other situations such as the bending of epithelia involve pulling against the volume of the cell or extracellular structures such as bone. Once a blood vessel is formed, the flow of blood past the endothelial cell surface, as well as the

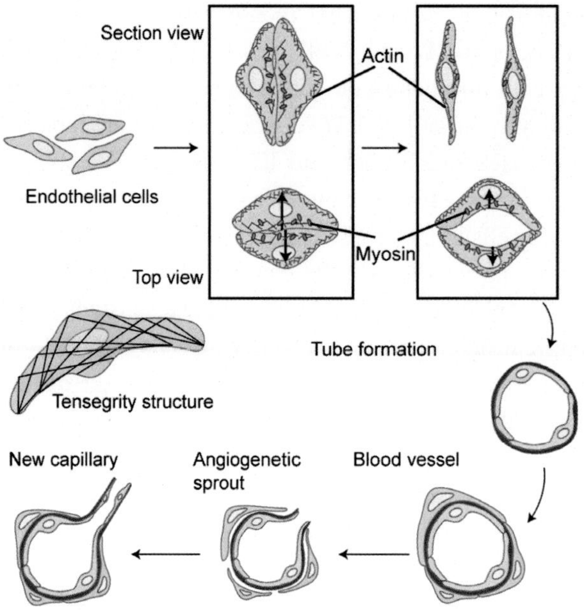

Figure 12.8

Endothelial cells and
tensegrity.

pressure pulse, causes the cells to change the diameter or character of the blood vessel wall. In the process of invasive movement, the lead cell must maintain contacts with the following cells and when it encounters a venous endothelial cell, it will form contacts with it to enable continuity of the circulation. There are many aspects of this process that are not understood at a subcellular much less a molecular level.

Blood vessels allow for the flow of blood cells and some of serum proteins from tissues. The strength of the cell–cell junctions are critical in its function, and also prevent the unwanted transfer of proteins into sensitive organs such as the brain. In this case the endothelial cells form the tight barrier, or blood–brain barrier, whose selective permeability is maintained by tight junctions. In contrast, there are occasions where multiple holes that can be up to 20 or so nm in diameter are formed through the endothelial cell enabling direct movement from the blood stream to the organ. These holes, which form across the endothelial cells, are often observed in the liver. Although the molecular mechanism that drives this process is poorly understood, certain conditions have been described that enable the transcellular holes to form. For example, they will form from the local collapse of the cytoplasm and the subsequent fusion of two plasma membranes. This is analogous to a nuclear pore. Because endothelial cells are typically very thin, even minor gaps in the cytoskeleton network can allow the two membranes to come into contact. Still, the fusion of the membranes is mysterious.

12.18   Curving or Reshaping Epithelia in Development

The embryo develops from a ball of cells to a sack formed by a monolayer of cells and that monolayer often changes morphology to form the primitive gut. In the fly embryo the epithelial cells form around the yolk and then expand as they invaginate to form the gut. Forces developed in the monolayer drive the curvature change and can also reshape the embryo by compressing the tissue on one axis and expanding in the orthogonal axis. These different types of shape changes involve different motile processes and we are just starting to understand the important subcellular steps in the shape changes.

Critical for both of these shape changes are axial differences in the composition of the egg that direct the morphological changes. Early segregation of factors in one pole of the egg is sufficient to define a polarity, particularly if that asymmetry drives asymmetric contraction in the actin cortex. At later multicellular stages, the axes formed earlier would direct the contractions that occur. This is particularly true for the changes in embryo shape that involve rearrangements of the cell boundaries, such as in convergent extension where the embryo elongates along the body axis and shrinks on the off-axis. As the tissue morphology changes, cell borders are altered in characteristic patterns (see Figure 12.9). Furthermore, in some cases, cell–cell junctions may be modified independent of morphology changes, indicating that cells can change partners and still know which axis is to be contracted. As in intracellular cohesion, the multicellular cohesion requires tension from myosin contraction and actin dynamics and that tension will direct the axes of contraction and elongation. In some cases, relatively continuous lines of actin bundles are observed on the major force-bearing axis, but often there is no obvious basis for the asymmetric forces in the tissue. Changing the shape of

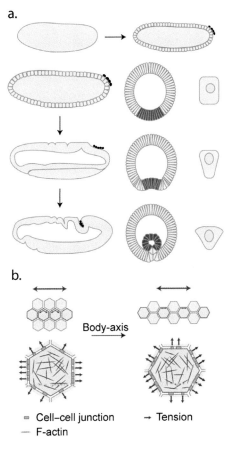

Figure 12.9

Cell borders are altered as tissue morphology changes in development.

a.

b.

Body-axis

▭ Cell–cell junction   → Tension
— F-actin

the tissue is activated at the appropriate developmental stage and is directed by asymmetric tensions in the organism.

In the case of tissue bending, the area of the apical and basolateral surfaces of the cell must be carefully controlled. Again, the overall process of movement involves a series of short contractile events followed by pauses. In these movements there are brief contractions of the apical surface with corresponding expansion of the basolateral surface followed by consolidation of the contracted area and these steps are repeated until the final shape is reached. Clues to understanding the process come from a number of mutations that alter tissue morphogenesis in *Drosophila*. For example, mutations in two transcription factors, snail and twist, will alter the ability of the embryo to contract the apical surface (snail) and to consolidate the contractions (prevent slipping back to the original area, twist). Because these are critical movements in development that create the gut, there must be robust mechanosensory mechanisms in place to assure that the right shape changes occur. Thus, although the shape changes of epithelia are not fully understood, they follow characteristic patterns from *Drosophila* to humans and are performed through steps similar to other robust functions.

12.19 SUMMARY We have described a number of motile processes that we believe involve multiple linked steps to create the emergent properties of cell migration or tissue shape change. The different motility processes such as cell migration rely upon standard motile functions such as lamellipodial or filopodial extensions (i.e. they are driven by common molecular mechanisms in many cells and understanding the detailed mechanism in one cell type will greatly enlighten the understanding of the same process in other cells). Furthermore, motile processes are typically digital in that they are either fully on or off (differences in rate occur primarily because of mechanical factors such as load or factors affecting the on state, such as the availability of actin subunits for polymerization). Motile processes are also modular in that they occur over a finite distance (0.2 μm for the diameter of filopodia versus 1–5 μm of the edge for lamellipodia) and the overall motility of a cell can be described as the number of active modules per unit time. In terms of their kinetics, motile processes automatically turn off after a period. The typical 'on periods' for filopodial or lamellipodial extensions range up to 3–5 minutes, but most are on for less than a minute. Each motile process can generate similar mechanosensing signals, but those signals are often processed differently by the cells. Thus, the view is that the cell applies a number of mechanical tests to its environment through multiple motile functions. Based upon the physical and chemical nature of the microenvironment, those motile functions will then generate appropriate signals that direct the cellular response to its microenvironment, creating its shape.

For cells in tissues to adopt the proper shape and to shape the tissue properly, the correct set of motile functions should be used. This will primarily involve developing proper cell–cell and cell–matrix adhesions and the correct tension on those adhesions. During development and regeneration, exquisitely controlled morphological changes occur to create the proper tissue morphology. Again, the motile functions are turned on multiple times until the proper tissue morphology change is attained and then the sensory systems stop that motility and start the next process. Thus, the motile processes are tools for morphological changes but are under fine mechanosensory controls.

12.20 PROBLEMS

Please add a step to an existing complex function in the MBInfo or a review. Think about the complex function of interest as a process, which involves progressive steps with checks for relevant parameters, such as environment rigidity, ATP concentration, etc. If obvious steps are underdeveloped in the literature, please expand on those steps in the larger context of the function.

12.21 REFERENCES

Luo, W., Yu, C.H., Lieu, Z.Z., et al. 2013. Analysis of the local organization and dynamics of cellular actin networks. *J Cell Biol* 202: 1057–1073.

Machacek, M., Hodgson, L., Danuser, G., et al. 2009. Coordination of Rho GTPase activities during cell protrusion. *Nature* 461: 99–103.

Watanabe, T.M., Tokuo, H., Gonda, K., Higuchi, H. & Ikebe, M. 2010. Myosin-X induces filopodia by multiple elongation mechanism. *J Biol Chem* 285(25): 19605–19614.

Welch, C.M., Elliott, H., Danuser, G. & Hahn, K.M. 2011. Imaging the coordination of multiple signalling activities in living cells. *Nat Rev Mol Cell Biol* 12(11): 749–756.

Yu, C.H., Law, J.B.K., Suryana, M., Low, H.Y. & Sheetz, M.P. 2011. Early integrin binding to Arg–Gly–Asp peptide activates actin polymerization and contractile movement that stimulates outward translocation. *Proc Natl Acad Sci USA* 108(51): 20585–20590.

12.22 FURTHER READING

Actin polymerization: www.mechanobio.info/topics/cytoskeleton-dynamics/actin-filament-polymerization/

Bleb assembly: www.mechanobio.info/topics/cytoskeleton-dynamics/bleb/bleb-assembly/

Cell polarity: www.mechanobio.info/topics/cellular-organization/establishment-of-cell-polarity/

Cytoplasmic network contraction: www.mechanobio.info/topics/cytoskeleton-dynamics/contractile-fiber/

Filopodia assembly: www.mechanobio.info/topics/cytoskeleton-dynamics/filopodium/filopodium-assembly/

Focal adhesion disassembly: www.mechanobio.info/topics/mechanosignaling/cell–matrix-adhesion/focal-adhesion/focal-adhesion-assembly/#disassembly

Invadopodia: www.mechanobio.info/topics/cytoskeleton-dynamics/invadopodium/

Lamellipodia assembly: www.mechanobio.info/topics/cytoskeleton-dynamics/lamellipodium/lamellipodium-assembly/

Podosomes: www.mechanobio.info/topics/cytoskeleton-dynamics/podosome/

Rho GTPases: www.mechanobio.info/topics/mechanosignaling/rho-gtpases/

Small GTPases: www.mechanobio.info/topics/mechanosignaling/small-gtpases/

Stress fibers: www.mechanobio.info/topics/cytoskeleton-dynamics/stress-fiber/

13 DNA Packaging for Information Retrieval and Propagation

If stretched out, the double stranded DNA in a human cell nucleus would be almost 2 m in length. The complete DNA is packaged in a nucleus of 7–9 μm in diameter with histones and associated proteins. During the cell cycle, the DNA is available for transcription and replication, which will cause supercoiling and strand crossovers. In mammalian nuclei, the negatively charged DNA strands are first wound around a 6 nm, positively charged histone core. This forms a basic structure known as a nucleosome. Further compaction sees nucleosomes organized sometimes into 30 nm solenoids, which are subsequently condensed into looped chromatin structures. Dynamics in the solenoids allows DNA binding proteins to access the DNA strands on their surface. Despite the lack of observable nuclear movements in the light microscope, the nucleus is a hotbed of activity. For example, there is considerable evidence to indicate that DNA strands move through polymerizing enzymes and are scanned by other proteins. The nucleosomes are actively positioned by mechanoenzymes and although not fully understood, chromatin condensation at the core of chromosomes is known to be an energy-dependent process. In surface loop regions of chromatin, there is active transcription and recent studies indicate that chromatin is organized as flexible strings of uncondensed nucleosomes. Thus, inactive regions are hidden inside and active regions are on the outside of chromosomes.

13.1 Much More DNA than in Genes

The DNA sequence is often described as being the blueprint for the organism. However, this blueprint includes plans for self-repair and the mechanisms for avoiding damage from environmental challenges such as high temperature. As a result, there is considerable complexity in how the genetic information is propagated, as well as how the necessary portions are robustly read and transcribed. Some aspects of DNA structure and processing are understood, such as the checks that ensure only a single copy of the DNA is made in the S phase of the cell cycle. However, the roles of many features are not clear. For example, the majority of human DNA does not encode human genes that can be translated into proteins. Although normally silent, nearly 25% of the genome encodes for viruses that may be activated. In addition, there are large stretches of repetitive DNA sequences with no known purpose. Only about 2% of the DNA encodes for proteins and there is perhaps an equal amount involved in encoding for other RNA species and for regulation of transcription. There is not a clear function for the majority of the DNA in the human genome, although there is much speculation about what might be its role either in structuring the nucleus or in some undetermined aspect.

13.2   Retaining Accessibility of Condensed DNA

It is clearly important for the chromatin to be organized in a manner such that the regions of the DNA to be transcribed are accessible. To ensure DNA is accessible to the transcription or replication machinery, the cell packages DNA through a multistage condensation process. Here, DNA fibers are wound around histones in nucleosomes that can form solenoids of 30 nm in diameter, which are further condensed into loops that extend out of a highly compact, and inaccessible, core. The loops ensure that the transcription machinery does not need to penetrate into the condensed DNA core to access the genetic code. Furthermore, accessibility of the DNA is also increased through the larger surface area of the chromosomes. It has been argued that DNA is separated into individual chromosomes to ensure the required genes are accessible when DNA is wound up in a condensed state.

Excess DNA that does not code for protein and is not transcribed could provide a natural core for the chromatin, and ensure that the regions to be transcribed remain on the surface of the chromatids.

Figure 13.1

Chromosome condensation.

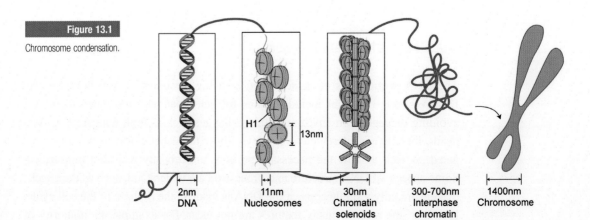

13.3   DNA Accessibility in Specific Cell Types

It is important to consider the problem of DNA packaging and genome accessibility in the context of specific cell types. Transcription patterns in stem cells, for example, are vastly different than in fibroblasts or endothelial cells. These differences come from the cellular environment, and the fact that stem cells need to respond to the local matrix and hormone patterns in order to start differentiation programs. Once stem cells go down a given path, they either receive additional cues from the environment, or in some cases, start to produce their own microenvironment. Differentiation involves shifts in the profile of transcription, and as differentiation progresses, the genes that are no longer required will be silenced by enzymatic modifications. The chromatin containing these genes will be more tightly packed and is known as heterochromatin. The chromatin-containing regions of actively transcribed DNA are classified as euchromatin, and these domains are often located on the surface of chromatids, where they are more accessible to the transcription machinery. These regions frequently associate with similar active regions of neighboring chromatids in what are called transcription factories. Thus, the problem of genetic information retrieval is made easier as a cell differentiates and locks into a defined transcription program.

13.4   Nuclear Envelope

The DNA is packaged within the nuclear envelope and that creates a separate compartment where specialized DNA-handling processes can proceed without interference from cytoplasmic filaments or organelles. Not only are the processes of replication and chromatin condensation potentially sensitive to interference by cytoplasmic elements, but also the molecular level processes of supercoiling and strand–strand collisions. The topoisomerases I and II that perform those functions, respectively, are mechanically activated enzymes and isolation in the nucleus increases the effective concentration, enabling greater dynamics. Further, there are many signaling processes that rely upon a transcription factor moving from a peripheral complex like an adhesion site to DNA-binding sites in the nucleus. This increases the fidelity of those signals because the proteins must be both activated for DNA binding and be transported to the nucleus. This allows peripheral motility and adhesion processes to enable release of the transcription factors, but they would need the additional signal to move to the nucleus. Thus, the nuclear envelope fits well with the desired features of a robust device in that it enables separation of many functions that could otherwise interfere with one another.

An additional feature of the nucleus of mammalian cells is that the nuclear envelope is resorbed into the endoplasmic reticulum in mitosis and then reforms around the chromatin after cytokinesis. Fusion of the vesicles after cytokinesis creates a double-membrane envelope surrounding the chromatin with occasional nuclear pore complexes that regulate the traffic between the cytoplasm and the nucleus. The supporting nuclear lamins (members of the intermediate filament family of proteins) are disassembled during nuclear envelope breakdown and then assist in ordering the membrane around the chromosomes during cytokinesis. Mechanical stability of the nuclear envelope is much greater in fully differentiated, stiffer cells because of the increased production of lamin A. Conversely, stem cells, certain cancers and immune cells have a softer nuclear envelope with folds and undulations that changes morphology readily upon mechanical manipulation of the cells. Those nuclei have low levels of lamin A. Thus, the structure of the nuclear envelope and its rigidity are diagnostic of the degree of differentiation and/or the stiffness of the cells.

13.5   Nuclear Pore Complexes

A nuclear envelope structure that has a significant impact on nuclear organization and function is the nuclear pore complex. These complexes are large, multiprotein assemblies that span the two nuclear membranes and provide a conduit for small molecules to freely diffuse between the nucleus and cytoplasm. Proteins larger than about 50 kDa have a difficult time entering the nucleus unless they contain a nuclear localization signal (typically a stretch of 6–10 cationic amino acids), which enables their active transport along with importins, through the nuclear pore complex, using a gradient of RAN GDP (higher in the cytoplasm than the nucleus). Fibers extend from the pores into the cytoplasm and these fibers are thought to have multiple weak binding sites for the RAN–GDP–importin–protein complexes. Binding of the importin complex to a filament is thought to condense the filament, bringing the transported protein closer to the nuclear pore, where it will enter the nucleus. Once in the nucleus, the RAN–GDP is converted to RAN–GTP that can serve to power nuclear export in a similar system that couples the RAN–GTP gradient to the transport of proteins out of the nucleus.

There is a lot of traffic across the nuclear pores of proteins, nucleic acids, and some suggest even lipid molecules. Many proteins are moving simultaneously in opposite directions (estimates indicate that ~1000 proteins/s are moving both ways through the pores). Not all movements depend upon RAN proteins, but the other mechanisms are not well characterized. Even large structures such as

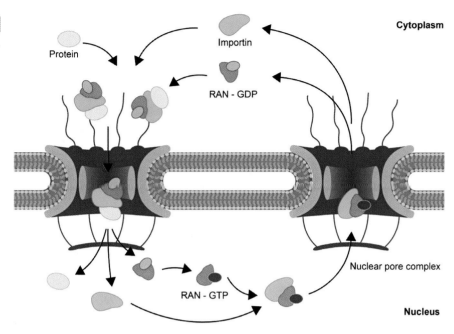

Figure 13.2

Nuclear pore complex.

ribosomes and mRNA–protein complexes move rapidly through the pores, but these movements must involve significant inputs of energy, more than a single nucleotide could provide. Once inside the nucleus, many proteins like the histones bind to the DNA directly or indirectly and that keeps them in the nucleus. Some proteins like NF-kappaB move to the nucleus for several minutes, perform a function such as activating specific gene transcription, and then move out. Others are constantly shuttling into and out of the nucleus. Recent studies show that inositol lipids are carried into the nucleus by steroid-binding proteins and they may participate in the assembly of actin filaments at sites of DNA damage to recruit repair complexes. Thus, the nuclear pore complexes are important regulators of the nuclear traffic and many feel that the nuclear concentrations of transcription factors and their modifiers are critically related to the activities of the chromatin.

13.6   Nucleolus

A very specialized region of the nucleus is the nucleolus that is evident in the light microscope as a nuclear region with a different density of protein. It is the site of ribosome synthesis and contains the polymerase (Pol III) for ribosomal RNA (rRNA) production. In addition, the proteins of the ribosomes are all synthesized in the cytoplasm and transported into the nucleus where they join up with the rRNA to form the complete small or large ribosome subunits in the nucleolus and then the ribosome subunits are transported to the cytoplasm. This seems like a convoluted way to produce ribosomes; however, they are a critical component of the cell and must be made in large numbers for the cells to grow rapidly. There are many questions to be answered about their synthesis and assembly. Likewise, transport appears to be a complex process, as many nucleoli are quite centrally located in the nucleus.

13.7 Nucleosome Condensation and Organization

The cationic histones serve not only to condense the DNA into a more compact structure but also to neutralize many of the negative charges of the DNA polyphosphate backbone. There is about 50 mM of negative phosphate charges in the 8 μm nucleus (volume 200 μm^3 or 2×10^{-13} liter), as each base has a negative charge and there are 6×10^9 bases (1×10^{-14} moles). Thus, a high concentration of cationic histone proteins is required to balance the charge. This charge balance provides stability to the higher-order, compacted chromatin structure. DNA is wrapped around histones in nucleosome discs that are 13 nm in diameter and 11 nm in height (Figure 13.1). Each nucleosome is then separated by about 100 base pairs. This organization is generated by SWI/SNF protein complexes that move the DNA over the nucleosome to achieve the desired spacing, a process that first involves unwinding portions of the DNA from the nucleosome (Cairns, 2007). In mechanical unwinding experiments, a force of 3–9 pN is required for this partial unwinding. Thus, the spacing of nucleosomes is regulated and will probably have an effect on the stability of higher-order structures such as potential solenoids and the location of loops.

Following formation of the histone core, an additional 'linker histone' interacts with the DNA. This promotes further condensation and provides further charge neutralization. The H1 histones are the major linkers in the organization of the solenoids and they help to condense the length of the DNA a further 25- to 35-fold. Compaction continues as the solenoid structure forms loops. Recent studies of the organization of DNA in cell nuclei in G1 phase through high-resolution cryo-electronmicroscopy did not find solenoids. Thus, although solenoids can be found *in vitro*, there is concern about whether they are physiological. During interphase when there is active transcription, the growing consensus is that the nucleosomes are freely mobile at the surface of the chromosomes, enabling the access of the transcription machinery, whereas the more condensed core of the chromosomes is largely the inactive chromatin.

13.8 Euchromatin and Heterochromatin

As cells differentiate, protein expression patterns stabilize. This means that the expression of particular genes is reduced, and the chromatin regions containing those genes become less active. This is useful to the organism as a whole as it ensures that a cell in a tissue is unlikely to change into another cell type and grow inappropriately. Furthermore, although the microenvironment controls the cell behavior, there are circumstances where the cell can fall out of that environment, such as an inappropriate axis of division for an epithelial cell or trauma. In such circumstances, the cells should keep their phenotype, which often means that those cells will apoptose because they are in an inappropriate environment. Such stabilization of the cell phenotype occurs during adulthood and serves to make the cells less adaptive for other functions. Thus, it is useful for the cell to alter the chromatin in those silent regions and stabilize them as heterochromatin.

The silencing of genes as heterochromatin following differentiation often results from covalent modifications of the histones and the DNA within the heterochromatin regions. These modifications include histone methylation as well as DNA methylation that result in the binding of additional proteins to help to condense the DNA into an inactive structure. The DNA methylation will serve to inhibit transcription factor binding that will decrease the likelihood of transcription of those genes. To aid in the condensation, the spacing of the heterochromatin nucleosomes is decreased relative to active regions of the chromatin. The process of condensation is also reflected in a decrease in the local mobility of the nucleosomes. As cells enter mitosis, the chromatin is condensed further with the binding of condensin complexes. After mitosis, the heterochromatin regions are only partially decondensed and then tend to be sequestered in the core of the chromosomes. This means that not only are they less motile but also less accessible, which would further assure that the genes in those regions will not be transcribed.

In contrast to heterochromatin, the more transcriptionally active euchromatin is characterized by a general pattern of histone acetylation and a larger spacing between nucleosomes. As discussed above, the structure of euchromatin will be nucleosomes on a string that gives more mobility to the nucleosomes and greater access to DNA-binding proteins such as transcription factors, some of which may already be assembled in complexes on the DNA. Actively transcribed regions are generally organized with a looser nucleosome structure, and are much more likely to be at the outer ends of loops on the chromosome surface, rather than condensed inside a chromosome. This certainly would make the dynamics of searching for transcription sites easier. The details of

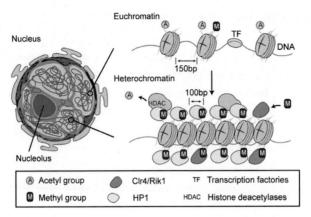

Figure 13.3

Euchromatin and heterochromatin.

the transcription process are not well understood and those details raise a lot of questions about the organization of the chromatin. For example, if many of the transcription factors in an active cell are already bound to the DNA, then there is no need to search extensively for the active regions. Secondary factors that are needed for activation of the transcriptional machinery can then bind to the attached transcription factors to activate the process. On the other hand, if the DNA sequence needs to be scanned to find the transcription factor binding sites, then the physical organization of the DNA must be dynamic to allow the scanning process. Considerable energy is needed to move the DNA in the nucleosomes to be able to scan for specific sequences, but if the transcription factors stay bound for long periods, then the need for scanning is less. Further, the binding of transcription factors to the DNA will help to keep it from being condensed into heterochromatin. Thus, for many reasons the euchromatin will be on the surface of chromosomes and will likely be organized as nucleosome beads on a string with some transcription factors bound.

One way in which the chromosomes are organized in differentiated cells is that during mitosis the mechanism of condensation of the chromosomes does not alter heterochromatin or euchromatin organization. During replication, the methylation of regions the heterochromatin DNA are preserved and in G2 those regions are condensed. During mitosis, the condensin complexes form the chromosomes and aid the in the division process. Upon reformation of the nucleus and the decondensation of the chromatin, the heterochromatin regions remain condensed and will naturally stay at the core of the chromosome while the more active euchromatin regions expand and take up the largest volume on the surface. Thus, simple diffusional mechanisms can aid in the sorting of the active euchromatin to the chromosome surface where transcription factories can then be assembled.

13.9   **Chromosome Territory Organization**

Distribution of interphase chromosomes in daughter cell nuclei occurs in an ordered manner that is generally observed throughout a cell population. This can be seen from mapping of the chromosome distributions in interphase cells. Perhaps the most impressive issue is the common alignment of the most active chromosomes as chromosome pairs. Further, the order is preserved through mitosis. If the histones are fluorescent and a pattern is bleached on the nucleus, then the daughter cells will have a similar pattern of fluorescent histones. This indicates that the chromosome order is preserved through condensation, alignment, separation, and decondensation. Several mechanisms could explain the similarities in chromosome distribution, but a recent hypothesis is based upon the organization of nearest neighbors through transcriptional activity. Nearest-neighbor contacts may then persist through the operations of mitosis and could account for the similar order in the daughter nuclei.

Studies correlating gene expression patterns with chromosome organization show that those chromosomes with the highest levels of expressed proteins often associate with each other, whereas chromosomes that have low protein expression localize to other parts of the nucleus. This fits with the fact that most of the

Figure 13.4

Chromosome organization.

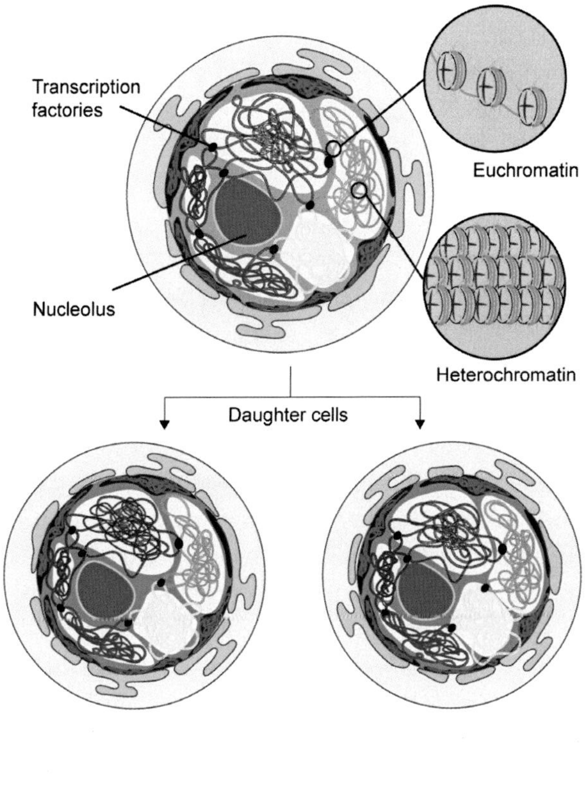

transcription is occurring on the surface of the chromosomes. These regions where high levels of transcription occur have been called 'transcription factories' and it is proposed that these could link neighboring chromosomes. There is a definite economy of scale with the localization of the transcription and the processing of the mRNA. Particular chromosome regions may therefore be arranged to ensure transcription factories can simultaneously express proteins encoded by different chromosomes. Over time, chromosomes containing highly expressed proteins would be organized so that they remain in close proximity, thus sorting them from chromosomes encoding lowly expressed proteins. It should be noted that this is not a very well-defined order, but rather a tendency that can help to understand the process of transcription.

13.10   Organization of Transcription Factories in between Chromosomes

In the process of transcription there are many events that must occur, such as intron splicing (introns are sequences cut out from mRNAs that enable several alternatively spliced proteins to be made from a single gene), or the priming of mRNA to facilitate its movement to the cytoplasm for translation. Previous studies have shown that this processing occurs during transcription, and fluorescence imaging of the splicing proteins shows that they are concentrated in speckles within the nucleus. Within the same regions are found many other proteins involved with transcription, including the Pol II polymerase. Hence these speckles appear to be the places where DNA is being read into mRNA. The number of speckles (hundreds) is considerably less than the calculated number of genes being transcribed at any given time for a rapidly growing cell. This means that several genes are being transcribed in the area of a single speckle (within ~500 nm). These sites where multiple genes are transcribed are termed transcription factories. It is believed that these sites form through the assembly of supermolecular complexes around active polymerases in active chromatin regions. The transcription factories are normally located at the chromosome surface where the chromatin is most accessible, and where there is access to the channels that normally form between chromosomes. Because mRNA diffuses to the cytoplasm through these channels, the newly formed mRNAs are able to start their journey out of the nucleus immediately after synthesis and processing.

As the factories often involve two or more chromosomes, there are logical questions about whether active chromatin regions are moved to an organized complex. This is unclear, because large displacements in the nucleus are observed only occasionally. Still, occasional chromatin movements are seen over micrometer lengths that could recruit activated genes to the transcription factories. Another possibility is that chromosome loops, which are ready for transcription,

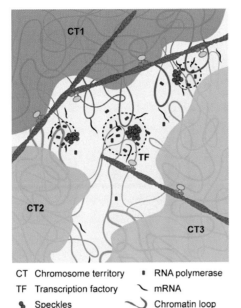

Figure 13.5

Supermolecular complexes in active chromatin regions.

CT Chromosome territory
TF Transcription factory
      Speckles
      Motor

RNA polymerase
mRNA
Chromatin loop
Nuclear actin

can diffuse to the neighborhood of a factory where they will be captured and retained. This begs the question of how new factories form when hormone signals activate new regions of DNA for transcription? It is likely that there is a pool of transcription factory-related proteins that can then form around active regions of chromatin and recruit the other transcription-related components, such as those required for splicing and RNA processing. In the future, it will be useful to follow the induction of transcription from the initial stimulus to the complete mRNA, but it is not clear now how to follow those different events in a living cell.

13.11   Nuclear Envelope and Chromatin Organization

Chromosome organization and chromatin condensation depends not only on the nucleosome components and associated proteins, but also on the wider nuclear structures, including the nuclear envelope and the nucleoskeleton. Like the cytoskeleton, the nucleoskeleton plays a role in structural support and the transduction of mechanical signals. A major component of the nucleoskeleton is a type of intermediate filament composed of lamin proteins. Although initially thought to coat the surface of the nuclear membrane, there is increasing evidence that the lamins extend into the nucleus or at least interact with chromatin and affect transcription. Their importance in nuclear dynamics is highlighted by the dramatic changes in nuclear plasticity that occur after increased expression of lamin A. The nucleus becomes more regular in shape and more rigid (measured as how much force is needed to deform the nuclear membrane by a set amount) with increased levels of lamin A. Furthermore, there is a direct correlation between the level of lamin A expression and tissue rigidity. How such a connection is established is not clear, but the sensing of rigidity may simply produce a signal that activates expression of lamin A. It should be noted that stem cells and cancer cells have very little lamin A and this likely contributes to the deformable nature of nuclei in those cells. Having a deformable nucleus is beneficial because such cells will often

Figure 13.6

Nuclear lamin A.

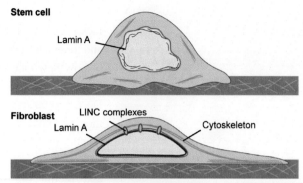

need to change shape in order to undergo differentiation, invasion or other motile processes. Fibroblasts, on the other hand, have very rigid nuclei, and a very tight organization of the nuclear membrane complexes. Thus, major differences in the organization and mechanical characteristics of the nuclei can differentiate between different cell types, and this presumably relates to the differences in nuclear functions as well.

Importantly, the nucleoskeleton and the dynamics of chromatin condensation do not function independently of the cytoskeleton. Numerous links exist between the lamins and the cytoskeleton that can carry force across the double nuclear membrane. These protein complexes, which are known as the LINC (linker of nucleoskeleton and cytoskeleton) complexes, extend from the nuclear lamins through the nuclear envelope to the SUN-domain proteins that then interact with the nesprins (KASH-domain proteins) at the outer nuclear membrane, and the nesprins bind to the actin and intermediate filament cytoskeletons. Deletions of the nesprins and SUN proteins have severe consequences on cell and nuclear morphology as well as the expression patterns of the cells. Several observations indicate that the tension in the cytoskeleton has a significant impact on chromatin organization through the development of tension in the nuclear cytoskeleton. For example, highly spread cells will develop stress fibers on their dorsal surface that indent the nucleus, and this clearly distorts chromatin. The tension on the nucleus could also alter the nuclear volume by compressing it (the major factors determining the volume of the nucleus would be volume of the chromosomes and the balance of cytoskeletal tensions through the LINC complexes). Because the chromatin occupies a large fraction of the space inside the nucleus, small changes in overall nuclear volume can cause major changes in the volume surrounding the chromatin. This is the space where mRNAs must diffuse. Diffusion rates of the mRNAs are extremely slow (two orders of magnitude slower than expected in free solution). Altered diffusion of mRNA can have major effects on cell function, and thus mechanical changes that result in a decreased nuclear volume, or the degree of condensation of the chromatin, can alter mRNA assembly and transport rates dramatically. From many recent findings, however, it is clear that the nuclear cytoskeleton–actin cytoskeleton interactions are critical for nuclear function, not only in terms of the transport of mRNA but also in the control of transcription. This is an exciting area for future research.

13.12 Replication

Cell DNA replication is essential for organism development, growth, and tissue repair. Several factors determine when a cell will undergo replication, including the nutrients available in the cell environment, the presence of biochemical signals and mechanical factors of the microenvironment. Replication commences at the end of the first growth phase of the cell cycle (G1), typically when the cell reaches the appropriate size and the environmental signals all support growth. Cell size is controlled by both the activation of mitotic cyclin-dependent kinases (Cdks) and the inhibition of an opposing phosphatase, protein phosphatase 2A (PP2A). Here, a checkpoint mechanism is initiated to survey the DNA for damage so that several factors all need to be positive (size, no DNA damage, environmental growth signals) for the cell to commit to S phase. Recent findings implicate mechanical factors of the microenvironment in the activation of cell division. For example, the stretching of tissues following internal growth (swelling) in an organ, or the loss of cells in a surface epithelial monolayer, can activate epithelial growth. A question that arises in these situations is how some cells become activated while their neighbors do not. Presumably, all of the factors that go into the checkpoint for S phase participate and once the decision is made, the expansion of a cell in an epithelium will discourage neighboring cells from also transitioning to S phase through mechanical effects.

The mTOR and YAP/TAZ systems integrate the inputs from forces on a cell, the presence of proper nutrients and circulating hormones that all contribute to the activation of a signal for replication. Conversely, DNA damage or the periodic relaxation of force provide negative inputs to the same integrators. For the cellular levels of the signal to exceed the threshold for activation of replication there must be a combination of factors. For example, recent studies have shown a correlation between cell growth and YAP movement to the nucleus in response to severe stretch of an epithelium that also involved beta-catenin movement through a Wnt signaling pathway. Studies of large cell populations have not determined the critical parameters for making these decisions or their relative effects, but now microfluidics and single-cell assays can focus on single parameters such as local forces to determine how differences in magnitude of force through an otherwise

Figure 13.7

Stretch of the epithelium
and G1-cyclin positive
feedback loop.

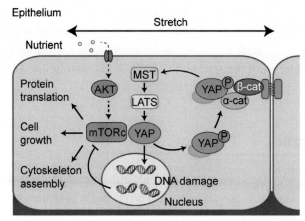

homogeneous population in an epithelium will affect replication rates. Thus, the decision to replicate the genome is perhaps the most critical decision for cell growth and many factors need to be integrated through the mTOR and YAP/TAZ systems to make that decision.

Once replication is activated, the process of making a single complete copy of DNA starts. The pre-replication complexes are first disassembled by the phosphorylation of components, which flags them for degradation. This ensures that no new replication initiation sites will form until the existing sites have been copied. Although many sites of replication are activated, the DNA polymerase is slow (~30 bp/s) and therefore S phase can last up to 8 hours. Presumably, any pre-replication complexes that are activated but have not been copied will maintain a signal for continued replication until the process is complete.

13.13   Mitosis

Chromatin organization is particularly important during mitosis. In this process there is a major reorganization of the chromatin into paired daughter chromosomes and a condensation process that orders the condensed loops perpendicular to the long axis of the chromosome. Although several proteins, including the condensins, cohesins, and topoisomerases, have been identified as playing a role in DNA condensation during mitosis, these only represent a small fraction of the total proteins involved.

Nearly 2000 proteins have been linked to the chromatin complex through mass spectrometric analyses, and even if only a small fraction are important for the condensation process, it will take a long time to define the critical events and the molecular bases of those events. There clearly are many steps in the process of condensation and ordering and the shear number of proteins involved belies the axiom of robust systems that the simplest mechanisms are generally most robust. From the analyses of gene deletions or mutations in model organisms such as *Drosophila*, the loss of single proteins rarely causes a block to mitosis and this indicates that there are many overlapping systems to assure that the daughter chromosomes are packaged into the daughter cells. Mutations in important proteins for mitosis often cause an increase in the number of chromosome-sorting defects, but the majority of cells still get the proper complement of DNA.

Critical steps in the process of mitosis include the condensation of the daughter chromosomes into a complex separated from other chromosomes but still linked to the other daughter chromosome so that they can be moved to the metaphase plate together. Movement to the metaphase plate means that the two daughter chromosomes have kinetochores attached at the proper locations. When the kinetochores become aligned after a tug of war, the chromosomes are then separated, moved to the daughter cells and the decondensation process begins. Because this complex mechanochemical process must occur for the formation of every nucleated cell in the body, it must be robust and include mechanical as well as biochemical feedback controls.

13.14 Reversal of Differentiation

The finding that fibroblasts can be converted back into stem cells was a major discovery because it showed that heterochromatin can be reactivated. Although the initial experiments were inefficient (only about 1 in 10,000 fibroblasts became pluripotent stem cells or IPS cells), more recent studies have reported methods to increase yield. This means that there is no loss of genetic information upon differentiation, and raises the question of whether an organism can activate de-differentiation programs, and under what conditions de-differentiation will occur. Severe trauma to an organ, which may occur through disease or injury, is the obvious type of signal that might cause de-differentiation. Many researchers are looking for de-differentiation and there are many papers on the effect of hypoxia (high CO_2 and acid conditions) on the differentiation state. One high-profile paper purporting to have found pluripotent stem cell formation in anoxic, acidic environments has been retracted, but there are many papers showing partial activation of some stem cell markers. The release of hormones and other molecular signals will likely augment the de-differentiation response and this is an area that is currently being actively researched. Thus, although it is possible to de-differentiate fully committed cells by transfection with chromatin-modifying enzymes, there is little evidence of how de-differentiation might occur under physiological conditions to help in repair after trauma.

13.15 SUMMARY Although 2m of DNA are condensed to fit inside the nucleus, which is just 6–9 μm in diameter, the DNA is still accessible for mRNA production and a variety of other activities. Condensation involves the formation of nucleosome complexes properly spaced in a chain of nucleosomes that potentially coils into 30 nm solenoids to make loop structures from the core to the surface of the chromosome. Surface loops of chromatids are highly accessible to the transcription machinery as a string of nucleosomes. In regions that are not being transcribed, the DNA is acetylated and packaged as heterochromatin in the core of the chromosome, which is a hallmark of differentiation. The organization of interphase chromosomes will also influence transcription programs, with transcription factories forming at the boundaries between chromosomes through recruitment of active chromatin regions even from different chromosomes. These factories contain enzymes for the processing of the mRNA (splicing, GTP processing, and polyadenylation) as well as the synthesis (Pol II). Once formed and assembled into mRNA–protein complexes, the mRNAs move to the cytoplasm by initially diffusing through channels between the chromosomes, and are actively transported through the nuclear pore. During the S phase of the cell cycle, a single copy of DNA is made, following the activation of multiple initiation sites. As cells move into mitosis, the chromosomes become highly condensed and ordered such that the two daughter cells can obtain a copy of each chromosome that will decondense to preserve the pattern of heterochromatin. Silencing of the genome in heterochromatin can be reversed *in vitro* to produce induced pluripotent cells and similar processes may also occur *in vivo* after trauma. Because many of these processes occur in the concentrated environment of the nucleus, it has been difficult to follow the step-by-step progression in these complex functions that involve many mechanical as well as biochemical functions.

13.16   PROBLEMS

Please describe additional steps in nuclear functions such as replication, translation in specific conditions, or chromatin condensation in mitosis.

13.17   REFERENCES

Cairns, B.R. 2007. Chromatin remodeling: insights and intrigue from single-molecule studies. *Nat Struct Mol Biol* 14(11): 989–996.

13.18 FURTHER READING

Chromatin: www.mechanobio.info/topics/genome-regulation/chromatin/
DNA packaging: www.mechanobio.info/topics/genome-regulation/dna-packaging/
DNA replication: www.mechanobio.info/topics/genome-regulation/dna-replication/
Genome regulation: www.mechanobio.info/topics/genome-regulation/
Nucleus: www.mechanobio.info/topics/cellular-organization/nucleus/
RAN GTPases: www.mechanobio.info/topics/mechanosignaling/ran-gtpases/

14 Transcribing the Right Information and Packaging for Delivery

The production of RNA (mRNA, rRNA, miRNA, etc.) occurs through a process known as transcription. This process occurs in a well-defined sequence, starting with the selection of one or two sites on the DNA that encode for the appropriate RNA. A polymerase complex will then assemble, and begin translocating along the DNA, producing the RNA as it moves. Finally, the RNA must be processed and transported out of the nucleus and processing is different for mRNA versus the other types. In differentiated mammalian cells, transcription initiation sites are partially determined by the silencing of major portions of the chromatin through post-translational modifications. In addition, there are tissue-specific aspects of mRNA production that include alternative splicing of the mRNA, and micro-RNAs, which decrease the lifetime of mRNA subsets. There has been considerable interest in how the cell transcribes the right DNA, processes and then packages the transcript for transport to the cytoplasm. To fully understand the primary signals that govern gene expression, the expression levels of various transcription factors must be defined, and more comprehensive data on the changes in transcription patterns attained. For sufficient specificity, and regulation of the transcription process, we would expect several factors to interact with the DNA or transcription machinery. Thus, the definition of the site of transcription is the critical aspect and the subsequent process of mRNA production and processing follows as expected.

Steps in mRNA production and processing:

(A) Identify regions of the genome to be transcribed.
(B) Activate transcription.
(C) Process mRNA transcript.
(D) Recruit RNA binding proteins and release from transcription site.
(E) Diffuse to nuclear pore and move to cytoplasm.
(F) Move and/or diffuse to translation site.
(G) Degradation.

To introduce the topic of RNA production and processing, we will first consider how proteins are produced in cells. Cells transcribe DNA into RNA as a first step towards producing functional protein complexes. Once again, the principles of manufacturing a car are analogous in many ways to this important cellular process. In a well-run automobile factory, each car is produced from a set number of parts, each of which is incorporated in a set sequence. Highly specialized machines are required to both produce the parts and incorporate them into the final product. It would be wasteful to produce an excess of any one part, and incorporating a part out of sequence will render the car unusable, and slow down further production. Furthermore, a quality control system is required to check the functionality of each part as it is made. As they are needed for the completion of production, a number of different parts must be produced in parallel for both the assembly of the car and the daughter cells. In the case of the cell, there is a cell cycle that requires many different molecular machines to be activated in proper sequence so that at each phase the cell will have the required components to perform the functions needed in that phase. One subset of proteins is needed for mitosis, while others are active only in the DNA synthesis phase, or during cytokinesis. In a multicellular organism, there are specialized machines that will be produced in one cell type but not another. As we mentioned previously, cells are primarily concerned with robustness and not with efficiency. This is often accomplished by having many proteins around that will only be needed for certain stresses that the cell may experience. Thus, of the 4000 or more different proteins present in many cells, only a small fraction will be active at any given time. The rest are available upon demand, i.e. when the cell changes phase or receives an environmental stimulus, such as a hormone or exposure to cold.

Changes in diet or environmental stresses can challenge an organism in a matter of minutes to hours. For cells to meet those challenges there must be protein systems ready to be activated or at least to be readily transcribed. If a cell is too highly specialized, it will lose its ability to rapidly respond to changes, i.e. its robustness. Many recent studies have shown that differentiated cells are capable of responding rapidly to unusual environmental conditions in unexpected ways. Thus, cells appear to have components for multiple activities, but the environment only activates a subset. For the organism, the ability to survive multiple challenges is more important than efficiency of energy utilization.

Although it takes a lot of energy to produce a protein (at least 2 ATP molecules will be hydrolyzed for every amino acid in a mature protein), having a required protein on standby for when conditions change is well worth the energy lost in its production. For example, extensive studies have been performed on the expression of heat-shock proteins when cells are exposed to increased temperature (heat shock). The general finding is that a previous mild heat shock enables the organism to survive much higher heat stress than a naïve control cell. This is because the first heat shock triggers the expression of heat shock proteins that

Figure 14.1

Heat shock and expression of heat-shock proteins.

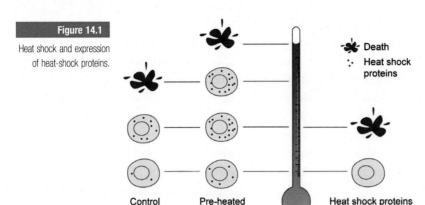

persist long after the first challenge. Furthermore, the knockout of heat-shock proteins make the organism much more sensitive to heat. Thus, the organism strikes a balance and makes some heat-shock protein even in optimal environments, but upregulates production when challenged with heat stress. This flexible compromise is made in many different protein systems that deal with potentially damaging conditions.

14.1 RNA Transcription Factors (Specificity Two Sites per Nucleus or 30 Picomolar)

To ensure that proteins are produced as needed, or in a limited quantity when on standby, the process of transcription is highly regulated. Certain criteria must be fulfilled in order for each step in the process to be initiated. The working model is that initially transcription factors bind to the DNA and enable transcription; however, secondary and perhaps tertiary factors are needed to activate the actual transcription process. As a first step in the activation of transcription, activation complexes scan sequences upstream of the genes that need to be transcribed in order to identify transcription factor binding sites and facilitate binding. In many cases, transcription factors remain bound to the DNA, even when the gene is not being transcribed. This may obviate the need for rescanning the same regions if the gene is required later in the cell cycle, or in the event of cell damage. Then, the secondary transcription factors can bind to a subset of primary factors and activate the process. This increases efficiency because activation is less dependent on finding the binding site, and only depends on the ability of the cell to recruit the secondary transcription factors to the initiation site. Further, the transcription of several proteins are often linked such as in the cases of multimeric proteins or proteins in an enzymatic pathway.

In considering the problem of transcription factor binding in the nucleus and the accuracy of transcription, it is useful to look at the important elements of protein–protein binding because the concentration of the two sites of transcription initiation for any gene is two sites in a billion base pairs or approximately 30 picomolar in a cell. In physical terms, the implications of having just two initiation sites in the nucleus are radically different if you have one or multiple transcription factors involved in binding to that site. If a single transcription factor protein was used to find that sequence, it would need to bind to a sequence of at least 15 nucleotides to find a single site in a billion base pairs (a sequence of 15 base pairs with four different bases could have roughly a billion different possible sequences). Typically, however, mammalian transcription factors bind to only 6–10 bases, meaning that multiple transcription factors are needed to specifically locate the two initiation sites for each gene. This is in contrast to bacteria, which rely upon transcription factor proteins that bind to 12–14 base pairs and don't need multiple factors. Further, bacteria have a much smaller cell volume. Thus, they have higher concentrations of transcription factors (~300-fold smaller than the nucleus or 150-fold higher concentration of initiation sites). For the mammalian systems it is possible then to have multiple steps in the activation of transcription that will enable different but overlapping sets of proteins to be activated in the same cell. The linking of multiple transcription factors could occur in multiple ways to enable detailed control of which subsets of genes are being transcribed.

14.2   Timescales and Transcription Factors

One question that arises from the need for transcription factors to scan DNA for their binding sites is how the whole nucleus can be sampled given the majority of DNA is condensed into chromatin. This can be approached by measuring lifetimes of transcription factors in the nucleus. In general, the recovery rates in fluorescence recovery after photobleaching (FRAP) studies are relatively rapid (with diffusion coefficients of 1–0.01 $\mu m^2/s$) and this would correspond to a transcription factor binding to DNA for seconds rather than minutes, as has been observed for the histone proteins in the chromatin. The actual lifetime of a transcription factor binding to a specific sequence site in the nucleus has not been measured, and the rapid dynamics may involve weak binding with non-specific DNA sites particularly as many of the factors bind to similar sequences at only slightly lower affinities. For multiple transcription factors to activate RNA polymerase complexes at the transcription start site, a relatively stable complex is required. Therefore, it is expected that the activation complex, which can contain multiple transcription factors, stays intact for minutes, allowing hundreds of mRNAs to be produced. Transcription factories or concentrations of RNA polymerase molecules are commonly observed as localized spots in the nucleus and they last for many minutes. Thus, we suggest that the scanning of DNA by individual transcription factors is rapid; however, when there is binding of additional factors to create a larger complex, the factors are stabilized such that they recruit the transcription machinery, including the RNA polymerase.

With a very rapid production of protein in certain circumstances, there may be a need to modify our thinking about how rapid-response genes are activated. One case in particular is after serum is added to serum-starved cells. In that case, mRNA encoding the rapid-response genes is produced within 15–30 minutes and protein is measurable shortly thereafter. A hypothesis to explain this rapid transcription is that a subset of transcription factors is stably bound to the activation sites, and only the activation of the secondary serum response elements is needed to produce full activation. If the transcription factors are diffusing rapidly along the DNA, and nucleosomes are continually being remodeled, then a better hypothesis would be that the chromatin regions containing the rapid-response genes are particularly open and more dynamic. This serum response mechanism has been studied extensively in differentiated cells. In those studies, the cells have extensive portions of chromatin stabilized as heterochromatin that is silenced for transcription. This means that large regions of chromatin are, in fact, much less dynamic because they are modified as heterochromatin. Much less of the genome then needs to be scanned because the rapid-response genes are most likely contained in the euchromatin. For the later-response genes, there is more time to scan the less-dynamic regions of the chromatin, and a normal activation mechanism can work.

Localization of components in the nucleus is commonly observed as speckles of fluorescent antibodies directed against mRNA processing factors in active cells and may serve as a critical step in the activation of transcription. For the speckles to occur, the mRNA processing proteins must be locally concentrated through their binding to nuclear matrix or scaffolding proteins. In fully differentiated cells, there are reproducible sets of genes that are to be transcribed. For this to occur, a general activation of the relevant regions of chromatin may result in the concentration of transcription factors, polymerases and other components required for the transcription of genes within those local nuclear regions. Biochemical signals from the cytoplasm which can include the LIM proteins, nFκB, MRTF-A, or YAP would then activate the transcription machinery to start transcription in those regions using partially assembled complexes.

In addition to early serum response genes, there are intermediate and late response genes that are transcribed up to hours later. The degree of chromatin dynamics could distinguish these genes from the early response genes; however, it is also possible that the product of the early response genes could also contribute

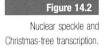

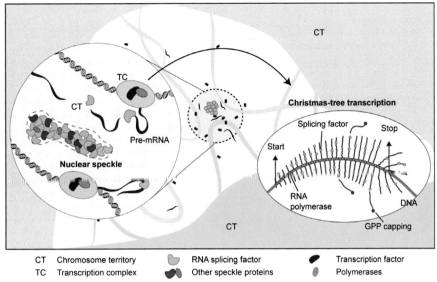

Figure 14.2

Nuclear speckle and Christmas-tree transcription.

| CT | Chromosome territory | | RNA splicing factor | | Transcription factor |
| TC | Transcription complex | | Other speckle proteins | | Polymerases |

in their activation, as those proteins are translated within 30 minutes of serum addition. Additional characterization of the dynamics of chromatin is required to fully understand the exact mechanisms that control transcription (Hager et al., 2009). If the implications from the photobleaching recovery data are correct, and large regions of chromatin are rapidly remodeled on a continual basis, then it follows that a multistep process for the assembly of the transcription complex is more likely than a simple two-step process. The additional steps are to enable robust controls of transcription, which would involve activation of transcription through multiple signals from the cytoplasm.

14.3 Signal Transduction from the Periphery to the Nucleus

For a cell to maintain its differentiated state and viability, it is important for the cell to receive communication from its microenvironment. If that microenvironment changes, then the cell should adapt and there are a number of adaptations that do not involve a change in cell type or apoptosis, but are rather brought about through altered gene expression profiles. Although we do not fully understand how the matrix– and cell–cell adhesion molecules regulate transcription, it is clear that the shuttling of adhesion proteins between the cytoplasm and the nucleus is involved. For example, the LIM-domain proteins will be shuttled from cell–matrix adhesions to the nucleus in response to changes in rigidity or reactive oxygen species (ROS), while β-catenin moves from cell–cell adhesions to the nucleus as part of the WNT signaling pathway. In both cases, the proteins will bind to transcription factors to cause changes in transcription. Researchers in the field are asking what is the number of genes that are activated and how activation of some will affect the level of transcription of others through competition or indirect effects. Overall cell growth clearly depends upon hormone activation as well as metabolic and tissue size control pathways that are linked to the mTORC (Target of Rapamycin) and the YAP/TAZ (tissue size-controlling genes discovered in *Drosophila*) pathways. Again, how many different signals are integrated into a single response pattern is difficult to understand. Assays for expression are difficult and only recently can single-cell responses be monitored reliably. Thus, much more study is needed to be able to predict changes in expression patterns from a simple set of inputs.

As an example of how signals from the environment might affect protein expression, we will consider the inhibition of cell growth by soft surfaces mediated by the four-and-a-half LIM-domain protein 2 (FHL2). Fibroblasts typically do not grow on soft fibronectin matrices. On rigid matrices, FHL2 is associated with focal adhesions and will bind to several adhesion proteins including FAK and GRB7. However, when fibroblasts are on soft surfaces or myosin contraction is inhibited, FHL2 moves to the nucleus in minutes. Movement requires FAK activity and the phosphorylation of a specific tyrosine in FHL2. In the nucleus, FHL2 is associated with active PolII sites, indicating that it is activating transcription. Further, nuclear FHL2 is crosslinked to upstream control sequences of the p21 gene and activates its expression (Nakazawa et al., 2016). Increased levels of p21 inhibit cell-cycle progression. Although the details of how sensing of soft surfaces is translated into FHL2 phosphorylation by FAK, this shows that the rigidity-sensing complex can rapidly change protein expression levels.

Many factors are clearly involved in the control of a cell's gene expression profile. The regulation of transcription can be understood at a number of different levels. At a simple local level, the loss of a neighboring cell can activate genes that cause the cell to re-enter the cell cycle and undergo replication, in order to replace the cell that died. Similarly, following the proteolysis of matrix proteins, genes may be expressed to increase the synthesis of matrix. At a broader level, the loss of multiple cells by an infectious agent (bacterial or viral) or a major activation of matrix degradation by toxins (e.g. certain spider venoms) can activate a significant wound response that involves many neighboring cells. There can be the recruitment of macrophages and other scavengers to remove debris and dead cells as well as fibroblasts to close a wound or lay down a fibrous network that maintains tissue integrity. Considering gene expression profiles at the organism level is also possible. Major trauma to an organism can trigger the recruitment of stem cells for repair and in some cases de-differentiation may also occur to aid repair. At each of these levels a measured cellular response would be expected. This could involve two to three neighboring cells, or even involve changes to the organism's metabolism and state. At the cell or tissue level, there can be a scaled proliferative response and repair mechanisms can proceed without major trauma to the organism. However, when the whole organism is involved, major trauma signals will alter the metabolism to keep the organism alive until major repair processes can be assembled. Thus, the processes of activation need to be scalable from a cell to tissue level.

Scaling the transcription process will often require factors that circulate throughout the whole organism. While local mechanical factors are sufficient to induce local repair processes, changes in the concentration of a major circulating hormone, or other soluble factor, would be required to initiate large-scale changes in gene expression patterns. The recruitment of macrophages or the fibroblast repair systems, for example, would alter gene expression profiles beyond the local cells. This recruitment is often driven through the secretion of chemotactic factors. The proliferative response that results is a major factor in the repair process. However, it is only when extensive damage occurs that the healing response involves a major change in the transcription profile. The major control of transcription comes from the environment, and even changing the shape of a matrix pattern can cause a dramatic change in the expression profile. For example, experiments have found that the mRNA levels for thousands of different proteins can be altered as a result of changes in matrix shape, while keeping matrix composition and area constant.

Figure 14.3

Three levels of control.

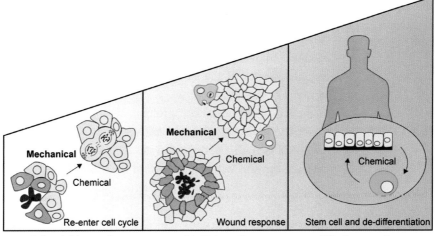

| 2-3 neighbor cells | Multiple cells | Whole organism |

14.4   Differentiation and Expression Patterns

The pattern of mRNAs produced appears to be very broad in pluripotent stem cells and becomes restricted as the cells differentiate. A rationale for this is that stem cells should respond to many external stimuli, and thus should have multiple systems in place. As cells differentiate, however, the pattern of expression, which is stimulated by the environment, becomes more stabilized. Over time, the modification of the chromatin increases and inactive regions become condensed into heterochromatin. However, this condensation is reversible, and is therefore only a partial block. Other factors such as the positioning of condensed chromatin at the periphery of the nucleus also can contribute to the inactivation.

14.5   Transcription Machinery

The synthesis of mRNA from a single strand of DNA is carried out by RNA polymerase II. This is the fundamental mechanism of transcription; however, it is important to note that RNA polymerase II requires several activation factors before it can locate the transcription start site and commence the transcription process. Like other cellular processes, a series of 'checks and balances' are undertaken before RNA polymerase II begins transcribing the DNA. For example, metabolic conditions of the cell are checked to ensure that the cell has proper nutrients. The influence of the cell cycle and the presence of hormones will also be checked.

Once a region of DNA is activated, multiple polymerases move through the gene. Each will start at the same site, and follow the preceding polymerase to create a Christmas tree-like array of gene transcripts. These transcripts will become progressively longer the further the polymerase moves away from the origination site. At the end of the coding sequence, a poly A tail and the GPP cap will be added. These are important steps for the mRNA to be transported to the cytoplasm, because proteins will bind to those modified regions and enable the transport of bound mRNAs to the cytoplasm. Without those signals, the mRNAs will be degraded in the nucleus. Once transcription has commenced, the site will remain active for a limited period before polymerization automatically stops as in term limits, usually after a period of minutes. Thus, the activation of the polymerization complex at the start site in the chromatin will synthesize a burst of mRNAs and they will then be processed in rapid sequence.

Methods of Measuring mRNA Production

Static Measurements
- Bromo RDU incorporation into the mRNA and then antibody localization.
- Hybridization methods.

Dynamic Measurement
- Incorporation of fluorescent ribonucleic acids.
- Recent studies have focused on the real-time observation of fluorescently tagged mRNA in the nucleus (tagged by the Ms2 protein that binds to RNA loop sequences (Tsukamoto et al., 2000; Kues et al., 2001)). The tags were found to initially localize to the site of transcription, which made them useful in identifying where the tagged mRNAs are transcribed. When coupled with targeted activation of transcription, this technology can help to better understand how related sites may be activated or spatially linked.

14.6 Transcription Factories in the Nucleus

For a proper mRNA transcript to be produced, a number of processing events are needed in addition to the poly A tail and GPP modifications. Of particular importance is the removal of portions of the RNA transcript called introns to create a string of exons that form the final mRNA. In different cells, there can be different exons formed from the same RNA transcript, which is called alternative splicing as the introns and some exons can be spliced out of an mRNA. This provides an economy of the genome in that the same portion of DNA can encode for several different alternatively splice mRNAs. Because the splicing machinery is tissue- or even cell-specific, the final mRNA can be customized for the cell type or cell state to enable the same basic protein function to be controlled or localized differently. Often several different RNA transcripts are processed by the same splicing machinery to enable several alternatively spliced proteins to be produced in concert. This processing has been shown to occur during transcription by proteins linked to the transcription process, and in many cases, is facilitated by 'transcription factories,' which are supermolecular complexes that assemble around active polymerases in active chromatin regions.

Fluorescence imaging of the splicing proteins as well as RNA polymerase II (produces mRNAs) reveals localized speckles within the nucleus. The number of speckles (100s) is considerably less than the calculated number of genes being transcribed at any given time for a rapidly growing cell. This means that several different genes are being transcribed in the area of a single speckle (~500 nm), which indicates the presence of transcription factories (Van Bortle & Corces, 2012). Because the factories form at boundaries between active regions on different chromosomes, there is likely a mechanical aspect to the formation of the factories. For example, there may be actin filaments associated the activated chromatin that enables it to be moved to a factory by a myosin; however, no evidence for this currently exists. An alternative model is that diffusion of active regions of chromatin enables them to contact other active chromatin, and then a complex of soluble transcription factory-related proteins would assemble around active regions of chromatin and recruit the other transcription-related components, such as those required for splicing and RNA processing. In either case, the transcription, splicing, and processing events occur in relatively few locations within the nucleus, indicating that multiple mRNAs are produced at a single transcription factory.

14.7   Processing of mRNA for Export

As the mRNA is produced, it associates with RNA binding proteins in an mRNA–protein (mRNP) complex that condenses the mRNA into a much smaller mRNP (100–500 nm) to facilitate the diffusion out of the nucleus. This is critical because the length of the mRNA can extend to micrometers (0.3 nm/base and dystrophin message is 25 kbase or 8 μm). Different mRNP complexes form on different mRNAs, and this helps to direct the mRNAs along one of four different export pathways. Thus, the binding of sequence-specific proteins to mRNAs can localize them to as many as four different regions of the cytoplasm for translation (Muller-McNicoll & Neugebauer, 2013).

In neurons, many of the messages are packaged for transport to the synaptic regions and that involves binding and activating microtubule motors in the cytoplasm, which will power the movement of the mRNP complex to the synaptic region. In fibroblasts, the beta-actin mRNA is transported to the lamellipodial region and blocking that transport alters cell migration. In general, there are zip code sequences (40–100 bases) in the mRNAs that determine the type of transport it will undertake, and this is based upon the proteins that bind to those sequences. In addition to the different transport functions that can be encoded, the mRNA can also contain sequences that flag it for sequestration in the cytoplasm, at least until specific signals are received. Finally, there are mRNAs like the tubulin mRNA that are only translated when the level of the encoded protein drops below a certain level. After taxol treatment to stabilize microtubules and prevent their disassembly, the tubulin mRNP is activated for the translation of tubulin mRNA. How many similar proteins are present in cytoplasm is unclear. Thus, the nature of the

Figure 14.4

mRNP is formed to structure mRNA.

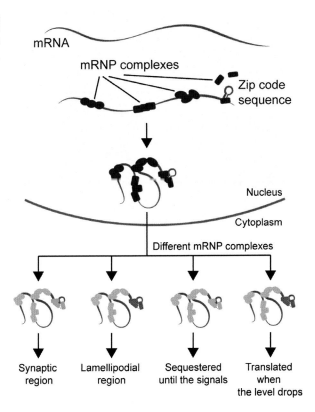

proteins in the mRNP complex is critical for the targeting of the mRNA to the right region of cytoplasm, and for the proper control of expression of the mRNA.

14.8 Quality Control of mRNA

The transcription error rate is one in 10^4 bases, which is much higher than DNA replication (one in 10^9 bases), and this results in numerous improper mRNA molecules. If the error is in the processing regions responsible for the poly A tail and GPP cap, then the mRNA will be unable to move out of the nucleus and will be degraded there. Many other premature truncations of the mRNA synthesis will cause the transcript to be rapidly degraded in the nucleus. If the mRNA exits the nucleus but is mistargeted to the wrong regions of the cytoplasm, there is a high probability that it will be degraded because the wrong protein in a region will potentially be very damaging to a cell. The signal for degradation is linked to the targeting signal and the proper targeting signal is needed for the mRNA to be synthesized in the proper place. If the error in the mRNA causes the premature truncation of protein synthesis, then the mRNA will be rapidly degraded. In addition, certain misreading errors can also cause rapid degradation of the mRNA presumably because the protein will not fold properly. Similarly, problems can arise from improper intron splicing and this will also cause degradation. Thus, there are several levels of quality control on mRNA production. This is to assure that only proper mRNPs will be transported to the cytoplasm and those that produce abnormal proteins will be rapidly degraded.

14.9 Movement Out of the Nucleus

Once the mRNP is processed, it is released from the transcription factory and will diffuse to the cytoplasm. Several properties of this process have been defined using Ms2 tags (see text box on methods to visualize mRNAs), to follow mRNP movement at the single molecule level with single particle tracking methods. For example, the movement of the mRNA in the nucleus is diffusional until the mRNA reaches the nuclear pore. The diffusion of many mRNPs is restricted to the channels between chromosomes, and this means that many mRNPs will follow the same paths when moving out of the nucleus. In addition, the diffusion coefficients are extremely low, indicating that the particles are interacting with material, most likely chromatin, as they diffuse. For example, the measured diffusion coefficients are about 4×10^{-11} cm^2/s whereas a 2 μm sphere (potentially the extended size of many mRNPs) would have a diffusion coefficient of 4×10^{-9} cm^2/s in water, which is 100-fold faster. It is most likely that the mRNP particles are less than 500 nm in diameter and would theoretically diffuse much faster. Decondensed chromatin at the surface of chromosomes will likely interfere with the diffusion of the mRNPs and there are potential interactions of the histones with the mRNA on the surface of the particles. Many weak interactions from the chromatin lining the channels are the best explanation for both the slow diffusion of mRNPs and the fast diffusion of small molecules in the nucleus. One factor that could aid the directional transport of mRNPs out of the nucleus is the assembly of new mRNPs at the transcription factories. Because new mRNPs are constantly being made at the factories, there will be a gradient of mRNPs from those sites that will increase diffusional transport, particularly in an essentially one-dimensional channel.

Once they reach the nuclear pores, the transport of the mRNP particles through the pores is extremely rapid (~1 s or less). This must be an active process, because the likelihood that a large (500+ nanometer) mRNP would diffuse through the 20–30-nm pore complex is extremely small. Even accounting for the fact that the fluorescent tags are on one end of the mRNP in the studies of transport, the speed is remarkable. It is attractive to think that condensation of the fibers that extend from the nuclear pore complex contribute to the transport; however, active transport of mRNP components across the pore itself is needed. This is made more efficient through the condensation of mRNA by proteins and could be further aided by some regularity in the structure of mRNPs. The nuclear pore acts as a sink for the mRNPs and this consequently increases the diffusional transport rate.

14.10   Micro-RNAs Can Decrease the Lifetime of mRNAs

Once the mRNP reaches the cytoplasm, its degradation can prevent proteins from forming and one of the major mechanisms for degrading mRNAs is through the action of micro-RNAs. Since the discovery that added small interfering RNAs could cause degradation of endogenous mRNAs, the role of endogenous micro-RNAs that can also cause the degradation of cytoplasmic mRNA has been of interest. It is now emerging that numerous endogenous small RNAs, micro-RNAs or other RNA species have been identified as playing key roles in the regulation of specific mRNAs. As the micro-RNA precursors are transcribed, they are processed in the nucleus to give a stem loop form that diffuses to the cytoplasm and then produces the micro-RNAs that target specific mRNAs. They will hybridize with specific mRNAs to form a double-stranded RNA that recruits a degrading enzyme

Figure 14.5

microRNA Mir-21.

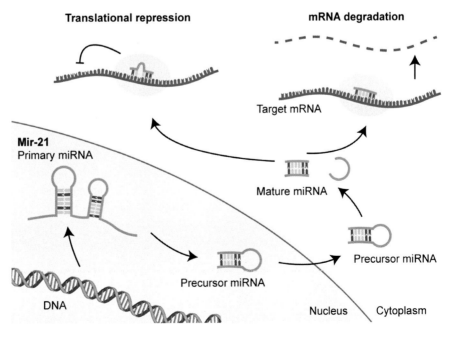

complex called dicer that chops up the mRNA. In a number of cancers and in a number of differentiated cells, there is a major role for micro-RNAs to decrease the concentration of mRNA for critical proteins. For example, Mir-21 is a micro-RNA that is commonly expressed in several cancers and correlates with severity of the disease. It has a major role in decreasing the level of the mRNA for cytoskeletal proteins that are tumor suppressors. Physiologically, it has been hypothesized that micro-RNAs are part of transcription/translation regulation pathways. This could explain how specific cells in a tissue could decrease the levels of a subset of genes that were activated at the chromatin level for a transient function. This would bring about a desired response of decreasing the level of the microRNA targeted mRNA and protein levels without changing the overall transcription profile for that cell type. For example, it is useful for the cell to activate growth mechanisms by knocking down the mRNA levels of critical growth control proteins, rather than changing the global transcription pattern of the cell.

14.11 SUMMARY The production of mRNAs by transcription is a critically important step in the life of any cell. Although the important events for activation of transcription initiation sites occur in the nucleus, the critical signals are coming from the microenvironment, including hormonal signals. Those signals typically cause the movement of transcription factors to the nucleus where they will activate particular transcription sites. The transcription process is then carried out within transcription factories that contain multiple copies of the relevant transcription-related proteins; both those that carry out the task and those that process the mRNA. Several different genes from potentially different chromosomes are transcribed at a single transcription factory site, and the congregation of the active genes may depend heavily upon the mobility of the euchromatin at the surface of the chromosomes. In the areas between the chromosomes there are channels adjacent to the transcription sites that enable the processed mRNP particles to diffuse to the nuclear pores, where they are actively transported out of the nucleus. By sequestering, in the nucleus, many of the critical steps in the identification of the transcription sites, the transcription process, mRNA processing, quality control as well as packaging mRNA for transport, the cell makes the overall process of mRNA production much more efficient. This enables the final product to be transported to the correct location in the cytoplasm, where the mRNA will be translated into a protein. Once the mRNA is in the cytoplasm, microRNAs that hybridize with the mRNA can cause its degradation and this provides another level of control of protein levels in the cell that can be responsive to specific signals where changes at the level of the overall transcription profile are not needed. Thus, both the production and the degradation of the mRNAs can be controlled from the nucleus.

14.12 PROBLEMS

Please describe additional steps in transcription-related functions such as mRNA processing, transport to the site of translation, or other aspects of translation.

14.13 REFERENCES

Hager, G.L., McNally, J.G. & Misteli, T. 2009. Transcription dynamics. *Molecular Cell* 35. 741–753.

Kues, T., Dickmanns, A., Luhrmann, R., Peters, R. & Kubitscheck, U. 2001. High intranuclear mobility and dynamic clustering of the splicing factor U1 snRNP observed by single particle tracking. *Proc Natl Acad Sci USA* 98: 12021–12026.

Muller-McNicoll, M. & Neugebauer, K.M. 2013. How cells get the message: dynamic assembly and function of mRNA-protein complexes. *Nat Rev Genet* 14(4): 275–287.

Nakazawa, N., Sathe, A.R., Shivashanker, G.V. & Sheetz, M.P. 2016. Matrix mechanics controls FHL2 movement to the nucleus to activate p21 expression. *Proc Natl Acad Sci USA* 113(44): E6813–E6822.

Tsukamoto, T., Hashiguchi, N., Janicki, S.M., et al. 2000. Visualization of gene activity in living cells. *Nat Cell Biol* 2: 871–878.

Van Bortle, K. & Corces, V.G. 2012. Nuclear organization and genome function. *Annu Rev Cell Dev Biol* 28: 163–187.

14.14 FURTHER READING

Chromatin: www.mechanobio.info/topics/genome-regulation/chromatin/

Chromosome territory dynamics and transcription factories in transcription: www.mechanobio.info/topics/genome-regulation/dna-templated transcription/

15 Turning RNA into Functional Proteins and Removing Unwanted Proteins

Most proteins in a cell are synthesized in close proximity to the nucleus through a process known as 'translation,' in which ribosomes translate mRNA into protein following its export from the nucleus. Proteins that will be secreted or integrated into membranes must be processed within the endoplasmic reticulum (ER). This requires a signal sequence that enables the docking of the ribosome to the ER. During translation, the protein is moved into the lumen of the ER, where it will be processed. Other mRNAs are targeted for transport to selected regions of the cytoplasm or for sequestration until proper signals are received. The number of proteins synthesized from a single mRNA is highly variable, but is usually in the range of 100–1000 copies. The subsequent lifetime of the protein depends largely upon whether it is properly synthesized. Defective, misfolded proteins, or proteins that are no longer required, will be actively removed from the cytoplasm. This is achieved through a process of ubiquitinylation, which flags the protein for degradation by a proteasome. This process is dependent upon the activation of ubiquitin ligases that are downstream in a number of signaling pathways. Both localized protein synthesis and targeted degradation can facilitate cellular functions by dramatically altering the concentrations of proteins at critical times.

Protein production is truly the hallmark of life as we know it, as proteins form the machinery that performs most cellular functions. As such, the translation of mRNA into protein is an extremely sophisticated process and one can imagine that virtually every mutation of the proteins in the ribosome and in the ribosomal RNA have been tested by cells or organisms. Any improvements would be shared with other organisms and, thus, the current ribosomes should be viewed as particularly robust. In this chapter, we will first discuss the actual process of protein production and the sophisticated adaptations for larger eukaryotes, including localized synthesis, protein folding, and processing. Protein degradation is the balance to synthesis and is used to rapidly block functions as well as to remove denatured proteins or recycle proteins in starvation conditions. Assuming proper nutrient levels, the goal of protein production and turnover is to provide the cell with the tools needed to perform the differentiated cell functions as well as face many environmental challenges from hyperthermia and trauma to various bacterial and viral challenges.

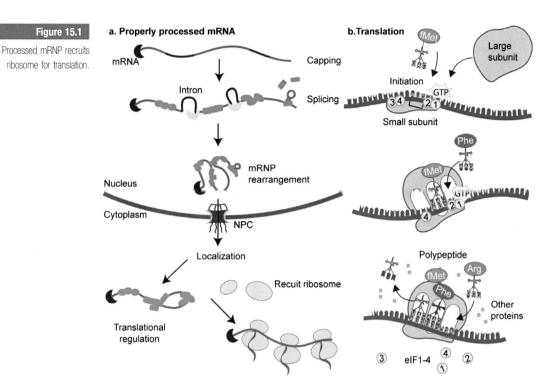

Figure 15.1

Processed mRNP recruits ribosome for translation.

15.1 The Complex Function of Protein Synthesis

1. Movement of mRNA, ribosomes and other components to the right location in cytoplasm.
2. Initiation and formation of the eIF complex in the right location.
3. Early translation and signal recognition (tRNAs, codon recognition and elongation as well as binding of signal peptides to ER).
4. Termination and recycling of the ribosome.
5. Protein folding and the role of chaperone proteins in quality control plus post-translational modification.

15.2 Translation by Ribosomes

Translation is primarily carried out by a highly complex molecular machine known as the ribosome. The atomic structure of this machine has been determined in a number of species, and despite several differences is generally composed of large and small subunits, each of which contains multiple proteins, and ribosomal RNA (rRNA) molecules. In humans, for example, the large 60S subunit contains 47 proteins and three rRNA molecules (5S, 5.8S, and 28S), while the small 40S subunit contains 33 proteins and a single 18S rRNA molecule. During translation, a large and small subunit assemble onto the mRNA with typically three tRNAs bonded to the appropriate amino acids that are marching through the ribosome as the polypeptide chain is assembled.

The ribosome has been studied extensively at the single molecule level. It has a number of control parameters including quality control and fidelity of the translation process. There are surprising features of the system that are emerging, such as the fact that the nature of the amino acid on a specific tRNA influences the synthesis process. This implies that there is a feedback between the tRNA structure and the amino acid binding site that influences the synthesis of the polypeptide. Such quality control mechanisms make it difficult to exploit the ribosome function in the lab for experimental purposes. For example, it has not been possible to use ribosomes to incorporate a variety of fluorescent amino acids into a protein of interest. Only certain modified amino acids can be used. The many features of this very large RNA–protein complex are quite remarkable both from the engineering as well as the biochemical viewpoint.

The rate of synthesis for a single ribosome is about 20 aa/s or 25 seconds to make one copy of a 500-aa protein (about 60 kDa), which means 144 proteins/hour or about 3400/day. For a cell that is rapidly growing (24 hour division time), there is the need to produce 4×10^{-9} g $\times 0.18 = 7.2 \times 10^{-10}$ g of protein or about 1.2×10^{-14} moles of a 60-kDa protein (about 10^{10} molecules). Thus, there must be at least 3×10^6 ribosomes per cell to produce the proteins. If each ribosome is 4.2 million daltons, then the mass of the ribosomes per cell is $0.5 \times 10^{-17} \times 4.2 \times 10^6 = 2 \times 10^{-11}$ g or 3% of the total cell mass. This is a significant burden on the cell because this is only a theoretical 'minimum' number of ribosomes needed. The real figure is more like >5% of the total cell mass, particularly if you take into account the ribosome assembly factories in the nucleoli. Growth places a significant burden on the cells, and cells in static conditions need far fewer ribosomes because the level of protein synthesis is much less.

15.3 Initiation of Translation

The first step in the formation of the translational complex is initiation. This is mediated by eukaryotic initiation factor (eIF) proteins and involves the formation of a pre-initiation complex, which serves to scan the 5′ untranslated region of mRNA to detect the initiation codon. Although there are up to 12 eIF proteins involved in the regulation of the initiation, eIF1–4 are the primary eIFs to interact with the small 40S subunit, mRNA and methionine-tRNA (the initial amino acid in proteins). Once the mRNA is activated by the relevant cofactors, a large ribosome subunit is recruited to the complex. Once the ribosome has assembled, the relevant loaded tRNAs, which are bound to their corresponding amino acids, are then bound and the synthesis of the protein proceeds.

15.4   tRNA Loading with an Amino Acid

Transfer RNAs are relatively short RNAs (76–90 bp) that form into a clover leaf shape through hybridization (see Figure 15.2). They extend roughly 10 nm from the anticodon to the 3′ end where the amino acid is attached. In the cytoplasm, an aminoacyl tRNA synthetase will add the appropriate amino acid to the tRNA and there is typically one synthetase per amino acid even though there are multiple tRNAs per amino acid.

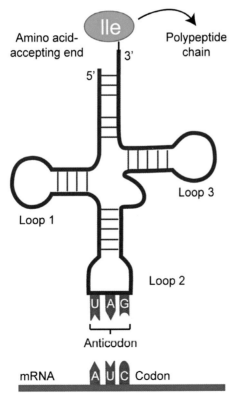

Figure 15.2

Clover leaf shape of tRNA.

15.5   Elongation of the Polypeptide

For elongation of the polypeptide from the initiation complex on the ribosome with the methionyl tRNA, there needs to be a supply of the tRNAs with the appropriate amino acids bound and GTP as an energy source. In the ribosome, there are three basic sites for tRNA binding: an entry, a transfer, and an exit site. After the ribosome assembles around the initiation complex that contains the *N*-terminal methionine with tRNA, the next tRNA binds with its amino acid to the entry site. The tRNA binding is aided by eIF1 and relies upon the proper anticodon–codon interaction. If there is a match, then the tRNA is physically moved to the transfer site where its amino acid is added to the growing polypeptide chain. After the next tRNA binds to the next codon, the last tRNA is moved from the transfer to the exit site as the next amino acid replaces its bond with the polypeptide chain. This cycle contains many sophisticated steps and the ribosome complex is basically a motor that moves upon the mRNA. There are a number of surprising elements to the process, such as the triggering of movement from the entry site upon proper pairing of the codon and anticodon. Further, the amino acid is about 10 nm away from the anticodon and there is communication between the sites because attempts to trick the system with tRNAs containing an unusual amino acid have met with limited success (the rate of incorporation of the wrong amino acid on a given tRNA is much less than when the proper amino acid is on the tRNA).

15.6   Termination and Ribosome Recycling

Termination of translation is caused by the entry of a mRNA stop codon (UAA, UGA, or UAG) into the ribosome entry site. The class 1 release factor (RF1) recognizes this codon because it is bound to the ribosome in eukaryotes in a pre-assembled ternary complex comprising eRF1, eRF3, and GTP. Once translation is terminated, the ribosome is released from the mRNA in a five-step process that first involves the binding of the ribosome release factor to the tRNA entry site and then the binding of the release factor–GTPase complex. The peptide is then released followed by the release factor that then enables the ribosome release factor 1, EF-G and ABC ATPase E1 complex to bind. They cause the separation of the ribosome subunits and the release of tRNA plus the mRNA. When the release factor complex leaves, the ribosome can be reinitiated for translation. If the translation is stalled, then this process occurs along with the rapid degradation of the mRNA. Thus, the robust nature of the process involves multiple triggers for release and recycling of the ribosome. Recycling is important because of the large size of the ribosome and the energy needed to produce it.

15.7 Protein Translation Occurs at Specific Sites

Translation often begins after the processed mRNP diffuses from the nucleus to the cytoplasm. In other cases, however, the location in the cytoplasm where the mRNA is translated is critical for the proper function of the proteins being produced. The reason for this is not always clear, but it is possible that functional complexes need to form immediately after their protein components are produced. In other words, proteins are produced at the site where they are required for a given process. For example, in neurons, mRNPs are transported down axons to synapses where they are then translated. This saves the cell time and energy that would otherwise be required to transport the proteins from the cytoplasm to the synapse. Similarly, proteins destined for secretion or integration into membranes are assembled and processed in the ER. This ensures their packaging into membrane vesicles, or integration into the membrane will occur as they are translated. In addition to localization, there are important metabolic controls on protein translation to stop the cell from using needed energy during periods of starvation.

15.8 Movement of mRNA in Cytoplasm (Specific Localization)

To ensure that translation of certain proteins is carried out in the correct region, 'transport sequences' of about 40–100 bases in length are incorporated into the appropriate mRNAs. These sequences are targeted by specific protein factors that then recruit the motors and other proteins needed to carry the mRNP to the proper region of the cytoplasm, prior to the initiation of protein translation. At least four different types of targeting signals have been identified.

In neurons, there are sequences that bind protein complexes involved in targeting mRNPs to the dendrites or axons; whereas in oligodendrocytes, the mRNPs are transported to the very periphery of the myelin sheaths. In fibroblasts, the mRNP for actin is transported to the periphery. The transport of mRNA in the cytoplasm has mainly been studied in oligodendrocytes (myelin-forming cells) and neuronal dendrites. In those systems relatively large complexes of mRNPs (as large as 0.5–0.8 μm) are transported along microtubule tracks by motors. Translation is repressed in the transport complexes by presumably blocking the initiation sequences until they reach the appropriate destination. Once at the correct peripheral sites, the transport complex is disassembled from mRNPs and the translation can start when the appropriate activation signal is received.

If the mRNPs are delivered to the wrong destination, then the mRNA must be degraded. This is facilitated by deadenylases and decapping enzymes that bind to the poly A tail and cap structures potentially because of the targeting protein complex. For proper cell function, it is more important to degrade mRNAs that end up in the wrong location than to risk the damage caused by the production of a protein in the wrong location.

In some cases, the localization of the mRNP to the correct area is not enough to activate translation. In *Drosophila* or *C. elegans* embryos, many mRNPs are deposited during oogenesis, but are not activated until later in the organism's development. Although many different protein mRNAs are similarly distributed, they are activated at different times, which indicates that not only is there a signal for the localization, but also there is a signal for the timing of activation that is specific for different species of mRNA. These mRNA transcripts typically encode transcription factors, which will activate the production of the protein components required for subsequent developmental processes. The major issue is how they get specifically transported to the proper location and activated for translation at the right time.

Figure 15.3

Movement of mRNA in cytoplasm.

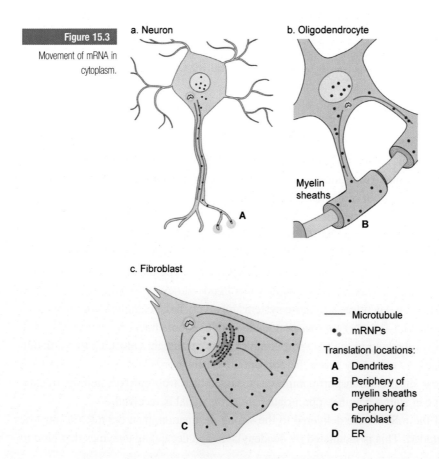

a. Neuron

b. Oligodendrocyte

Myelin sheaths

c. Fibroblast

—— Microtubule

•• mRNPs

Translation locations:

A Dendrites

B Periphery of myelin sheaths

C Periphery of fibroblast

D ER

15.9   Quality Control of Protein being Synthesized (Chaperones)

As the protein is being assembled and exiting the ribosome, it is folding into its tertiary structure. There is a significant possibility that the protein will fold improperly, which can produce dysfunctional proteins. Misfolded proteins tend to aggregate, and will be cleared by autophagosomal processes that package the aggregates for degradation.

To assist in protein folding, and ensure that a functional protein with correct configuration is produced, the cell employs chaperone proteins. Many of the chaperone proteins use ATP energy to massage the protein into the proper configuration. Because heat, organic solvents, mechanical stretching and other perturbations cause an increase in the fraction of misfolded proteins, they all increase the demand for chaperone proteins, and in particular, the activation of heat-shock protein (chaperone) expression. There are many cases where proteins are unfolded in the normal functioning of the cell and the basis for recruiting the chaperone proteins to refold the denatured proteins is not clear. One possible explanation is that the surfaces of improperly folded proteins have exposed hydrophobic regions that promote their aggregation/ degradation and/or the binding of chaperone proteins, which will induce unfolding and refolding of the protein. Because the concentration of protein is so high in the cytoplasm, the probability of aggregation is increased, and the presence of multiple hydrophobic domains could enhance the binding of the chaperone proteins.

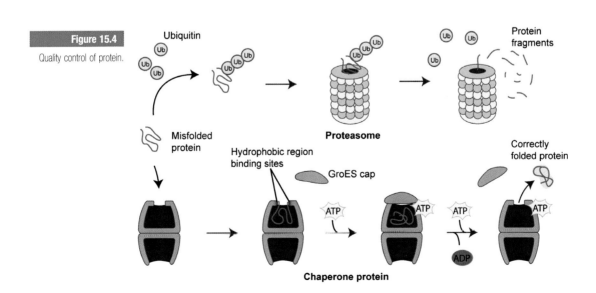

Figure 15.4

Quality control of protein.

15.10   Signal Sequences and Translation into the ER

A significant number of the mRNPs are destined to produce proteins that are processed in the endoplasmic reticulum (ER). As those mRNPs are activated and bind to ribosomes, they then synthesize a short signal sequence that binds the Signal Recognition Particle (an RNA–protein complex) that brings the mRNP to the ER surface. It then facilitates docking to a pore complex, which guides the polypeptide chain into the ER (Akopian et al., 2013). Once the protein is in the ER, it is modified by chaperone proteins, protein disulfide isomerase, and glycosylation enzymes. Modifications are critical for subsequent trafficking and there is a requirement for proper folding as well as proper glycosylation of the proteins. These are critical pathways for the cells and ensure that functional proteins are secreted or moved to the cell surface.

15.11   Control of Translation by Protein Concentration

In a few cases, an mRNP will remain in a quiescent state within the cytoplasm, but can be activated by the concentration of the same protein it encodes. In these cases, the protein synthesis machinery will respond relatively rapidly to changes in the free concentration of the protein. This is observed, for example, following taxol treatment, which promotes microtubule polymerization and stabilization. Here, the polymerization of tubulin will activate the synthesis of tubulin subunits from existing cytoplasmic mRNPs. This can cause major changes in cell morphology as the synthesized tubulin subunits incorporate into existing microtubules and cause their elongation over time.

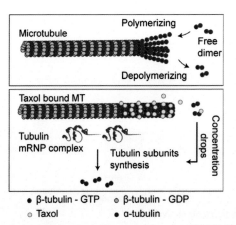

Figure 15.5

Tubulin concentration control.

15.12   Protein Degradation

The final concentration of a protein in a cell is determined not only by its rate of synthesis but also its rate of degradation. There is a wide range of protein degradation rates in cells, with some proteins, such as nuclear pore complexes, being very stable, while others, like p53, turn over constantly. Some proteins are degraded as part the cell cycle, and others degraded in response to the activation of signaling pathways.

The major reason for general protein degradation is that the protein is denatured or has lost activity. This is particularly relevant for muscle proteins, because inactive myosin can slow the contraction of active myosin. Because myosin is normally assembled into large filaments that are embedded in sarcomeres, it is very difficult to imagine how inactive myosin can be selectively removed from a muscle. Although we don't know how myosin turnover actually occurs, there are some intriguing findings that do provide some insight into the process. For example, highly ordered sarcomeres are formed in contracting muscles, but when the muscles stop contracting, the sarcomeres become disordered. This indicates that the sarcomeric proteins are being solubilized, or at least dissociated from normal filament assemblies. That would enable them to be tested for activity because inactive proteins often bind to actin irreversibly and the inactive proteins could be then targeted for degradation. This raises the possibility that periods of rest or sleep are indeed times for removal of inactive or denatured proteins.

15.13 Autophagy and Abnormal Protein Turnover

Phagocytosis is a process whereby macrophages help to clear bacteria and cellular debris from sites of infection and cell damage. A similar intracellular process, known as autophagy, exists to do basically the same thing for internal debris. If a cell is producing an excessive amount of an abnormal protein, or a localized subcellular trauma has resulted from infection, then the abnormal or damaged proteins may be isolated within a cytoplasmic membrane vesicle. This vesicle will then fuse with a lysosome, which contains an acidic interior and enzymes that will degrade any captured proteins or lipids. This process enables the cell to clear non-functional or damaged components. How the cell decides that proteins are abnormal, or that organelles are non-functional, is still a mystery. If the protein is denatured, it will often aggregate and then there will be a mechanical change in the protein's behavior. If the protein is bound to membranes, it can be tested by membrane tubulation and dynamics because the aggregate will potentially be less able to enter highly curved regions. Indeed, an accumulation of proteins at sites of high membrane curvature during ER movements potentially triggers the binding of components involved in the autophagy process. Similarly, the crosslinking of multiple filament systems or the binding of multiple chaperone proteins can trigger the recruitment of components involved in the autophagocytic response.

A mechanical component is also likely in the process of detecting protein aggregates, as well as in their subsequent processing. This may involve a procedure to test if the structure is dynamic. If the aggregate does not disperse, then it is considered abnormal and should be cleared. When the trauma or disease has led to unusually high levels of protein aggregates, the testing processes may be overwhelmed and will not be able to rescue all cells. Thus, trauma and significant inactivation of proteins can trigger major rearrangements and clearance processes that rely upon an active cell response, often including the new production of damage repair proteins.

15.14   Stimulated Degradation of Proteins by Ubiquitin Ligation

As we suggested earlier, the most rapid and reliable way to stop or prevent a specific cellular activity is to remove the proteins involved in that activity. This can be done by ubiquitinylation of the protein by E3 ligase, which is activated in response to a specific signal and will target the protein to be degraded. There are a number of E3 ligases that target different proteins in response to different signals. Some are involved in housekeeping functions, such as the removal of inactive protein complexes. However, many are part of specific signaling pathways, and play critical roles determining the outcome of a given signal. Once proteins are

Figure 15.6

Protein degradation.

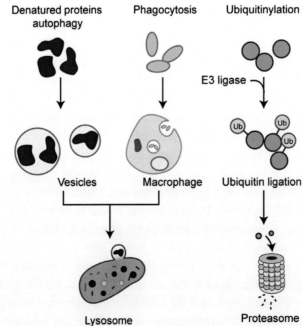

modified with ubiquitin, they will be targeted for degradation by the proteasome, which is a large, barrel-shaped complex of proteins that draws ubiquitinylated proteins into the barrel, where they will be proteolysed.

Targeted ubiquitinylation is critical for the cell cycle phase transitions and explains how the cell can reliably make abrupt transitions from one phase to another. It is, therefore, perhaps not surprising that some of the most effective anti-cancer drugs at present are compounds that inhibit E3 ligases or the proteasome complex itself.

15.15 Rationale for Overall Protein Levels

Cells produce subsets of the genome proteins to ensure they have the full complement of proteins to sustain their normal activities within the tissue as well as to address common challenges. There has been a lot of concern about how tightly protein levels are controlled in cells.

Even with only half the normal concentration of active protein complexes a cell should still function normally. This is because many of the essential functions are robust, with excess capacity or back up mechanisms able to take over when a particular protein is lacking or damaged. For example, if activation of a gene produces 200 mRNAs and they go to the cytoplasm and each produce 500 proteins, then the 100,000 proteins in the average cell of 2000 μm^3 will boost the concentration of that protein by 100 nanomolar. This is a large amount of protein and should therefore be sufficient for the protein to perform its function. Furthermore, if the protein has a half-life of days, it will not need to be synthesized again for days. Thus, cells can readily produce sufficient protein to restore normal function of the cell in a short period.

In most cases, the fine-tuning of protein expression level is not a critical issue for the cell. Even with less than optimal levels of protein, the cells will still function properly. This is because the important controls are exerted by the microenvironment, and it is the ability to respond to local matrix proteins and neighboring cells that is critical for cell function. Although a threshold level, or minimum amount, of many proteins is needed to support function, this tends to be far lower than the amount the cell produces. In some cases, however, the level of specific proteins is important. For example, the level of p53, which is involved DNA damage repair, is critical and yet it is continually turning over. For this reason the cell will continually transcribe the protein, and when DNA damage is detected, the p53 degradation pathway is inactivated. This in turn increases the level of the protein. As one might expect, it is better in this case to turn off degradation than turn on transcription, as those genes encoding DNA repair proteins such as p53 may be damaged. Other proteins are also turned over rapidly in cells under the appropriate conditions.

There are no iron-clad rules for the lifetimes of proteins and thus each situation should be approached individually.

There are some common properties that seem to apply to protein synthesis and degradation. First, protein degradation often serves as the signal to bring about a rapid change in cellular functions. This is because protein synthesis is much slower than degradation, and would therefore be ineffective when rapid changes are necessary. For example, the rapid degradation of cyclins, which can occur in seconds to minutes, causes abrupt changes in the cell cycle. In contrast, the activation of the serum response in serum-starved cells results in measurable protein production in about 15 minutes, but protein levels don't start to plateau until closer to an hour. Thus, it is simply easier to degrade a protein than to synthesize mRNA, process it, transport it out of the nucleus, and synthesize a protein.

The second property that is often observed relates to protein exchange from larger complexes. The exchange rate of proteins in functional complexes or large structures like focal adhesions is often much more rapid than the degradation rate (an exception may be some of the ribosomal proteins). Thus, the protein complexes are typically not degraded *en bloc*, but individual proteins are selectively exchanged. This is observed, for example, in axons and muscles, where it is critical to maintain active proteins and repair damage to cytoskeletal structures. In general, the repair appears to involve the exchange of active proteins for inactive proteins; however, we do not understand how that selection is made, particularly as microtubules in axons and myosin filaments in muscle are large structural complexes that are not easily remodeled for exchange.

15.16 Dynamic versus Stable

Because researchers typically work on cells that are growing in a lab, our prejudice is to think of all cells, including those in the organism, as being very dynamic and continually growing. However, this is not representative of many tissues, where cells are not growing and there is a very stable population of proteins. Most organisms have proteins that are very stable, while other proteins are always rapidly turning over. For example, extracellular matrix proteins are the most stable, often lasting for the life of an organism. Cells in a mature tissue are also relatively stable so long as they are not affected by trauma through injury or disease. However, even in these cases, excessive cell use, through stretching or starvation, can result in a dramatic increase in protein turnover. Even resting cells in a G0 or equivalent state (e.g. senescence or hibernation for certain mammals) will require some proteins to be replaced, and this means a basic level of turnover

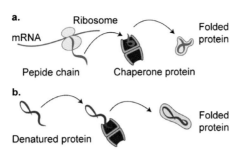

Figure 15.7
Chaperone proteins.

is maintained at all times. Generally, there are a variety of different cell states that are characterized by a relatively slow rate of protein synthesis rates.

The major activators of protein synthesis, even in stable tissues, are activity and trauma. For example, in analyses of cytoskeleton stretching, we found that many heat shock or chaperone proteins were bound to the stretched cytoskeletons. Furthermore, the stretching of endothelial tissues was shown to activate heat-shock responses. Thus, mechanical unfolding of proteins during normal tissue activity or stretching causes a trauma response similar to heat, or solvent exposure, both of which denature proteins. The heat-shock response presumably increases the expression of chaperone proteins to help maintain proper folding of proteins, even during stressful conditions. Because unfolded proteins are more readily degraded, the increased expression of chaperone proteins will reduce the loss of proteins caused by trauma. This allows for a reduced level of protein synthesis, and a higher percentage of functional proteins.

15.17 Starvation, Feasting, and Senescence

Even with normal changes in metabolism, which will occur during times of stress or as an organism ages, there are dramatic changes in the balance between protein production and protein degradation. Naturally, with starvation, the balance shifts from protein translation to protein degradation. Such changes will bring about changes in tissue mass, as is observed in muscle tissue, which can change in mass dramatically with aging and usage. Starvation is usually accompanied by inactivity. Because the muscles are being used less, they can serve as a reservoir for protein. At the same time, fatty tissues supply lipids and glycogen stores provide carbohydrates. When the starvation is reversed, the muscle mass is not necessarily restored unless there is exercise of those muscles. With age and inactivity it is very hard to maintain muscle mass, and a poor diet can exacerbate the problem.

In any tissue, growth versus catabolism is critical, and the balance can be tipped by any number of factors. One of the primary factors that determine which metabolic direction a cell takes is circulating hormones. Those hormones are in turn controlled by the overall state of the organism, the combination of food intake (type of food and amount) and exercise being the major factors under normal conditions. With cancer and other diseases, the normal balance is tipped and overproduction of growth hormones particularly aids tumor growth. This is an active area of study because hormonal issues are often associated with diabetes and a rich diet. On the other hand, cancer, and the growth of the tumor, may occur at the expense of other tissues.

An additional cell state exists with unique consequences for the translation of mRNA into protein. Cellular senescence is described as a state of 'irreversible cell cycle arrest' that is often stimulated by a cellular stress response. The fact that the cells can no longer divide was considered as playing a critical role in inhibiting cancer growth; however, recent studies find that senescent cells will often secrete growth-promoting molecules (cytokines, chemokines and growth factors), implicating them in the promotion of tumors in the rest of the body. From the viewpoint of protein translation, senescent cells will not continue in the cell growth programs, but can take on other programs that are not part of the normal tissue pathway, while still involving protein synthesis. Senescence is stimulated by the damage caused by ischemia–reperfusion injury wherein the tissue is deprived of oxygen due to temporary loss of blood or the absence of oxygen for the organism. Often the senescent cells will secrete factors that may be designed to aid the local recovery process because the damaged, senescent cells themselves are no longer able to aid in the repair process. Thus, senescent cells can create problems for the organism.

15.18 SUMMARY   Once an mRNP leaves the nucleus, it will be transported along one of many potential pathways to the site of translation. The simplest method is diffusion; however, in many cases motor proteins will carry the mRNPs to distant sites. This is particularly relevant to neurons. Once at the correct location the mRNA will be activated and translated by a polyribosome factory to produce 100–1000 copies of the protein. In the case of oocytes, this activation and translation may be delayed until a specific time in the development of the embryo. After the mRNA has been translated, it will then be degraded. Many mRNPs will produce a signal sequence upon translation that will bind the Signal Recognition Particle, targeting the complex to the ER where the new protein will be inserted into the ER for further processing. From the ER the protein will be transported to the plasma membrane, lysosome, or a secretory vesicle. Chaperones and other factors help the protein to assemble properly in the right complex. If proteins don't fold properly, they will often aggregate, which will promote the premature degradation of the protein through autophagy. Trauma signaling pathways can also flag the damaged protein for degradation.

Once produced, a functional protein can have a wide range of lifetimes. Some proteins are targeted for degradation by signaling pathways early in their lifetime. The cyclins, for example, regulate phases of the cell cycle, but are short-lived.

Other proteins are very long-lived (nuclear pore complexes and extracellular matrix, for example) and may have roles in aging because they are not produced in mature adults. It is critical for the cell to be able to remove inactive proteins, but we know relatively little about how inactive proteins are selected for degradation. Both the synthesis and the degradation rates define the final concentration of the protein. Thus, it is important to understand both the synthesis, and degradation, of a given protein to know how the concentration can change during cell functions.

15.19   PROBLEMS

Please describe additional steps in translation such as initiation of translation at nerve synapses, degradation of mRNA, translation of secreted proteins or other aspects of translation.

15.20   REFERENCES

Akopian, D., Shen, K., Zhang, X. & Shan, S. 2013. Signal recognition particle: an essential protein-targeting machine. *Annu Rev Biochem* 82: 693–721.

15.21   FURTHER READING

www.mechanobio.info/topics/genome-regulation/translation/

PART III

Coordination of Complex Functions

16 How to Approach a Coordinated Function: Cell Rigidity-sensing and Force Generation across Length Scales

Having discussed specific complex functions in a cell, it is good to understand how a cell integrates them when carrying out a specific process. This will be done in this chapter using fibroblast spreading and adhesion formation as an example. Before this, however, we will recap some of the fundamental concepts discussed in Chapters 1–4. At a whole-cell level, there are many complex functions that we would like to understand in more detail. However, it is difficult to isolate an individual function from the multitude of alternative functions that modify the function of interest. Isolating individual complex functions is made easier by standardizing the *in vitro* conditions in which cells are growing, so as to cut down on the number of uncontrolled variables. The cells should be synchronized to start from a well-defined state, rather than an array of states. The function of interest must then be followed, from initiation to a later stage, as this can enable the complete sequence of events to be analyzed. At present, this approach has only been used for a handful of complex functions in mammalian cells at the single cell level. By better defining these few functions, however, new paradigms have been provided in which to consider other established functions. Clathrin-dependent endocytosis and cell–matrix adhesion formation are two complex functions where many of the proteins involved are known, and a rough sequence of events has been described in some special circumstances. Those circumstances include specification of the cell environment, cell state, and several other factors that then enable reproducible observations of the cell functions. In the case of cell binding to matrix, there are rapid transitions between distinct cell states. This provides a good example of how cell state changes are needed for complex functions, but they complicate our understanding of the process. In a general context, the descriptions of complex cell functions read like the engineering descriptions of complex functions in factories.

At a practical level, the analysis of functions in single cells requires a number of conditions to be met and we will see in a special case how that can be done. In Chapter 3, complex cellular functions were described as multistep processes that employed many different functional modules. For complex cellular functions to operate under the wide variety of cellular conditions, they need to adapt to the changes in cell behavior that will follow environmental challenges. This often entails modifications to specific functional modules through signaling pathways, both biochemical and biophysical. For example, even the relatively simple function of clathrin-dependent endocytosis will be dramatically affected by changes in cell membrane tension. It is therefore extremely important to define the cellular state, and to start a function with a defined stimulus, so that all subsequent studies of the function have a reference point.

In studies that defined cell–matrix adhesion formation, the complex function began with the addition of suspended cells to a matrix-coated surface. This has worked well in terms of providing a reliable reference point, but fails to provide physiological relevance to the steps that are subsequently observed. We imagine that cell extensions to new matrix sites *in vivo* will activate analogous functions and spreading from suspension will provide a start to analyzing those processes. Nevertheless, spreading from suspension is reliably activated in almost every instance so that any changes to the cell state over time, can be defined, and attributed to the function under observation. Only a limited number of cell states are found in spreading, and defining them will enable their identification in other environments or cells. From there, it should be possible to move to later times that involve other cell states, and then to determine how different motile functions of interest are involved in those states. Thus, for all of the functions that have been discussed in earlier chapters, it is important to be able to study them in defined cell states and under conditions where the function can be abruptly activated to enable reproducible analyses of the steps involved.

Cell–matrix interactions occur at the cell surface, and are therefore relatively easy to address at the single cell level. What has emerged from nearly 20 years of studying this process is that our original thinking was very naïve and now many additional steps have been added that were not originally imagined. Most of the change has come from the discovery of new steps that weren't envisioned initially and new complexities that have emerged. Perhaps a suitable analogy is that we often travel from one city to another by airplane without considering all the steps that must be completed for that process to occur, including the fueling of the plane, loading and unloading of bags, air traffic control, security, etc. In approaching any complex cell function, it is important to start from the detailed observations of the process from the time that the cell starts the process until it reaches some quasi-steady state. For cell spreading on matrix-coated surfaces, it is important to follow the steps from when the cell first encounters the surface until it is fully spread and polarized or migrating. For example, just measuring the rate of spreading on different densities of fibronectin can tell us a lot about the process, as will be discussed below. Often the literature does not provide the detailed analyses that will identify the important subcellular steps involved in the process and many high-resolution movies are needed to see how different components might be involved in a given step. With knowledge of where molecules are during the process, a testable hypothesis can then be derived for at least one of the steps. Such an approach has been taken when considering the formation and function of matrix adhesions in fibroblasts.

16.1 A Systematic Approach

For an understanding of the process of cell spreading, we should describe the various steps that we believe to be involved and how they might be performed by the cell, i.e. what does the cell need for these tasks. Further, the timing and order of events are critical but often difficult to determine, particularly if you rely upon data from many cells that are in different parts of their cell cycle and often have different morphologies. With the careful definition of the initial, suspended cell state, the elements that must change for the cell to perform the function will become clearer, and the steps that occur can be determined along with the potential decision points. For example, cells in suspension are primed to test their environment and will activate those tests upon contact with a surface or a cell. Dependent upon the outcome of those tests, the cell will establish a new cell state that is typically dynamic, and will change depending on the outcome of additional tests. We suggested before that the tests form a nested set of if/then decisions that can cause the cell to end up in one of many different possible states. Thus, we want to experimentally determine the nature of those tests that are embedded in the steps and to define which are terminal or leading to growth.

To put this in a more operational form, we will move from the initial state to the final state of cell–matrix adhesion formation by using the following systematic approach:

1. Define the initial state.
2. Define the initial stimulus.
3. What are the steps that follow the stimulus (where are decisions made)?
4. Trace a logical decision tree to the final state.
5. Define the final state.

Because a lot is known about the process of cell spreading and force generation, we will focus on the functions that lead from the suspension state to a force generating state with stable focal adhesions.

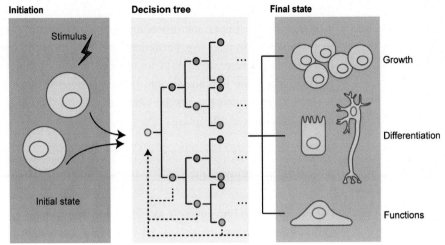

Figure 16.1

Systematic approach.

16.2 Description of Initial State

In considering the process of mouse embryo fibroblast spreading and force generation, the initial state is a suspension state. If we standardize the system by serum starving the cells overnight in a subconfluent culture, then cells that are suspended by trypsin/EDTA treatment will be relatively similar (additional standardization of the initial state in the cell cycle before starvation would further help). In the process of suspension, the cell will typically endocytose a significant fraction of the plasma membrane. Pre-incubation of the cells at 37°C will enable the cells to complete the endocytosis process as well as restore integrins that were cleaved by trypsin. We believe that the cells will be in a relatively stable state at this point. Although it is very difficult to completely define the states of cells, such measures will produce a relatively homogeneous cell population wherein a significant fraction of the cells will behave in a reproducible manner. Recent developments in the microfabrication of substrates with defined shapes and defined areas can potentially be used to put many cells into a reproducible state. The goal is to obtain consistent behavior that is critical for defining the subsequent steps as there is often significant variability in the duration of events due to the stochastic nature of most biological functions.

Once cells are in a defined initial state, it is worthwhile to consider the cell parameters that may be important in a given function. During cell spreading, for example, the parameters of cell morphology, volume, surface area, and cytoskeleton organization/dynamics are all critical and are coordinately controlled by the cell for the desired result. A preliminary list of the various parameters is provided in Table 16.1.

Table 16.1 Important parameters in cell spreading.

1. Cell volume.
2. Plasma membrane area (cells can increase area by 50% upon mechanical stress).
3. Actin filament assembly and disassembly.
4. Actin filament contraction.
5. Matrix binding.
6. Response to matrix rigidity.
7. Release of matrix binding.
8. Cell growth/apoptosis signals.

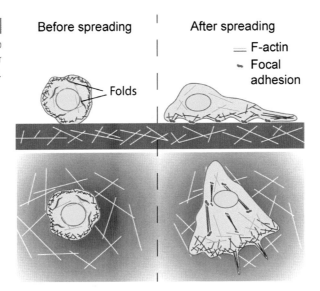

Figure 16.2

Cell morphology prior to spreading and after spreading.

To get a better understanding of the additional parameters that may be involved, we can look at the cell morphology prior to and after spreading (Figure 16.2). Before spreading there are many folds in the membrane, and the expected area of contact with the substrate is very small. It is not tenable for the cells to spread through substrate–adhesion binding with a diffusive expansion of the contact region, because the cytoskeleton will hold the membrane off the surface. Instead, what typically happens is based on the cellular IF/THEN decision-making process; IF the cell finds sufficient matrix within a certain time, THEN it will spread on the matrix. IF the concentration of matrix is too low, THEN the cell will send out unsupported outpocketings of membranes (blebs) that can increase the area of substrate contact. IF there is insufficient matrix for a long period, THEN the fibroblasts will apoptose. In all three cases, the cell changes from one distinct behavioral state to another. We can then consider how those changes might occur.

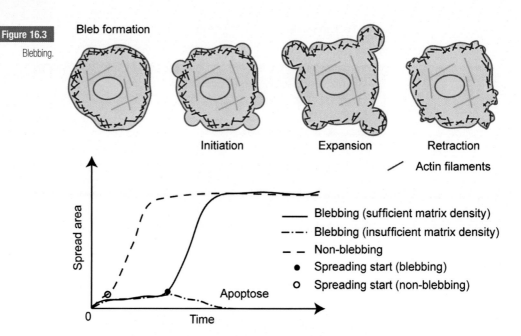

Figure 16.3

Blebbing.

As mentioned, if the matrix density on the surface is too low, then the cell will begin producing blebs. Blebbing represents a change in cell state where the cytoskeleton-based membrane support is decreased. The probability of a cell blebbing generally increases with the time it spends in suspension, because the loss of adhesion signals from the microenvironment causes a decline in the physical connection between the membrane and the cytoskeleton. Logically, without matrix adhesion signals, there will be decreased stimulus for actin polymerization, and therefore depolymerization will predominate. This will reduce the density of actin filaments supporting the membrane and the internal pressure of the cell from myosin contraction will cause blebs to form. Once blebs form, the diffusion rates of membrane glycoproteins are increased, promoting their rapid binding to matrices, which will then recruit the proper components to form adhesions. If sufficient matrix density is not found in time, then the rigidity-dependent cells will transition to an apoptotic program.

To understand the spreading process further, we need to alter many factors that affect actin and myosin function and to repeatedly analyze the process. It is important to know if the perturbations used will significantly change the cell state. For example, soluble growth hormones can activate signaling pathways that will alter the state of suspension cells and would logically alter the behavior of the cell upon interaction with the surface. Further minor aspects of the surface can vary from place to place such as roughness, curvature, or ligand density. Because those factors can influence the cell response, it is not trivial to develop experimental conditions that will give a reproducible cell response. Practically, careful cleaning of the glass coverslips and the use of highly reproducible matrix ligands are usually sufficient for reproducible spreading. Some matrix ligands like laminin or full-length fibronectin will aggregate after stirring or shear, which will cause variation in the density on the glass. A major reason for taking the care to control for variations in the conditions is that there are often a surprising number of steps that were not expected. When an unusual result is found, it could be explained as an unexpected step or a random uncontrolled event. Reducing the number of the latter makes progress easier. A large number of the steps in the process can be rationalized in retrospect, but are hard to predict without dissecting every step in a process from several different viewpoints.

16.3   The Process of Cell Spreading on Fibronectin

Many studies have described how mouse embryo fibroblasts spread on fibronectin or RGD ligand matrices. We will discuss some of these findings here; however, it should be noted that this will obviously give us a limited view of the cellular behaviors that can occur. The expectation is that once spreading processes are defined at a molecular level, then we can move on to define other complex functions and will have tools to determine when a given function is operating within a given cell as well as potentially ways to block that function. For most of these studies, subconfluent mouse embryonic fibroblasts are treated in the standard procedure described above to put them in a standard initial state (see the preceding two paragraphs). When such cells are placed on fibronectin-coated glass surfaces at 37°C, they typically spread and polarize in 20–40 minutes through $\alpha v \beta 3$ or $\alpha 5 \beta 1$ integrin binding to fibronectin. In that short period, the cell executes a series of steps involving binding, activation, spreading, and mechanical testing of the surface. What follows is our current understanding of those steps and some of the decision tree that is involved in this process.

16.4   Initial Matrix Binding and Activation of Spreading

Although standard fibroblasts spread almost instantaneously on fibronectin (or RGD, which is the three amino acid sequence in fibronectin that binds to the beta integrin)-coated coverslips (with 10 μg/ml of fibronectin), studies on different substrates indicate that several distinct events occur within seconds. For example, it is possible to attach cyclic RGD to supported lipid bilayers on a glass surface, and RGD will diffuse at the rate of the lipid, or about 100-fold faster than integrin diffusion. Without barriers in the bilayer to resist or restrict RGD movement, however, the high rate of diffusion means that no meaningful lateral force can be developed on the integrins. Fibroblasts will adhere to the RGD-conjugated lipid and form clusters of $\alpha v \beta 3$ integrins that are about 100 nm in diameter with about 50 integrin dimers (Changede et al., 2015; Schvartzman et al., 2011). Those clusters will then recruit a number of cytoplasmic proteins including talin1, FAK, and the formin, FHOD1 (Yu et al., 2011). Actin filaments will extend from the clusters if FHOD1 is active, and the clusters will move in a myosin-dependent manner. If barriers are present in the lipid bilayers, then the cells will spread rapidly, as barriers restrict RGD movement and enable force to be generated on the clusters. However, if there are no barriers, the force cannot be generated, and the cells do not spread. This suggests that several steps occur in very early activation of spreading. First, there is cluster formation, and then the recruitment of actin-polymerizing machinery to the cluster. Actin assembly from the cluster provides a handle for myosin to pull the cluster. If myosin pulls on the actin associated with the cluster and force develops, then spreading is activated. Although this seems like a logical series of events for cellular spreading, they were initially found through supported lipid bilayers studies because they occurred too rapidly on glass surfaces to be studied.

Now, we feel that the events observed on lipid bilayers also occur on glass and are critical for proper cell spreading. Despite these events occurring too quickly to measure, their effects were evident. For example, the 100 nm clusters form on RGD-coated glass and contain a similar number of integrins. FHOD1 associates with the clusters and the depletion of FHOD1 by siRNA expression causes both a dramatic loss of adhesion formation, and a highly abnormal motility (Iskratsch et al., 2013). The peak of FHOD1 activation is about 3–5 minutes after cells are added to fibronectin-coated glass and this fits with a very specialized role for FHOD1 in early adhesions. Thus, we believe that the very early activation of spreading depends upon integrin clustering and force on the integrins. The signal that is generated by the force on the integrin clusters causes a dramatic change in cell state, from largely a contractile behavior to a spreading behavior where there is very little contraction. Thus, early adhesion

Figure 16.4

Integrin clustering.

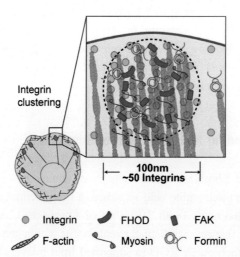

appears to be an important decision for the cell that is made upon the basis of a cluster of integrins promoting actin attachment, and then force generation.

16.5   Spreading is Dependent on Fibronectin Concentration

One interesting observation on early cell spreading that could tell us a lot about how the cell decides to spread comes from analyzing the effect of fibronectin concentration on the spreading process. Cells can be fixed after being in contact with fibronectin-coated glass for a given time to give a measure of the rate of spreading. In such experiments, the percentage of spread cells decreases as the concentration of fibronectin on the surface decreases. This implies that the cells spread slower on lower densities of fibronectin. However, when tracking individual cells, it becomes apparent that the time on the surface before spreading is initiated increases as fibronectin density is reduced. In some instances, it can take over 30 minutes for the cells on the surface to spread, yet once spreading begins, it will continue at a constant rate (Dubin-Thaler et al., 2004) (Figure 16.3). This indicates that there is either an integrated effect of fibronectin over time, or that the cell has a lower threshold for spreading with time. In either case, the spreading process is activated in a digital fashion, whereby it is either on or off, and the time to activation is inversely dependent upon the concentration of the activator, fibronectin. This appears to be an integration process that is similar to others we discussed earlier, as low levels of a signal over time can lead to a digital activation event. Thus, there are additional steps that we do not understand that enable single cells to average over minutes the concentration of matrix ligand on a surface.

Steps in RGD(FN)-dependent Adhesion Formation

1. Integrin Activation and Initial Cluster Formation: this does not involve force but may require talin, paxillin, FAK, etc., binding. Src kinase activation is seen very early as well.
2. Actin Polymerization from Cluster: actin polymerization occurs in the absence of myosin contraction but requires Src kinase activity for robust polymerization.
3. Contraction of Clusters: myosin II will contract clusters.
4. Force on Clusters Activates Rapid Spreading: we suggest that the stretching of a protein will result in G-protein activation and activation of spreading.
5. Actin Assembly both from the Edge and Clusters: during rapid spreading on lipid bilayers, there is an increase in the distance between clusters as well as a separation of the edge from the clusters, indicating a general activation of actin assembly.

(continued)

6. Membrane Tension-dependent Activation of Contraction: as the cell flattens with actin polymerization, it runs out of excess membrane area and there is a rise in membrane tension that activates contraction (and often exocytosis).

7. Rigidity Sensing by Local Contraction: adhesions can form on structures as a result of local contractions that sense rigidity.

8. Stabilization of Adhesions on Rigid Substrates, Reversal on Soft: adhesions couple to flow of actin to create inward force that matures adhesions. Myosin inhibition causes loss of coherent cytoskeleton and adhesion reversal.

9. Maturation of Adhesions.

16.6   Rapid Spreading of the Cells Following Force-dependent Activation

The rapid spreading of cells on supported RGD–lipid bilayers following activation by force is analogous to the process of rapid early spreading of cells on fibronectin coated glass, which has been described as phase 1 (P1). Studies exploring cell spreading in P1 have shown that once the cell edges begin to move, they will continue over adjacent non-adhesive coated surfaces at the same rate they spread on fibronectin coated glass (Rossier et al., 2010). This indicates that once activated, spreading does not require further integrin activation. When spreading occurs on a supported bilayer, the integrin clusters moved outward, almost as rapidly as the cell edge. Only at the final stages of spreading did the cell edge move away from the clusters (Yu et al., 2011). In general, the clusters moved away from each other on the supported bilayers. This is likely to be driven by actin polymerization from the clusters. As the cell spreads further, polymerization at the edge appears to increase relative to the internal polymerization. This causes

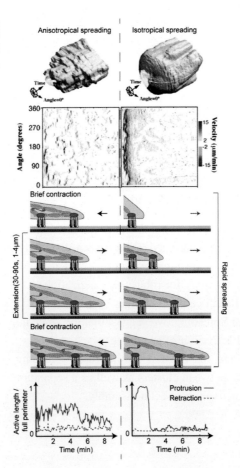

Figure 16.5

Isotropical and anisotropical spreading.

the gap between the outer clusters and the cell edge to expand. This is all consistent with a general activation of actin polymerization at multiple sites that ends when the cell reaches the limit of membrane area (see below). However, when cells spread on glass, only a fraction of fibroblasts will spread in this isotropic manner. The remainder will spread anisotropically through limited spreading events that occur for 30–90 seconds, and involve 1–4 μm lengths of the cell edge. These events are then followed by brief contraction events. It is not known how a cell decides to spread isotropically or anisotropically, although the fraction of isotropic spreading is increased by serum starvation (Dubin Thaler et al, 2004). Rapid isotropic spreading has been studied more extensively because there is a clear separation between the spreading and contraction steps over the whole cell. In anisotropic spreading, the spreading and contraction behaviors occur in local regions, asynchronously over the cell, making it difficult to study the steps separately. Although the spreading of cells can occur synchronously or asynchronously around the cell, the spreading events are generally followed by contractile events that involve the new integrin–matrix contacts.

16.7   Membrane Tension-driven Activation of Contraction

What ends isotropic P1 is a rise in plasma membrane tension, which occurs as the cell changes shape from a sphere to a flattened disc. As mentioned previously, the cell volume and plasma membrane surface area are critical in determining the spread area of a given cell. In the suspended state, the cell is spherical, and the plasma membrane has many folds over its surface. These folds will flatten as the cell spreads into a disc-like shape, eventually becoming depleted. It has been reported that a rise in membrane tension occurs at the end of P1, concomitantly with the loss of all membrane folds. This tension spike of twice the resting tension for 2–4 s presumably signals to increase myosin contractility, as well as to increase plasma membrane area (typically the increase is abrupt and by about 40%) (Gauthier et al., 2011). A hypo-osmotic shock will also cause a spike in membrane tension that will stop rapid spreading as well as activating both contraction of the cytoskeleton, and exocytosis (Gauthier et al., 2011). The purpose of exocytosis is to counteract the rise in membrane tension and it enables further spreading on stiff surfaces. If myosin II is inhibited during spreading, membrane tension rises much more dramatically at the end of P1 to sustained levels almost 10-fold greater than normal (still the rise in membrane area is only about 50%). How an abrupt change in membrane tension is converted into a biochemical signal that activates myosin II contraction and/or exocytosis is not understood. In anisotropic spreading,

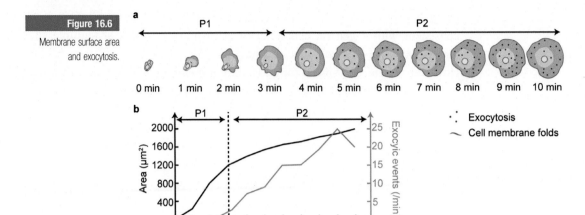

Figure 16.6

Membrane surface area and exocytosis.

there is intermittent myosin contraction during the whole spreading process, and a delay in the activation of myosin is believed to promote isotropic spreading. The simplest explanation for the rise in contraction at the end of the rapid P1 phase of spreading is a membrane tension-dependent activation of a GTPase exchange factor (GEF). This would cause a local rise in GTP-Rho, or other small G-protein, that then causes myosin activation and could also activate another protein for membrane exocytosis. Again, there is an abrupt change in cell state at the end of P1 due to a rise in membrane tension, and a variety of cell functions are active in P2 that are not observed in P1, and vice versa. In this case, a tension signal is converted to a biochemical signal that causes a major phase change.

16.8 Periodic Contractions and Rigidity-sensing

The cell will transition from P1 to phase 2 (P2) when myosin contraction is activated. This is often characterized by periodic contractions in cells that are growing on rigid surfaces (Giannone et al., 2004; Doebereiner et al., 2006) and there is good evidence to indicate that these contractions are involved in rigidity-sensing. If the surface is soft, for example, the cell will not exhibit periodic contractions and will stop spreading. Furthermore, on soft surfaces, there is a rapid flow of actin inward without the development of adhesions (Giannone et al., 2004). If the surface is rigid, however, the cell will continue to spread, and undergo several cycles of periodic contractions, occurring

Figure 16.7

Lamellipodial-actin periodic regeneration.

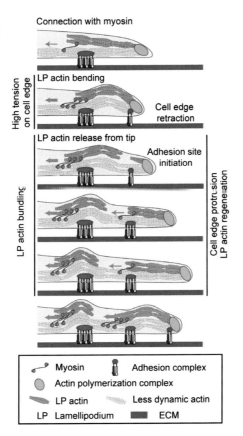

Connection with myosin

High tension on cell edge

LP actin bending

Cell edge retraction

LP actin release from tip

Adhesion site initiation

LP actin bundling

Cell edge protrusion
LP actin regeneration

Myosin   Adhesion complex

Actin polymerization complex

LP actin   Less dynamic actin

LP Lamellipodium   ECM

approximately every 25 s. This initiates the rigidity-sensing contractile cycle driven by the local contraction units described below (Giannone et al., 2007). These cycles involve the reinforcement of initial adhesions at the cell edge that then mature over time into more substantial adhesions. Several models have been proposed to describe the mechanical basis for the maturation of adhesions from a purely theoretical viewpoint. We suggest that maturation involves the movement of an upper layer of actin filaments over the adhesions driven by myosin filaments located behind the adhesions. Moving actin filaments transiently bind to adhesion molecules creating a stick–slip stretching and relaxation of talin and other adhesion molecules that then results in maturation.

16.9 Rigidity-sensing by Local Contractions of Sarcomere-like Units

How cells test the rigidity of their substrates during the periodic contractions can be understood using submicrometer pillars. Here, it is possible to see that there are local contractions (2–3 μm in area) at the edge of the cell, which are the hallmarks of rigidity-sensing. Recent studies show that local contractions near the cell edge displace the adhesion sites by about 50 nm for about 30 s, regardless of the rigidity of the substrate. This means that rigidity-sensing appears to involve sensing the force needed to produce molecular level displacements. For cells to produce such displacements, they need to assemble sarcomere-like contractile units that are capable of producing very high forces on the matrix (> 15 nN/μm^2). Those local contractile units are found in many different cell types including stem cells, and appear to be a general tool for sensing matrix rigidity. Detailed analyses of the pillar contractions, which are mediated by local contractile units, reveals step-wise movements of ~2.7 nm total displacement (the distance of movement from an actin on one side of an actin filament to the next actin on the other side). When the force on the pillars reaches about 25 pN, there is a pause in contraction as α-actinin, and potentially other proteins, are recruited to the adhesion sites (Wolfenson et al., 2016). Thus, the force is the signal to the cell to form an adhesion at that site. If the pillars are soft, the adhesive links from matrix to the cytoskeleton will often release and the pillars will jump back to a zero force position. This rapid disassembly of adhesions on soft surfaces can cause the cell to revert to a non-adherent state and then to reinitiate spreading. On the other hand, if the cell has many rigid contacts, there will be a cumulative signal to the cell that will direct it in terms of growth or differentiation. From many studies, it appears that the local rigidity-sensing contractions form on new matrix contacts to test the rigidity of those matrices.

16.10 Enzyme Pathways in Rigidity-sensing

Several enzyme systems are involved in rigidity-sensing. These are predominately involved in the assembly and disassembly of the contractile apparatus, and in the sensing of force generated during the local contractions. A rigidity-sensing screen involving the knockdown of all human receptor tyrosine kinases (RTKs) showed that at least 15 RTKs affect rigidity-sensing. Preliminary evidence indicates that those RTKs are involved in the mechanical regulation of the local contraction units. For example, the amount of displacement and the duration of displacement in the local rigidity-sensing contractions are regulated by different RTKs, AXL and ROR2 (Yang et al., 2016). AXL controls the length of contraction and interacts with both myosin filaments and tropomyosin on the actin filaments, putting it in a position to measure the relative displacement of the two filament systems in the local contractions. On the other hand, ROR2 controls the duration of contraction and binds to an actin-crosslinking protein, filamin A, that has been implicated in holding the actin network together, implying that the relaxation of crosslinks causes the cessation of contraction (Figure 16.8). The idea that RTKs might be mechanosensory proteins is novel but is consistent with the roles of many of those kinases in cancers and development. What is missing is a more complete understanding of how the RTKs can sense mechanical aspects of cell contractions or other cell mechanical processes and convert those measurements into the desired cell response. Thus, we propose that local contractions test the rigidity of the surface, and generate the rigid surface response. If the surface is rigid enough then adhesions can grow and participate in signaling to the rest of the cell. If the surface is soft, fibroblasts will revert to a non-adherent state and stem

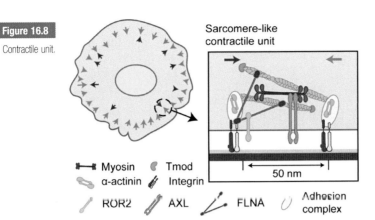

Figure 16.8

Contractile unit.

cells will adopt a soft surface state. Thus, the sensory systems appear modular (e.g. the rigidity-sensing module) and produce signals that are treated differently in different cell states (e.g. a fibroblast versus a neuron).

16.11 Stabilization and Maturation of Adhesions by Force

Once an adhesion is stabilized on a rigid surface, it will undergo maturation through the recruitment of various proteins, including vinculin, paxillin, and zyxin (see the adhesome article by Zaidel-Bar et al., 2007 for a longer list). This will allow the adhesion to sustain increased loads for increased periods of time. Under typical culture conditions, the adhesions will last for tens of minutes before turning over. However, photobleaching recovery experiments reveal that the individual proteins that make up the adhesion will turnover within tens of seconds. This means that the adhesion is a dynamic structure with some stable but many transient components. Consistent with that, the inhibition of myosin contractility after one hour on an adhesive surface will result in the rapid disassembly of adhesions, indicating that the continued contraction of the actin past the adhesion is needed for adhesion maintenance. Protein mass spectrum analyses of adhesion proteins from cells plated on fibronectin, with or without force, have shown that several hundred different proteins bind to the adhesions under contraction versus when myosin was inhibited. Thus, the association of many of these proteins to the adhesion complex is force-dependent, and most of their interactions with the complex are secondarily dependent upon protein force-dependent conformation changes. One idea is that the adhesion is a signaling center for the cell, and components will bind, be modified and then release from the adhesion as a means of facilitating this signaling. In other words, the adhesions will inform the cell through mechanical signals to confirm that the cell is in the correct environment for growth or differentiation. Adhesions are therefore important in maintaining cell viability.

16.12 Sensing by Actin Flow

In active cells, there is typically a flow of actin filaments from the cell edge to the central regions. Because many proteins in the matrix adhesions also bind actin, they will often attach transiently to flowing filaments. This will stretch the proteins and generate signals that then travel to other portions of the cell. Thus, the adhesions are mechanical signaling centers and all of the previous steps are to assure that the proper signaling apparatus is assembled at the adhesion site. In the case of fibroblasts on a rigid fibronectin matrix, these signals enable the cell to grow through both the activation of growth pathways and the inhibition of apoptotic pathways. The complex function of adhesion formation will be different depending on the cell type and the composition of matrix proteins. Nevertheless, looking at how fibroblasts respond to a rigid matrix provides a good example of how the cell can reliably respond to its microenvironment.

16.13 Analysis of Other Complex Functions

As we suggested at the beginning of the chapter, cell spreading and adhesion formation is only one of many complex functions that deserve a similar step-by-step analysis. Less than a decade ago, the description of this complex function would have been much simpler and with far fewer steps. New and different technologies have been applied to studying the spreading process, and these have provided many important insights into how the many aspects of this system are incorporated into the larger function. For example, the spreading of cells on RGD lipid surfaces with barriers provided an unexpected insight into the polymerization of actin from the early integrin clusters, and the subsequent force-dependent activation of the spreading process. Studies of early spreading on glass logically missed those steps because the RGD was immobilized and force was immediately developed on integrin clusters. Subsequent studies of the overall process of cell spreading showed that the formin involved in actin polymerization from the early integrin clusters was needed for normal early spreading on glass. Thus, a complex function like matrix-adhesion assembly needs to be analyzed at a single-cell level where the individual steps can be observed and the bases of those steps can be analyzed at a molecular level.

In terms of the general analysis of complex cell functions, many such functions are buried within the cytoplasm or even the nucleus and much more difficult to access than motility functions. Mutations in proteins that affect complex functions are often very hard to understand in the context of our current working models because they may be part of an unknown step in the complex function. Even with new technologies to follow individual fluorophores within a cell, single molecules are hard to isolate in a dense cytoplasm. The key when designing experiments to investigate a function involving steps, which cannot be observed or measured, is to have a rapid means of activating the function. In the past, that has been possible through temperature-sensitive mutation or washing out a specific inhibitor. New optogenetic technologies can increase the specificity, localization, and the speed of activation. This will make it possible address the major questions in the function within single cells and then build upon those results to address additional aspects of the function. As in a good mystery story, there are often obscure elements of complex functions, which once known, make it possible to understand what actually happens in the function.

16.14 SUMMARY To understand a complex function, it must be broken down into the many constituent steps, and it needs to be analyzed in a reproducible cell state. In this chapter we have discussed, as an example, the process of fibroblast spreading on matrix, and the development of matrix adhesions. This is an ideal example because the function starts reproducibly when cells in suspension contact a matrix surface. Fibroblasts typically test matrix-coated surfaces in several ways before they assemble adhesions and grow on those surfaces. Initial clusters of matrix-activated integrins recruit formins that polymerize actin from the integrin clusters to enable force generation on the clusters. If the matrix supports force, then the cell will spread rapidly in phase 1 and upon a rise in membrane tension it will transition to phase 2, which is characterized by the testing of surface rigidity through local contractions by sarcomere-like units. If sufficient force is generated by local contractions, then adhesion protein complexes will form. These will signal to fibroblasts to grow. At steady state, the frequency of the matrix tests will be governed by the rate of cell growth and mechanical stimuli. From a systems biology perspective, this means that the cell uses a number of emergent properties of local mechanical tools to test its environment. Each tool provides a different signal to the cell that in turn causes the cell to respond by testing another aspect of its environment. These tests are critical for the cell to know about its environment in a dynamic fashion. Thus, we have to think of cell processes as a series of decision points that then give rise to the final function. Just as in the assembly of a modular component for a car or other complex machine, there are a series of steps and tests that need to be successfully completed to make the final product.

16.15 PROBLEM

16.15.1 Paper and Discussion of Anonymous Hypotheses

Because an important aspect of research is the generation of testable hypotheses, it is useful for grad students to develop their skills through analysis of their peer's hypotheses and listening to their peer's analysis of their own hypotheses. This is best done anonymously; therefore, we recommend that the student hypotheses be presented anonymously to the whole class for discussion. Normally, we give the students 1–2 weeks to analyze the given paper and then generate a testable hypothesis to extend our understanding of a biological process in the paper. It is best to find another paper that presents an alternative view. All of the class should generate a one paragraph description of a testable hypothesis or proposed experiment based upon the paper. Everyone will serve as a jury to judge the proposed experiments that will be presented anonymously by the instructor.

To give you an idea of the types of hypotheses that are testable, the paper and hypothesis by Mor et al. (2011) serve as an example. Please read the paper and then consider how the hypothesis can test an important facet of the process.

16.15.2 Hypothesis

- Mor et al. (2011) reported corralled diffusion of mRNPs through the nucleoplasm occurred slowly. In addition, this diffusion occurred in a directional fashion down 'channelled pathways.'
- I propose a model by which directional flow of mRNPs from the site of transcription to export from the nucleus occurs via a 'source and sink' (the source being the continuous expression of RNA transcripts, and the sink being the nuclear pore complex (NPC)).
- By this model, newly transcribed RNA would yield a crowding pressure on previously transcribed RNA to move away from the site of transcription. The 'channelled pathways' described would serve as an escape route by which RNA could leave the crowding occurring at the transcription site.
- As Mor et al. (2011) also described, NPC export occurs rapidly. Thus, the NPC would act as the 'sink' preventing crowding of RNA at the site of exit from the nucleus.

- To test this proposal, I would perform an experiment similar to that done by Mor et al. (2011) with the additional step of adding a transcription inhibitor (such as actinomycin-D) and then measure the rate of directional flux of RNAs in the nucleoplasm that were transcribed before inhibiting transcription.
- With this experiment, one could determine if continued transcription at the 'source' is necessary to encourage the directional flux of RNA towards the NPC 'sink.'

16.15.3 Sample Paper for Student-Generated Hypothesis

See Yang et al. (2016).

16.16 REFERENCES

Changede, R., Xu, X., Margadant, F. & Sheetz, M.P. 2015. Nascent integrin adhesions form on all rigidities after integrin activation. *Dev Cell* 35: 614–621.

Dobereiner, H.G., Dubin-Thaler, B.J., Hofman, J.M., et al. 2006. Lateral membrane waves constitute a universal dynamic pattern of motile cells. *Phys Rev Lett* 97(3): 038102.

Dubin-Thaler, B.J., Giannone, G., Döbereiner, H.-G. & Sheetz, M.P. 2004. Nanometer analysis of cell spreading on matrix-coated surfaces reveals two distinct cell states and STEPs. *Biophys J* 86(3): 1794–1806.

Gauthier, N.C., Fardin, M.A., Roca-Cusachs, P. & Sheetz, M.P. 2011. Temporary increase in plasma membrane tension coordinates the activation of exocytosis and contraction during cell spreading. *Proc Natl Acad Sci USA* 108: 14467–14472.

Giannone, G., Dubin-Thaler, B.J., Dobereiner, H.G., et al. (2004). Periodic lamellipodial contractions correlate with rearward actin waves. *Cell* 116: 431–443.

Giannone, G., Dubin-Thaler, B.J., Rossier, O., et al. 2007. Lamellipodial actin mechanically links myosin activity with adhesion-site formation. *Cell* 128: 561–575.

Iskratsch, T., Wolfenson, H. & Sheetz, M.P. 2014. Appreciating force and shape – the rise of mechanotransduction in cell biology. *Nat Rev Mol Cell Biol* 15(12): 825–833.

Jiang, G., Giannone, G., Critchley, D.R., Fukumoto, E. & Sheetz, M.P. 2003. Two-piconewton slip bond between fibronectin and the cytoskeleton depends on talin. *Nature* 424: 334–337.

Mor, A., Suliman, S., Ben-Yishay, R., et al. 2011. Dynamics of single mRNP nucleocytoplasmic transport and export through the nuclear pore in living cells. *Nat Cell Biol* 12: 543–552.

Rossier, O.M., Gauthier, N., Biais, N., et al. 2010. Force generated by actomyosin contraction builds bridges between adhesive contacts. *EMBO J* 29: 1055–1068.

Schvartzman, M., Palma, M., Sable, J., et al. 2011. Nanolithographic control of the spatial organization of cellular adhesion receptors at the single-molecule level. *Nano Lett* 11: 1306–1312.

Wolfenson, H., Meacci, G., Liu, S., et al. 2016. Tropomyosin controls sarcomere-like contractions for rigidity sensing and suppressing growth on soft matrices. *Nat Cell Biol* 18(1): 33–42.

Yang, B., Lieu, Z.Z., Wolfenson, H., et al. 2016. Mechanosensing controlled directly by tyrosine kinases. *Nano Lett* 16: 5951–5961.

Yu, C., Law, J., Suryana, M., Low, H. & Sheetz, M.P. 2011. Early integrin binding to RGD activates actin polymerization and contractile movement that stimulates outward translocation. *Proc Natl Acad Sci USA* 108: 20585–20590.

Zaidel-Bar, R., Itzkovitz, S., Ma'ayan, A., Iyengar, R. & Geiger, B. 2007. Functional atlas of the integrin adhesome. *Nat Cell Biol* 9(8): 858–867.

16.17 FURTHER READING

Actin filament assembly: www.mechanobio.info/functional-cell-modules/actin-filament-polymerization/

Actin filament disassembly: www.mechanobio.info/topics/cytoskeleton-dynamics/actin-filament-depolymerization/

Actomyosin contraction: www.mechanobio.info/topics/cytoskeleton-dynamics/contractile-fiber/

Blebs: www.mechanobio.info/topics/cytoskeleton-dynamics/bleb/

Cell–matrix adhesion formation including initial matrix binding and activation of spreading/maturation; integrin activation and clustering: www.mechanobio.info/topics/mechanosignaling/cell–matrix-adhesion/focal-adhesion/focal-adhesion-assembly/

Cell–matrix interactions: www.mechanobio.info/topics/mechanosignaling/cell–matrix-adhesion/

Clathrin-dependent endocytosis: www.mechanobio.info/topics/cellular-organization/membrane/membrane-trafficking/clathrin-mediated-endocytosis/

Exocytosis/membrane tension and exocytosis: www.mechanobio.info/topics/cellular-organization/membrane/exocytosis/

Initial matrix binding and activation of spreading/maturation of focal adhesions: www.mechanobio.info/topics/mechanosignaling/cell–matrix-adhesion/focal-adhesion/focal-adhesion-assembly/

Integrin clustering and signaling: www.mechanobio.info/topics/mechanosignaling/cell–matrix-adhesion/integrin-mediated-signalling-pathway/

Nucleation of actin filaments: www.mechanobio.info/topics/cytoskeleton-dynamics/actin-nucleation/

Release of matrix binding: www.mechanobio.info/topics/mechanosignaling/cell–matrix-adhesion/focal-adhesion/focal-adhesion-assembly/#disassembly

Rigidity-sensing and myosin contraction: www.mechanobio.info/topics/mechanosignaling/cell–matrix-adhesion/focal-adhesion/regulation-of-focal-adhesion-assembly/

17 Integration of Cellular Functions for Decision Making

In the design of a robust device, it is common to specify all of the ancillary functions needed, and then determine how they interface with other functions in the physical device. The previous chapter discussed how to analyze a complex function in the cellular context by controlling cell phases, so that one can reproducibly analyze the same function in a known cellular context. In the design of a robust device, the designer determines what is needed; however, if the device (cell) was developed over many generations through robust selection processes, many ancillary functions or features may be hidden. The treatment of biological systems as robust devices should include an appreciation for all of the needed functions, the necessary links between them and the dependent parameters, such as ATP, which contribute to larger cellular activity. This complete treatment of the problem can enable one to analyze biological functions with a new perspective. The major difficulty in this approach is that we do not appreciate the complexity of most biological systems in that dietary changes, exercise levels, startle reflexes, and environmental factors such as temperature, bacteria, or viruses can all alter the normal balance of cellular homeostasis. In many of these cases, the organism has compensatory or adaptive mechanisms to minimize the trauma of an abrupt environmental change. This is part of the definition of a robust device. In this chapter, we will discuss how to dissect a primary function into a series of dependent functions and their governing parameters. Although many of those parameters are automatically controlled in cells, knowing that they are important may help to explain why certain perturbations cause unexpected changes in a given function.

'Systems Biology' has been defined as the study of how interactions between specific components of *biological systems* give rise to that system's function and behavior. For example, the proteins and the cell phase in clathrin-mediated endocytosis. Operationally, many have approached these questions by using protein expression, and interactomics data to generate models with a number of experimentally determined reaction constants. The problem that occurs in many cases is that the system is too poorly constrained and therefore the models are built with too many adjustable parameters. A rule of thumb in mathematical modeling is that you can model the shape of an elephant with four adjustable parameters and get it to walk with a fifth. Thus, it is important that any systems biology approach is grounded in the quantitative analyses of a constrained biological system wherein many of the parameters can be experimentally determined. In most cases, however, the quantitative measurement of one parameter leaves many unknown parameters. These gaps can occasionally be filled by considering the system as a grouping of functional modules, wherein the stoichiometry of components, the modules' response to signals, and their outputs to other functional modules, are standard. This is similar to the way we now think about computers or cars as composed of functional modules, where modular units can be analyzed separately for performance and replaced individually.

If we were designing a robust, self-replicating machine, we would use a systems engineering approach to develop a machine that performs a desired function in the best way possible. However, when considering a cellular system it is best to take a reverse engineering approach, as many principles of robust devices, such as the independence of major functions, are already incorporated into the cellular system (see Thomas et al., 2004 for more details). In the case of cells, we are merely trying to understand how an existing robust device actually works rather than design one from the ground up. Such an approach is possible because the many functional units in the cell are integrated through relatively simple signals. For example, the small G-proteins, hormone-activated tyrosine kinases or other general activators. Activation or deactivation of these signals will often constitute a change in cell phase. This is because many functions are altered coordinately, and can revert back coordinately, in response to the signal. Further, there are often standard signals generated by the functions themselves that modify other functions and constitute a modification of activities within a given phase.

Developing a quantitative understanding of cellular functions requires data acquired from a single-cell level. Analyzing the responses of many cells leads to averaged values that often do not reflect what actually happens in the cell. For example, stimuli may cause a function to be fully activated in 50% of the cell population. However, analyzing all cells at once may give the false indication that 100% of the cells were activated at 50% of their normal level. The output is the same, but the process or steps that go into achieving that output is different. When reverse engineering the cell to understand the steps of a function, it is actually the process that is important rather than the overall output. A relevant analogy in the marketing world is that when you would like to understand the best sales techniques, averaging the total sales is a poor indicator even when it is the total sales for a regional office. It is much better to know individual performance figures and how individuals perform in several different environments before trying to

Figure 17.1

This figure describes individual cellular responses to TNF-α that involve pulses of NF-κB transport to the nucleus. The enzyme cascade that links TNF-α binding to the receptor to NF-κB transport to the nucleus is described in (a). The fraction of active cells over four logs of TNF-α concentration is noted in (b). In (c) and (d), the concentration of NF-κB in the nucleus is reported for individual cells versus time for 10 ng/ml in (c) and 0.01 ng/ml in (d) of TNF-α. The average number of molecules of NF-κB transported to the nucleus in each transport cycle is shown in (e) as a function of the TNF-α concentration. The average response time is plotted as a function of the TNF-α concentration in (f). This is adapted from Tay et al. (2010).

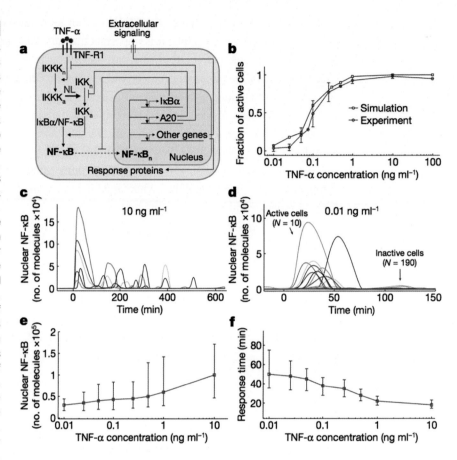

understand the method that they used. Similarly, knowing the level of activity within individual cells under a variety of conditions is critical for understanding how the system actually works. Such an approach is becoming increasingly achievable through newer methodologies, such as microfluidics and fluorescence assays at the single-cell level, which make it possible to monitor a variety of activities in many individual cells in a population.

In this chapter these principles will be applied to consider the relatively simple, single-hormone activation system, of tumor necrosis factor alpha (TNF-α) signaling. Here, TNF-α activates pulses of NF-κB transport to the nucleus and this subsequently causes a number of downstream effects, including the possible induction of a change in cellular phase, i.e. apoptosis. At high TNF-α concentrations of 10 ng/ml, the pulses of NF-kB in the nucleus last for 20–40 minutes before transport back to the cytoplasm and often repeat several times (see Figure 17.1). At a thousand-fold lower concentration of TNF-α (0.01 ng/ml), the pulses of NF-kB concentration in the nucleus are only slightly smaller than at the high concentration, but they are much less frequent. This means that NF-κB movement to the nucleus can be considered as a digital event, where it either will occur, or will not occur, but the amount transported is relatively constant. If the average rate of transport of NF-kB to the nucleus is measured for a large population of cells over several hours, then the transport rate is linearly related to the TNF-α concentration (over four orders of magnitude). The reason for this is that although the amount of NF-κB moved in each pulse remains the same, the frequency at which pulses occur is proportional to the TNF-α concentration. When the concentration of TNF-α is low, the pulses of NF-κB transport are very infrequent; however, when TNF-α concentration is high, these pulses are much more frequent, thereby producing a higher cumulative rate of movement of NF-κB to the nucleus. With a low TNF-α concentration, the interval between pulses may be hours, and over time this means less NF-κB is moving into the nucleus. When the expression of proteins downstream of NF-κB stimulation was analyzed, two distinct gene expression events were noted. First, a set of early genes was activated by the single pulses that occurred when TNF-α concentration was low, whereas a set of late genes was expressed only after multiple cycles of NF-κB transport occurred, which would produce a higher average concentration in the nucleus. Large amounts of quantitative data have been acquired from studies using mutations or knockdowns of important proteins in this pathway. Thus, using a quantitative modeling approach, it should be possible to address the question of how this hormone affects cell function.

17.1 Quantitative Analyses and Modeling

Although this signaling pathway is well described and a quantitative modeling approach can be applied, there are several factors that make it difficult to develop models with experimentally determined parameters. These difficulties arise due to the integration of upstream or regulatory signals from the environment, such as other hormones or movement of the cell toward a serum-starved phase. For example, at low concentrations of TNF-α, there is often a delay of 2–4 hours before a standard pulse of NF-κB transport to the nucleus is activated, and many other changes could occur in that time period that could affect TNF-α signaling. Further, the signaling pathway noted in Figure 17.2 has several components that receive inputs from other signaling pathways. There are many mathematical models describing the movement of NF-κB to the nucleus, and at the multicellular level there are very good fits of experimental data for high concentrations of TNF-α, but often they include rate constants that were adjusted to fit the data

Figure 17.2

TNF-α signaling pathway.

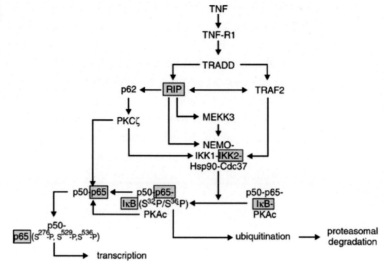

TNF-α acts by binding to its recepter TNF-R1. TRADD (TNFR-Associated Death Domain), RIP (Receptor-Interacting Protein), TRAF2 (TNFR-associated factor 2), PKC (Protein kinase C), p62 (PKC-binding protein), MEKK3 (Mitogen-activated protein kinase 3), NEMO (Nuclear factor-κB (NF-κB) essential modulator), IKK (Inhibitor of κB kinase), Hsp (Heat shock protein), Cdc (Cell division cycle), PKAc (Protein kinase A catalytic subunit), p50, p65 (Subunits of NF-κB transcription complex)

(Tay et al., 2010). The stochastic nature of responses in single cells, particularly at low levels of TNF-α, can be explained by possible variations in the concentrations of TNF-α receptor, and early components in the NF-κB signaling pathway. However, that postulate is very difficult to test and does not explain an average linear response with concentration. Furthermore, a highly non-linear amplification in the transport to the nucleus is postulated through the MEKK3 and IKK pathways. Even these postulates are insufficient to predict the amplitude and timing of the slow activation at low TNF-α concentrations with adjusted parameters. Thus, a simple system of one hormone and one major receptor cannot be modeled in detail because we have no understanding of the mechanism of integration of the TNF-α activation pathway at low TNF-a concentrations.

What this example serves to highlight is the general problem of understanding robust signaling phenomena in a physiological context. In this case, the cellular response is digital in that NF-kB is concentrated in the nucleus or not. Because the concentration of NF-kB in the nucleus will cause the activation of some transcription, those cells will be altered. Further, the lifetime of the proteins made as a result of the activation will be perhaps days; therefore, TNFα can have a long-term effect even at a very low concentration that would fit with many physiological conditions. At higher concentrations of TNFα, there will be multiple nuclear concentration events that cause transcription of additional genes and that will give a greater response. From the viewpoint of the tissue under severe stress, the high concentrations of the TNFα released can signal a toxic shock response that will preferentially kill growing cells. Thus, the signals from a single hormone interacting with a single receptor can cause different outcomes depending upon both the concentration of the hormone and the cell phase. This is a good example of a case where it is important to know how the cell can respond to both low and high hormone concentrations as a function of the cell state.

In thinking about how the nuclear transportation events linearly respond to the hormone over four orders of magnitude in concentration, it is particularly difficult to imagine a mechanism that would be linear over this range. Experimentally, the average level of NF-κB in the nucleus is linearly proportional to the concentration of TNFα, and because the nuclear concentration events are essentially the same, this means that either the fraction of cells that have a nuclear transport event drops linearly or that the time between concentration events increases linearly. It appears that the time between concentration events increases dramatically with decreasing concentration. Thus, there must be a mechanism for the cell to integrate the signal from very low concentrations of hormone over very long periods of time exceeding an hour.

In an attempt to better understand the molecular bases of biological processes, we will look at the established models of TNFα signaling, and some of the alternative approaches that can be proposed. In the model described so far (Tay et al., 2010), the responses that occurred at early times with high concentrations of TNFα were well-explained using the 16 adjustable parameters that were within expected ranges. To explain the low-concentration responses over hours, it was postulated that there was a variation in the number and phase of the TNFα receptor that accounted for the stochastic nature of the activation events. An alternative model would be that the activation of the receptor produces a relatively long-lived intermediate that can accumulate over time and produce a standard response. If low levels of TNFα gradually altered the number and state of the receptors through an integrative process that affected the inhibitors of the receptors, then there could be a cumulative effect that would lead to delayed activation. Thus, it is important to understand if the delays are part of a robust behavior that is able to integrate weak signals over long periods or are part of a purely stochastic process that is not really under cellular control. A more detailed mechanistic understanding could direct pharmacological intervention to affect the low-level responses that may be very significant in the organism because of the long-term effect of activation of transcription and therefore, it is important to test different models wherever possible. At a physiological level, the longer-term responses are important because a low level of the hormone may be present for hours and that could be important for the organism.

17.2   Modeling in a Cellular Context

To better understand these systems at the cellular level, new ways of thinking need to be applied. For example, the transport of NF-κB and many other factors such as YAP/TAZ or MRTFA (both involved in cell growth responses) to the nucleus will often involve proteins that are part of the cytoskeletal scaffolding complexes and these could be released upon a change in cell motility phase that is induced by other hormones or an internal cell clock that will activate motility on occasions. Understanding the full array of components that play a role in the given process can help reveal where signals are integrated into the system, and highlight other factors that could have major effects. Many of the all-or-none decisions that cells make require positive feedback systems for their regulation. These may be activated downstream of where signals are integrated, or involve positive feedback to the same signal over time. In either case, this adds another layer of complexity, as the feedback system may also depend on the accumulation of intermediates, which would need to reach a threshold level for the feedback to be propagated.

As in the case of TNF-α signaling, the idea of using linear modeling parameters with positive feedback and a stochastic input is not necessarily valid. This is because the integrative aspects of this, and similar systems, are more robust if there is an actual step in the process that allows the cell to control how the signal is integrated into process. For example, if IKK or one of the other early kinases would upon TNFα binding phosphorylate a scaffolding protein at low levels and multiple TNFα binding events would cause further phosphorylation that then would concentrate elements of the signaling cascade to lower the activation threshold over time. This would be instead of relying on stochastic variations in the concentration of TNF-α, or similar signaling proteins, alone. These or other possible methods of signal integration can only be approached if the system is well-controlled and there are single-cell assays of function.

17.3   Timescale and Integration

The time taken to integrate a signal and for the response to be carried out is another important parameter to consider when quantitatively modeling a cellular function. For example, it is important for growth signals to be integrated over time, at both the cell and organism level, so that reliable decisions can be made whether to grow or not. As we have emphasized before, most biochemical reactions occur rapidly. Thus, the time taken for these reactions to activate a process will be very rapid if there are no intermediate controls or additional steps.

For example, we know that many cellular proteins are stretched when a cell is stretched and stretching will result in growth. However, activation of growth by stretching is slow and appears to involve temporal integration of the stretch signal. Consequently, the time taken for the signal to be integrated in the cell system can be over hours and such timescales fit with a biological system that wants to make the right decision whether grow or not and has days to weeks to make that decision in the adult. It is therefore not obvious how mechanical signals can be converted to biochemical signals for integration over time. This is highlighted by the NF-κB system, where integration of the TNFα signal appears to occur over a period of hours.

Biological systems can integrate signals through cyclic processes that then are coupled to a cumulative parameter. Examples of this can be found in some of the growth hormone response systems, where there is usually a robust initial response, followed by a recovery period that involves receptor inactivation by endocytosis. Recycling of the receptors will give rise to a subsequent and more sustained

Figure 17.3

Timescale and integration in synapse.

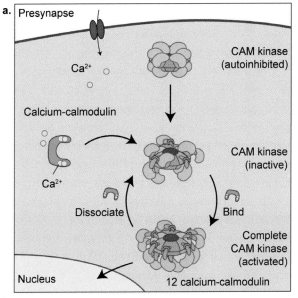

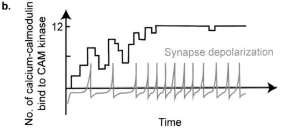

activation at a lower level. However, downstream enzymatic systems are often activated in parallel, and it is not always known if the signals from these systems are integrated as part of a feedback loop that regulates hormone response over time. Another factor could be that endocytosis of some hormone receptors causes them to be cleaved and the cytoplasmic domain of the receptor moves to the nucleus. This could result in a change in the cell phase such that a second round of activation of other receptors would cause a much different effect. Whether through this mechanism or another, it is clear that in most cases sustained hormone exposure produces different responses from a single burst of hormone.

One paradigm that may explain how a signal can be integrated over time is where the product of a cyclic process gradually accumulates, until a threshold is reached, and the signal is propagated. For example, depolarization of a synapse causes calcium entry, with a subsequent rise in both intracellular calcium as well as the calcium–calmodulin complex. In this case, the Cam Kinase needs to bind 12 calcium–calmodulin molecules before the complete complex can travel to the nucleus and stimulate the transcription and transport events that will cause consolidation of the synapse. Because the calcium–calmodulins slowly dissociate from the complex between depolarization events, the formation of a complete Cam Kinase:12calcium–calmodulin complex will only occur when events are frequent. Similarly, we know that the lifetime of many molecules in focal adhesions are very short but change with force and actin flow. If those molecules are modified in the adhesions (e.g. phosphorylated), then the concentration of the modified adhesion protein will increase over time as they bind, become modified and release. When many adhesions are present and the cell is actively contracting, the concentration of the phosphorylated species could be sufficient to trigger downstream processes, e.g. growth of fibroblasts, which is known to depend on adhesions and force generation. In general, many different cycles (binding and release with modification, phosphorylation/dephosphorylation, or nucleotide binding–hydrolysis–release) could be used to integrate a signal. For such schemes it is not the cycle per se that is important, but the frequency that molecules move through the cycle and become modified. With a rapid activation and slow turnover, there can be sufficient accumulation of a product that can trigger an important cell decision such as growth. If the frequency of activation events in a given cell is low, such that the degradation rate of the product is greater than the rate of formation, then the product concentration will never exceed the required threshold. This can explain why the levels of components in a given cycle do not change even though the cycle is fully activated and downstream effects are occurring. For example, the amount of protein in an adhesion complex is constant and yet the lifetime is very short and correlates inversely with growth.

17.4   Modeling Based on Composition vs. Environment

Given the problems that arise when modeling the comparatively simple scenario of the cellular response to a single hormone in a controlled environment, it is daunting to consider larger questions with the same approach. Yet, this is necessary in order to develop strategies that deal with cancer metastasis, regenerative medicine, and aging. Indeed, these scenarios are influenced by a vast array of factors, many of which play a role in the composition of cells, or changes therein.

One can begin to dissect the relevant processes by exploring the large amount of DNA/RNA sequencing data that has been acquired over the years. In the case of metastasis, for example, the identification of diagnostic mutations in critical oncogenes or tumor suppressors can provide a target for novel therapies. Much less instructive has been the effort to understand cell behavior based upon cell composition, and changes in behavior based upon changes in composition. The problem with understanding cell behavior based upon cell composition is that dramatic changes in behavior can occur during transitions in development or after hormone activation before changes in cell composition can occur. Thus, composition can reflect aspects of the prior cell history or previous trauma such as heat shock that have little to do with the current behavioral phase.

For example, when fibroblasts were grown on differently shaped fibronectin patches (triangular or circular) for 4 hours, over 4000 proteins were differentially expressed. In this case, the cells assumed a triangular or circular shape that corresponded to the shape of the fibronectin patch, but if replated on the same surfaces after several hours, they would probably be very similar in a number of assays. After 1–2 days on the different shapes, the cells from different shapes would show more differences because of the changes in protein content. Thus, it is a matter of degree in the sense that major modifications in composition will result in differences in behavior; however, many common functions will probably be preserved for a long time.

At least a few of the differentially expressed proteins, such as integrins, would be expected to cause changes in cell behavior. These are the proteins to look at initially to explain behavioral differences. Because both triangular and circular cells were of the same cell type, they should exhibit similar behaviors in a third environment after 1–2 days, unless one of the cellular behaviors involves modification of the environment itself. For example, those cells displaying a circular morphology after the first 2 days may secrete more matrix proteins, like laminin, when placed in the new environment. Epigenetic changes in the heterochromatin vs. euchromatin may also contribute to stable changes in composition when cells are placed in a new environment; however, these are typically slow to occur and to reverse. Cancer fibroblasts may form from tumors, and these can clearly be differentiated from normal tissue fibroblasts even after multiple passages *in vitro*. In most cases, after a relatively short period (1–3 days), changes in the microenvironment can produce dramatic behavioral changes in the cells due to changes in cell composition. Thus, differences in cell composition and cell behavior can over the long term cause changes in the cell responses to new environments, but similar cell states can be activated by defined stimuli in many different cell backgrounds.

Once a sensory protein is known to affect a given function, it is possible to design mechanistic studies that explore the steps in a function with or without the protein. A further way for a protein to affect a function is to alter the inputs to the function. As we will discuss below, many of the complex functions of cells need to have multiple inputs before the function will be activated, and in an otherwise homogeneous population of cells, only a fraction of the cells will activate the function. In the case of the cell cycle, which will be discussed in detail below, the decisions to move from one step to the next are not just dependent upon cell composition, but also upon other factors including nutrient level and mechanical forces. These factors will often influence the function in a stochastic fashion so that only a small fraction of the cells are activated despite most cells receiving essentially the same input. Thus, to be able to understand the roles of different stimuli in complex functions, it is important to control as many variables as possible and then follow a large number of single cells that are given the stimulus in close proximity to single cells that don't respond to the stimulus.

In organ regeneration, the ultimate goal is to control how cells differentiate. In the aging process, we want to restore how a young individual responds to a given stimulus or change the hormone composition to restore youthful function. In each case, we will need to alter the cellular response to its microenvironment, or control the microenvironment itself. However, it is particularly difficult to achieve this *in vivo*, especially when the properties of the microenvironment are already perturbed compared to normal, as is often the case in disease states or during aging. In some cases, hormone therapy can be used to alter hormone signaling *in vivo*, but even then the problems of altering matrix parameters, and the control by neighboring cells, remain. To be able to reverse a slow alteration in cell behavior is very difficult because of the many steps that have occurred prior to that point.

One issue worth considering is how bioinformatics can aid in plans to correct long-term alterations by providing knowledge and understanding of a cell's protein composition. Aside from the obvious benefit of knowing which diagnostic mutations may be present in an individual's genome, cell mRNA composition can help to understand how protein expression is altered. This would involve comparing the pattern of proteins expressed in the diseased cells relative to normal cells in a similar environment. Another bioinformatics approach that could be diagnostic of how factors in the microenvironment have changed in a given condition is to assess changes in the expression of protein isoforms. This of course assumes that the role of the protein in a given function is known. A problem with relying on bioinformatics to understand changes in the microenvironment of the cell is that cells can rapidly alter their own microenvironment. This means that after 1–2 days, the local environment can be totally remodeled by the cells, which would mean that the protein expression profiles could change as well. Therefore, although protein expression profiles can be diagnostic of certain situations that will respond to specific treatments, it remains that the information provided is diagnostic, and cannot form the basis on which to develop models for understanding cellular functions. Genetics has taught us that mutations of certain proteins can have dramatic effects on functions; however, this can occur when the protein is involved indirectly. Any correlation between the genetic information about a specific protein and a complex function can, however, stimulate interest in undertaking relevant mechanistic studies. This is analogous to having a critical part of a device in your hand and being able to identify individual defects in that part, but being unable to see how it actually fits and functions in the working device. For that, you would still need to disassemble, or reverse engineer, a working device.

17.5   Control of the Cell Cycle and Checkpoints

A critical question for cells in the adult is when to activate growth. In the case of a tissue at steady state, cells will die and others will need to grow to maintain tissue mass. Many organs maintain their size over the life of the adult organism. Outside of situations where there is major trauma, the loss of cells in these tissues is very slow and only causes a minor change in tissue size over time. However, some organs, like the liver, are fully regenerated after severe loss of mass. This means that some cells will need to be activated for growth before the tissue shrinks significantly. Similarly, in a regenerating tissue, some cells cease growing before the organ is fully regenerated.

In the case of cell growth, the critical point on whether to grow or not is at the G1/S boundary. At this point, the cell must decide whether to replicate the DNA during the S phase of the cell cycle. There are many conditions that must be met by the cell before it will commit the energy and resources into duplicating the DNA strands. Once the commitment is made to duplicate DNA, the machinery is activated and works for about 8 hours to complete the duplication and assemble the new chromatin into fibers.

One of the basic conditions that must be met for a cell to enter S phase is that the cell must first be in a growth phase, and in mature tissues most cells are in late G1 but not growing. There are other phases such as G0, senescence, and early stages of apoptosis where the cell will not be able to activate the replication machinery. Because G0 is often caused by serum depletion, it is a holding place in the cell cycle that can be readily escaped by simply adding the major serum hormones. Senescence is less well understood and, as the name implies, it increases in older cell populations or with stress from the environment such as disease. Thus, for cells to progress through the cell cycle, they will need the correct balance of hormones and nutrients, and to be in a relatively stress-free environment. Furthermore, mechanical factors, such as matrix geometry and forces from neighboring cells, are also important in the regulation of the cell cycle. Thus, it is possible to mechanically activate a population of cells to enter the growth phase, but otherwise the cells will normally be in a static situation.

A list of factors that will influence or stimulate the cell to enter the cell cycle (see Figure 17.4) can be developed; however, the problem of determining which of these factors contributes most in a crowded tissue environment must also be taken into account. For fibroblasts, the presence of the correct matrix and serum proteins, as well as a sufficient area in which to spread and develop force are important. This latter factor requires an absence of cell crowding. In tissues like skin, there are likely exocrine factors from neighboring cells that inhibit growth to prevent overcrowding. Furthermore, issues like tissue activity are also important, and can be manifested in cell or matrix stretching. Obvious issues, such as nutrient levels, infection or trauma to other regions of the organism, can also influence growth.

Mechanical factors at the cellular level, particularly cell shape, are critical for defining growth versus stasis in tissues. For example, in an epithelium, the loss of cells will cause the thinning of the epithelium as neighbors spread to fill the gaps. Because the cytoskeleton is dynamic, the cells can change to a thinner form and increase in area without changing tension; however, the cells typically respond by increasing the tension in the epithelium. Because contractile forces are the basis for altered tensions in tissues and forces are generated by the cells, the active response of increasing contractility is triggered by a change in cell morphology. As forces are increased on both cell–cell and cell–matrix adhesions, they are converted into biochemical signals at the adhesions. Some proteins like vinculin are common to both types of adhesions and are linked to growth. Thus, the homeostatic mechanisms must include a sensitive measure of changes in cell geometry (thinning in this case) as the cells would otherwise simply allow the change in morphology.

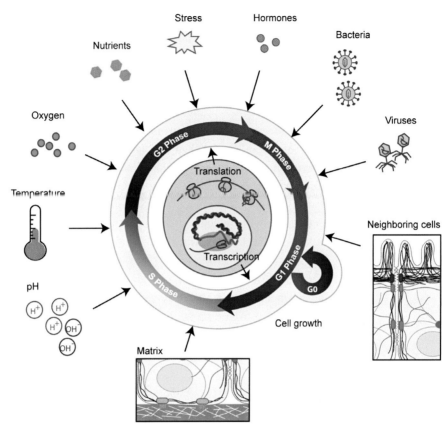

Figure 17.4

Factors that control cell growth.

To be able to predict what will happen in a given tissue during wound repair or other processes where growth is critical, experiments need to be performed in an *in vivo* context. Tissues like skin are the most accessible, and many aspects of wound repair have been investigated, but relatively few studies have been done to address the relative strength of signals that activate or inhibit growth. Although cells are robust, the relative influence of factors that control cell growth can change depending on age, temperature, nutrient concentrations and disease factors. Controlling for many factors in an animal context is very difficult but possible. Using a standardized trauma method, such as laser ablation, where a standard number of cells in a given tissue are removed, the rate of recovery can be measured under a variety of conditions to determine the relative importance of different factors in the repair process.

17.6 Integration in Cell Growth

There seem to be multiple proteins that act as integrators for the control of growth rates such as mTOR, Ras, AKT/PTEN, and YAP/TAZ. These proteins, which have inputs from many different sensory pathways such as metabolism or stress and certain cell-cycle regulators, are often linked to the growth of cancer cells and to regeneration processes. Ultimately, however, they must primarily be linked to morphological parameters, because they control the size and shape of tissues. It is not obvious how these proteins integrate mechanical forces into the biochemical signaling pathways to promote cell growth. The Ras, TOR, YAP/TAZ, and AKT/PTEN pathways have all been implicated in different cancers as well as in the control of organ size. When we look at these pathways, and the schematic diagrams that have been developed to explain them, we see that they show multiple inputs and a simple phosphorylation or movement to the nucleus as an output. Given the importance of these signaling events in the decision to grow or apoptose, and the reliance of cells on them to determine organ growth, there should be a number of checks and balances in place. However, studies investigating the mechanical regulatory factors are limited. This is because designing experimental conditions that provide the cells with a microenvironment that reflects a mechanically altered *in vivo* matrix has proven difficult. Proper controls are also needed to investigate how mechanical factors might be integrated in the tissue. Although many of the important scaffolding or integrative proteins are

Figure 17.5

Growth/size proteins.

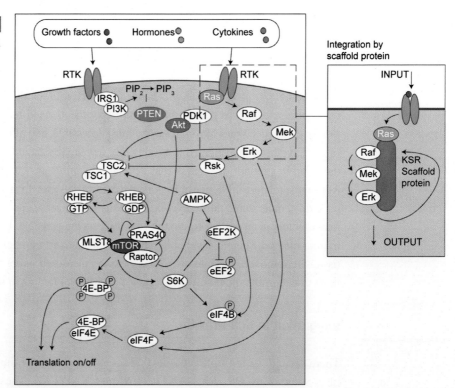

known, there is little evidence at the single-cell level to know the relative weight of different factors that could influence the decision to commit to growth.

17.7 Coordination of Multiple Functions

Another level of the systems engineering approach is to understand the dependent or linked functions that are activated or inactivated in concert with a given function. As the cell enters the S phase of the cell cycle, for example, the metabolic functions that produce deoxyribonucleic acids, as well as histones to complex with the DNA, need to be activated. Many events must be coordinated in order for these functions to be completed, and there will naturally need to be physical or biochemical links between these functions. As suggested in the design principles for robust devices, the simplest linkages are the best. For example, a fall in the level of thymidine triphosphate in the S phase is believed to be sufficient to increase thymidine kinase activity and synthesis. Metabolic pathways are linked in similar ways, such that the abundance or deficiency of specific metabolites will trigger the storage or production of another metabolite or enzyme such as in the glucose/glycogen pathway of metabolism. In S phase, there also needs to be a dramatic increase in histone synthesis that is coordinated with DNA synthesis and that can likewise be triggered by the excess uncomplexed DNA. More complex is the mechanism by which the cell completes the synthesis of a complete copy of the DNA. This requires that the synthesis is monitored and will not stop until all of the priming complexes are degraded. Thus, linkage of different functions for the completion of complex tasks can be by simple mass action as in the case of thymidine synthesis, but often more complex biochemical signaling mechanisms are required to develop the needed components.

Functions can also be robustly coordinated through physical force or other intensive parameters, such as concentration, that will influence many different functions. A simple example is membrane tension, which is an intensive parameter that can directly affect the coordination of endocytosis and exocytosis rates, as well as cell motility, and membrane resealing. It is relatively easy to understand how membrane tension can regulate exo- and endocytosis rates to keep the membrane surface area relatively constant. However, the very complex process of cell migration with cell polarization appears to be coordinated through membrane tension. Thus, the simple tension signal is transduced rapidly into changes in cell motile activity in very localized regions that then directs migration. This implies a very rapid feedback between the physical parameters and biochemical responses in motility. The BAR proteins that curve membranes can also sense membrane tension and they can also activate small G-proteins. With a transient increase in membrane tension, a BAR protein could rapidly cause a local increase in small G-protein activity. At another level, the physical size of cells is relatively constant but can be controlled, as in the case of liver regeneration. We currently don't understand the basis for this physical feedback, but it is fundamental for our understanding of many cell behaviors. Physical feedback mechanisms can be particularly effective in that they involve simple mechanical parameters and there are very finite boundaries to the forces and pressures that biological systems can generate. Thus, we need a much better understanding of the mechanisms whereby physical parameters are sensed in biological systems because that can enable a better understanding of the mechanisms of coordination.

17.7.1   Epithelial Wound-healing and Cell Proliferation

Another example of physical sensing is cell growth in an epithelial monolayer: the death of some cells naturally activates other cells to grow to fill in the gaps. In cases where cells apoptose or are acutely lysed, the neighboring cells close the gap rapidly by two modes of motility: (1) formation of a contractile ring near the apical surface, and (2) extension of lamellipodia into the gap at the basolateral region. Closing the gap is clearly important for maintaining the epithelial barrier function, but it is only a short-term solution. In addition, the cells in the monolayer should grow and divide to replace the dead cells potentially as a result of an increase in tension in the monolayer. This has been studied *in vitro* with monolayers of epithelial cells grown on elastic substrates that have been stretched (see Benhan-Pyle et al., 2015). In such situations, many but not all of the cells in the tissue will grow. In correlation with the stretch, YAP moves to the nucleus occurs and YAP is part of the Hippo/YAP signaling pathway that is involved in controlling the size of organs in flies and mice. Force on the cytoskeleton in the apical region of the cell can induce the movement of YAP to the nucleus, but growth is also related to metabolism through the protein complex that is sensitive to rapamycin (hence the name is Target of Rapamycin or TOR). This further emphasizes that many factors need to be integrated before the cell can enter the cell cycle and divide. Clearly, mechanical stretching of the cell is an important factor, because compression of the monolayer will cause cells to be shed that will result in their death. However, critical decisions like growth or death require many different inputs.

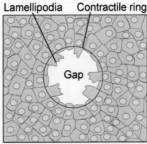

Figure 17.6

Epithelial wound-healing.

Lamellipodia   Contractile ring

Gap

Epithelial monolayer

17.8 Linkage to Protein Content

Because many cell functions are robust, and there is often excess capacity to fulfill their requirements, any given function will usually proceed without an optimal complement of proteins in the cell. There are cases, however, where this is problematic should the cell need to stop the activity of a relevant protein. As was discussed in Chapter 15 on protein synthesis and degradation, cells often use ubiquitinylation to target proteins for degradation. This serves as a reliable means to inactivate the protein. Degradation can be faster than synthesis of a new protein, and it is more reliable than post-translational inactivation of a protein, i.e. the most robust way to inactivate a critical protein is to simply degrade it. This is important, for example, in how the cell cycle progresses, because the targeted degradation of the cyclins is critical for the cell cycle to progress into each subsequent phase. The large number of E3 ligases that target specific proteins for degradation and the efficacy of E3 ligase inhibitors in treating numerous cancers highlights the importance of protein degradation in the control of cell functions.

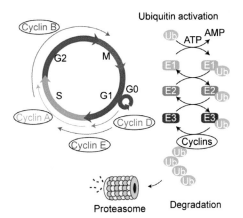

Figure 17.7

Linkage to protein content.

17.9 SUMMARY

In this chapter we have discussed how cellular functions are integrated into the larger context of the cell. Initially, we considered the simple case of a single hormone, TNF-α, which triggers pulses of reversible NF-κB transport to the nucleus. With lower levels of TNF-α, the frequency of these transport events decreases to levels of a single event every several hours, but always with a constant duration and amount of transport. This highlights the need for integrative mechanisms and the difficulty in modeling slow processes. Cycles of post-translational modifications and reversal can integrate through the buildup of a modified protein over time, to sense small changes in average forces, average calcium concentration, or other intensive parameters. Such integration is critical for deciding how much growth should occur to replace random cell death. Often only a small percentage of essentially identical cells will actually enter the cell cycle and grow. Similarly in cancer metastasis, many cells spread throughout the body but only a few become tumors. This highlights that the critical understanding of cell growth or other functional activities is not in the protein composition but in the essentially stochastic cellular response to environmental and physical factors. In the small cell space, there must be coordination (communication) between functions. Some of those coordination mechanisms are physical, and intensive parameters such as force or volume can help to link functions that would otherwise alter the parameters in opposite ways. Biochemical linkages are important when there are physically separated functions that need to be coordinately activated or inactivated simultaneously or in sequence. In the case of metabolism, the concentration of metabolites can be used to coordinate production and usage. Because the final morphology of the organism is tightly regulated, the coordination of many functions must be linked to sensors of morphology. To understand many important cell functions, we need to understand the mechanisms of integration over time, the digital responses to environmental stimuli including targeted degradation of proteins and how cell phase changes are controlled. Necessarily, these studies must follow critical transitions in cell state at the single-cell level. By treating these systems as complex machines with a finite number of complex, modular functions, an understanding of cell behavior is possible, albeit difficult.

17.10   PROBLEM

Take the Benham-Pyle et al. (2015) paper and develop a hypothesis to be tested, or use another more recent paper.

17.11   REFERENCES

Benham-Pyle, B.W., Pruitt, W.L. & Nelson, W.J. 2015. Mechanical strain induces E-cadherin-dependent Yap1 and beta-catenin activation to drive cell cycle entry. *Science* 348(6238): 1024–1027.

Tay, S., Hughey, J.J., Lee, T.K., et al. 2010. Single-cell NF-kappaB dynamics reveal digital activation and analogue information processing. *Nature* 466(7303): 267–271.

Thomas, J.D., Lee, T. & Suh, N.P. 2004. A function-based framework for understanding biological systems. *Annu Rev Biophys Biomol Struct* 33: 75–93.

17.12   FURTHER READING

DNA replication: www.mechanobio.info/topics/genome-regulation/dna-replication/

Fluorescence assays: www.mechanobio.info/topics/methods/super-resolution-microscopy/

Focal adhesions: www.mechanobio.info/topics/mechanosignaling/cell–matrix-adhesion/focal-adhesion/

Heterochromatin vs. euchromatin: www.mechanobio.info/topics/genome-regulation/chromatin/#heterovseu

Hippo-Yap/Taz pathway: www.mechanobio.info/topics/mechanosignaling/signaling-pathways/hippo-signaling/

Membrane tension, endocytosis and exocytosis: www.mechanobio.info/topics/cellular-organization/membrane/membrane-trafficking/

Microfluidics: www.mechanobio.info/topics/methods/microfabricated-technologies/#microfluidic

Receptor internalization by endocytosis: www.mechanobio.info/topics/cellular-organization/membrane/membrane-trafficking/clathrin-mediated-endocytosis/

Ras GTPases: www.mechanobio.info/topics/mechanosignaling/ras-gtpases/

Signaling pathways: www.mechanobio.info/topics/mechanosignaling/

18 Moving from Omnipotency to Death

Although the blueprint for the body is built into the DNA sequence, it is still mysterious how a single fertilized egg cell develops into the mature organism. In mammals, the microenvironment of cells is critical for their proper differentiation. This comes from the simple fact that killing cells in a tissue will not seriously alter the development of that tissue, i.e. neighboring cells rapidly fill the gaps both physically and functionally. In other words, the plan for differentiation of the cell is not cell-autonomous but comes from dynamic interactions of the cell with its neighboring cells and matrix. As the single cell becomes a multicellular structure, there are a number of very standard changes in shape that take place. These involve a dynamic mechanical feedback between individual cells and the surrounding tissue. At a protein level, stem cells express many different proteins, which allow them to respond to many different stimuli, and define their differentiation program accordingly. These stimuli will be received from the microenvironment as well as circulating hormones, and will activate a specific set of complex functions that allow the cell to progress to the next step in development. At later times the microenvironment provides mechanical as well as biochemical signals, making it very difficult to take cells out of the embryo and expect them to differentiate properly. Furthermore, mechanical changes in the tissue are critical for proper development, and can only occur through dynamic feedback between the cells, and an ill-defined organ shape parameter. There are a variety of tyrosine kinases that when mutated or deleted will cause major changes in the shape of an organ. Recent findings indicate that many tyrosine kinases are directly linked to mechanosensing, which can explain how they have been genetically linked to the pathways that define shape. As cells differentiate into one of the over 300 different cell types, the set of proteins that are expressed becomes limited. Further, there are major epigenetic changes in nuclei that involve the modification of chromatin to silence genes that are not needed for the specific cell type. Generally, the regularly transcribed genes will not be silenced, whereas those that are not expressed become silenced in heterochromatin. We will discuss how the differentiation process is controlled and how differentiation is manifested both at the cellular and nuclear level.

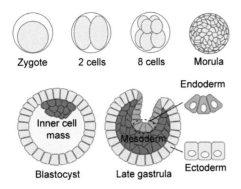

Figure 18.1

Development.

The earliest steps in the development process of many organisms are common. For example, in most organisms, after formation of the blastocyst, involution takes place to create a gut. It is not until relatively late in development that embryos differentiate from one another such that individual species can be recognized. Within the early steps of development, the ectoderm, mesoderm, and endoderm are created. These are the three major layers of tissue and are standard to many organisms. Once those layers form and seed the development of the critical organs, the detailed pattern of the organism begins to appear. Specific matrices are needed to support and define specific tissues and it is through the generation of the matrix that a true differentiation pattern is created. The relatively long life of the extracellular matrix (ECM) is likely to contribute to its ability to dictate many of the characteristics of a given tissue. The ECM will also define the size of an organ as well as the types of cells that can form in the organ or tissue. Thus, when stem cells are grown on a specific matrix, they will adopt the characteristics of the tissue from which the matrix was derived. Circulating hormones also play a major role in the determination of organ size (along with nutrients and other factors such as disease or trauma) and will even aid in aging after maturity. Aging is generally programmed and part of the plan. Development ultimately requires an orchestration of events that occur in multiple tissues. Coordination of these events involves factors such as hormones and nutrients, and this allows for development to be coordinated across the whole organism while individual tissues follow a local program that is dictated by the microenvironment.

18.1 Objectives of Development

The ultimate objective of development is to create an adult organism that can propagate both the species, and its DNA. However, there are also numerous intermediate objectives that relate to the steps needed to create a viable organism. Cell differentiation, for example, is important to create specialized cells that will perform certain functions in a dedicated manner. Because there are always environmental challenges, including disease and the presence of other predatory species, a need to adapt to meet those challenges exists. Adaptation can occur through mating and natural selection over multiple generations. However, once an organism is beyond the age for mating, it is a drain on resources needed by younger individuals. From an evolutionary perspective, the death of an aged member helps the species to survive by providing more resources for successive generations. Thus, it is generally believed that aging and death are part of the developmental program and aid in the ultimate propagation of the species.

18.2   Cell Differentiation

For a cell to move from being pluripotent to one of at least 300 different mammalian cell types, a major change in the expression pattern of the cell is required. This is a gradual process because there are several intermediate states before cells become terminally differentiated. How a cell progresses through a differentiation program depends on the signals received. These may be mechanical in nature and originate from the matrix environment or the neighboring cells. Alternatively, they can be biochemical and transmitted in the form of circulating molecules such as hormones. All of these factors contribute to the control of expression that will be manifest as a progressive change over time. Once a given path is chosen, there is a sense of momentum that enables a cell on a given path to continue along that path. Until recently, differentiation was believed to be essentially irreversible; however, many studies are now challenging that view as more methods are found to switch a cell from a fully differentiated cell type to a pluripotent cell type.

Naturally, this raises the question of why mammals cannot regenerate tissues, organs, or limbs to the degree that is observed with some amphibians, such as the axolotl. It has long been believed that the regeneration capacity of humans is limited to a few organs, such as the liver, skin, and gut; however, recent studies have indicated that kidney tubules can also be regenerated. Furthermore, until the age of 1 year, humans have the ability to regenerate the last knuckle in a finger if it is cut off, but that is relatively feeble compared to the axolotl, which can regenerate whole limbs throughout its life. It is often argued that the ability to regenerate limbs and whole organs would encourage more cancers; however, this has not been observed in amphibians.

The differences between mammalian and amphibian regeneration capabilities are difficult to understand. It is possible, therefore, that the low level of regeneration in mammals is a consequence of the stabilization of differentiation, which occurs within the nuclei of cells and is difficult to reverse. Nevertheless, there is turnover of cells in tissues and those dead cells are typically replaced rapidly. Practical issues such as scarring through the overproduction of collagen often inhibit regeneration in mammals.

In terms of the issue of why young tissues can regenerate better than old, it may have to do with the fact that the younger tissues have had less time to fully differentiate the chromatin for that tissue. If the chromatin is less stably differentiated, then it is easier for the cell to revert to a less differentiated state and contribute to the repair process when trauma occurs. This also means that younger cells will normally express a wider array of proteins and have many more functions that are typically active in them. Further, there is usually less matrix in younger tissues. Thus, a combination of less stable differentiation which will mean a more diverse set of proteins expressed and less matrix will make it easier for young cells to aid in tissue repair processes.

18.3 Early Development

As the organism moves to the blastula and later stages of development, there are more sophisticated morphological changes that are needed to create the body plan. The bending of the early embryonic epithelia into various shapes is an important early step. In *Drosophila* embryos, it is clear that the bending occurs over a series of small steps rather than through a continuous process (Figure 18.2). These steps will continue only until the tissue has bent sufficiently. During the process, the cells pull on their neighbors so that one surface (typically the apical surface) contracts, and then stops. The cell can then test if the contraction has been sufficient and, if not, another round of contraction and consolidation events will take place. Mutations that alter the overall process of bending appear to modulate specific steps in the process. Mutants of the *Drosophila* gene *snail* block contractile movements of individual cells and the epithelium doesn't bend. With mutation of the *twist* gene, there are contractions followed by pauses but during the pauses the apical area relaxes back to the original area. Thus, *snail* is involved in the contractile function whereas *twist* is involved in consolidation of the contractions. In addition, the curvature of the tissue is being tested by unknown components that will determine when the overall curvature is sufficient. The genetics has provided valuable insights into the steps in the process and clues about which proteins are involved in those steps. Still, much work is needed to understand how axes are determined in the process of tissue bending and how the emergent property of tissue bending is controlled at the single-cell level.

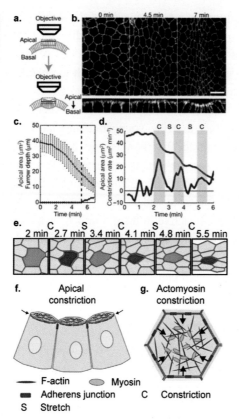

Figure 18.2

Apical actomyosin
constriction.

Several different paradigms involving contractions of boundaries along specific axes or geometrical rearrangement have been observed that can explain where the major cycles of contraction and testing could occur. For example, as discussed in the previous paragraph (Figure 18.2), a sparse actomyosin network on the apical surface pulls the edges on opposite sides of the cell together in steps (Martin et al., 2010). This is distinct from other systems where the major actomyosin complexes are aligned along cell–cell boundaries like a purse string and contract the boundaries thereby decreasing the apical area. Another mechanism is seen during dorsal closure in *Drosophila* where the apoptosis of cells in the contracting region and the contraction of the surrounding cells to close the gap is important (Teng et al., 2017). Despite all of these different paradigms, the contractions are intermittent, and there must therefore be a sensory mechanism that determines when the level of contraction is sufficient.

The establishment of embryo shape through bending is just one phase in development. As in the case of cell spreading (Chapter 16), the start of each new phase usually awaits the completion of the previous phase. In the case of development, however, each phase transition involves a coordination of functions in a monolayer that is made up of many cells. This coordination must take place within a reasonably short period so that there is no disruption to the development program. Each transition, then, constitutes a check-point similar to those in the cell cycle. Only upon completion of the previous step, or previous set of tasks, will the system be released to proceed to the next phase. Without such transitions, some processes would get out of synchrony with others when differences in temperature, metabolites, or other factors arise. Such transition points are useful because many functions commence specifically at the time of the transition and can therefore be studied reproducibly from that point. For proper development, it is important that many different organs coordinately progress from one phase to another and this can be accomplished with circulating factors.

18.4 Formation of the Three Tissue Layers

One of the earliest phases of development is the formation of three basic tissue layers, namely the ectoderm, mesoderm, and endoderm. This is a critical step in the formation of the organism as it is from within the mesoderm that cells will differentiate to form many organ precursors and the endoderm will go on to form the vasculature including the heart. During gastrulation, some of the cells break away from the two layers that become the endothelium and the outer epithelium to become mesenchymal cells. This epithelial to mesenchymal transition is not well understood in that there must be a signal to extrude the cells from the epithelium and then the cells must adopt a new behavior in the region between the cell monolayers. Because the behavior of some cancers is similar, there has been a lot of interest in characterizing the epithelial to mesenchymal transition. Both the transition to a mesenchymal cell phenotype and rapid growth characterize the cancer cell and the developing mesenchymal cells. Thus, one way to think about the cancer is to say that the cancer activates the path of normal mesenchymal development. Within the three layers, the patterns that develop will influence the neighboring tissues to enable coordinate differentiation. Thus, the characteristic behaviors in development that result in the formation of the body plan involve characteristic behaviors that can reappear in disease or regeneration states in the adult.

18.5 Heart Development

In the formation of the body, the formation of the vasulature is a relatively well-studied component. We will look at this developmental process in more detail to emphasize the many features that must be controlled for the system to actually work. The process of cardiac development has been studied in zebrafish embryos because of their transparency and ease of manipulation. For example, fluorescent cells can be incorporated into specific regions of the zebrafish heart which enables fate mapping. Here, stem cells expressing fluorescent proteins are injected into a defined region of the early blastomere, and will remain in the heart as it develops. Following the movement of these cells during the developmental process has allowed researchers to assess their location at each phase of cardiac development. Such phenomenological studies define the pattern of normal movements and cell divisions; however, if one wishes to determine the molecular factors that control movements, perturbations must be introduced into the system. An understanding of some of the steps in cardiac development can be derived from single-cell studies of the morphogenic movements during heart formation.

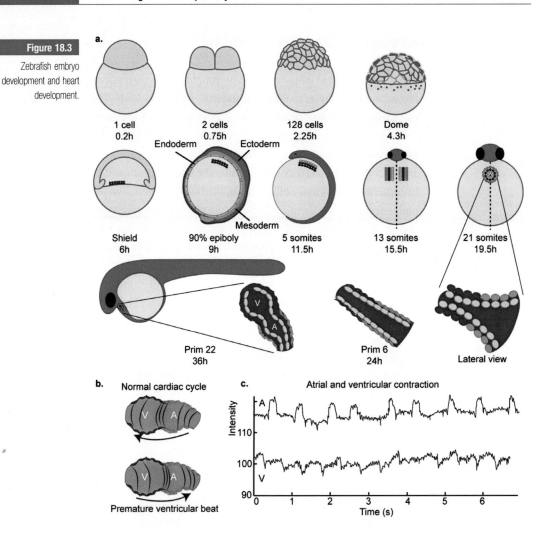

Steps of cardiac development have been revealed by careful characterization of the movements of the cell boundaries. Initially, cells receive cues from both neighboring cells in the monolayer and from the surrounding environment. A tubular arrangement of heart cells at each side of the organism will then fuse together and form a precursor organ that is beating and pumping fluid. From that point the tube lengthens and distorts so that it curves back upon itself and the four chambers of the heart start to differentiate from each other. These characteristic movements should be directed by cues in the neighboring heart tissue, including gradients of autocrine factors and the matrix produced by neighboring cells. By removing or mutating critical proteins in the process, it should be possible to define which proteins control the morphological changes, but the physical aspects of the matrices and neighboring cells must be considered. These shape changes require contractility and growth so that there must be coordination of these events such that there is feedback between the shape and the cellular behavior. It is logical to suggest that the initial signal to form the tube is critical and then the early vasculature will extend with growth while the heart region will expand rather than extend. Activation of the vascular growth is likely a global signal, but separate local programs are needed for extension versus expansion that are governed by feedback processes at a local level. In the heart, growth will expand the chambers and position-dependent mechanical signals can direct the cells to form the immature heart. Once the final shape is reached, further expansion is needed to develop sufficient force for flow. At an early stage, the ongoing process of feedback on heart growth from the vascular pressure must start and that will help to coordinate the system for the long term. Despite the fact that a dizzying number of events are coordinately reshaping the heart during development, the cells are logically using a standard set of functions to test their microenvironment and to properly sense signals from the environment as they dynamically form the organ.

18.6 Pattern Formation

A phase of development that has received considerable attention is the formation of anterior–posterior and dorsal–ventral patterns. In general, gradients of transcription factors or signaling molecules are needed to establish these patterns. Mutations in the relevant proteins, or alterations in the protein levels, will alter the patterns. However, determining how the gradients aid in the formation of distinct boundaries has not been a simple matter. *Drosophila* is an excellent example because initially it is a syncytium where soluble factors could diffuse readily between nuclei, and yet definite patterns of stripes are formed in the embryo.

A large amount of data on the *Drosophila* system has been acquired, and this has enabled various pattern formation models to be proposed. In 2013 alone, 13 papers were published that proposed various models. As there are many possible adjustable parameters, several of these models do fit the data. However, none of the current models fully explain how the abrupt changes in expression that often occur between neighboring cells can be predicted, when the system is perturbed in defined ways. Further, many of the adjustable parameters in the models do not have clear molecular mechanisms that underlie them. In most cases, differential equations are used to model the production of the intermediates that produce the final condition; however, most of the models exclude mechanical factors or other parameters that may play important roles in embryo development.

Figure 18.4

Drosophila embryo pattern.

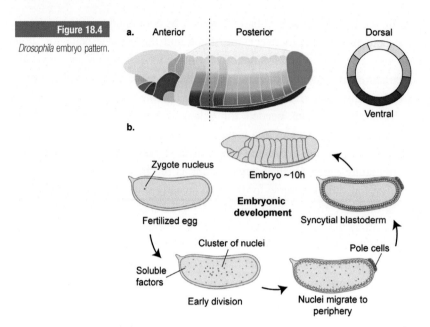

18.7   Size and Shape Matters

The size of organs as well as the shape is a critical issue for the survival of the organism. There are intrinsic factors in the genome that determine the size of the organs when extrinsic factors such as the availability of nutrients, the climate, and infections do not interfere. Further influences on the size of the organs come from feedback mechanisms in the whole organism, such as growth hormones or the needs of the organism. We can see the influence of intrinsic factors from experiments where cells are killed in a tissue. It is possible, with a few exceptions such as the pancreas, to remove 10–30% of the cells from early embryonic tissue using laser ablation and still have the tissue develop normally. Conceptually, this means that cells in a tissue are responding to signals that relate to the final size and shape of the tissue in an intrinsic process. Genetic screens have identified mutations in proteins that alter the size of organs through intrinsic processes. There appear to be two major signaling pathways involved in the control of tissue size, namely the TOR and the Hippo pathways. The TOR (target of rapamycin) pathway is generally involved in coordinating the cell-cycle growth pathways with metabolism and hence is receiving extrinsic cues. The TOR pathway involves a relatively large protein complex that integrates multiple signals to cause increases or decreases in organ and/or cell size. In contrast, the Hippo pathway, which includes several cytoskeletal proteins, appears to coordinate other factors from the cytoskeleton (perhaps including signals from the TOR pathway) to regulate growth, apoptotic rates and to some extent cell size. These pathways have related sets of protein components in both *Drosophila* and humans. In *Drosophila* the Hpo protein and Wts signal to Yki to send it to the nucleus, whereas in mice and humans the orthologs, MST and LATS kinases, signal to YAP. Thus, it is likely that similar transduction mechanisms are involved; however, we don't understand how cells integrate the intrinsic (cellular level) cues with extrinsic cues from the organism level such as hormones or nutrients. The Hippo pathway includes cytoskeletal proteins, and Yap's nuclear transport has been linked to changes in matrix rigidity, cell–cell strain as well as cytoskeletal actin assembly. Thus, it appears to be largely controlled by mechanical factors. TOR, on the other hand, is linked to metabolism and ribosome function and may have secondary effects on the mechanical factors or may act more directly on the Hippo pathway. Many studies are underway to determine how these pathways are controlled by physical parameters of the tissues including cell–cell junction tensions.

At the level of the whole animal, the growth hormones can increase overall size and cause proportional increases in the size of many organs, assuming sufficient nutrients are available. An extrinsic factor of dietary restriction will cause decreases in the size of many organs. At the level of individual organs, the size may be regulated by feedback from the whole organism that needs the organ to function at a certain level that will be dependent upon its size. The organism will not survive if the heart is too small to pump enough blood or, as in the case of liver, its metabolic functions are needed and there is regular feedback from the organism and the extrinsic cues determine the final size. In the case of the spleen, multiple spleens can grow to a total mass equal to a normal spleen. Organ mass can be increased by increasing cell size or cell number through increasing growth rates or decreasing rates of apoptosis. Thus, there are many control factors and a variety of feedback systems that enable the organs to maintain the proper size for the life of the organism.

At present, there are only models of how mechanosensation can occur on a cellular scale. As you increase in size from mouse to elephant the general pattern of the major organs does not change dramatically. Thus, there must at least a 100-fold range in the mechanisms to determine organ size. Models that describe the mechanisms behind the determination of organ size include many parameters, including gradients of surface protein expression, secretion of diffusible factors, and cytoskeletal protein mechanics. In all cases, there is a major question about the mechanism whereby physical parameters related to organ size can affect the molecular processes that control growth.

Figure 18.5

Different size and shape of
the tissues in humans and
Drosophila.

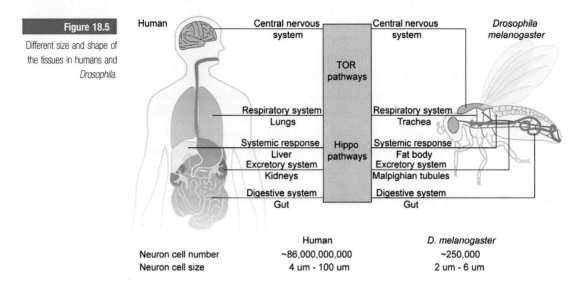

	Human	D. melanogaster
Neuron cell number	~86,000,000,000	~250,000
Neuron cell size	4 um - 100 um	2 um - 6 um

18.8   Manifestation of Differentiation State in Nuclear State

In many cases the differentiation state is manifested in the physical state of the nucleus. For pluripotent stem cells, the nucleus is very flexible and changes shape rapidly. This is also a hallmark of cancer cells and both cell types are able to express many different proteins, and consequently to respond to many different microenvironments. As discussed in Chapter 11, the major factor controlling the pattern of protein expression is the local environment. Typically, when there are only 2–16 cells growing in the embryo, there are minimal cues being generated in that microenvironment. However, as the embryo takes shape, more matrix is deposited and the cells receive more environmental cues. Multipotent cell nuclei have greater diffusional dynamics because the chromatin is not condensed, and this enables the rapid switching of expression patterns. Practical consequences of having more dynamic chromatin include the ability to easily express many new proteins as and when the developmental program changes. In the case of cancer cells, the ability to change expression patterns often enables them to differentiate to resist cancer treatment regimes.

As a tissue develops, the neighboring cells become more differentiated. At the same time, more matrix deposition occurs, and this produces greater stability in the signals received by the cells. With this stability, the nucleus is then modified to further inhibit the expression of proteins that are not required in the differentiation program or tissue function. Continued methylation causes chromatin condensation into heterochromatin. This is a gradual process and may take many cellular generations before it is complete.

18.9   Commitment in Differentiation

Depending upon the stage of human differentiation that is set by hormones as well as the cell's neighbors, a cell will proceed to differentiate toward one of the over 300 different cells types in the human body. As in the classic analogy from Waddington of a ball rolling down hill into one of many possible valleys, the cellular differentiation process involves a steady commitment to a finer and finer differentiated cell type. Movement to another type of cell as in moving to an adjacent valley is difficult and there should be many factors to discourage random or unwanted changes in cell type. As the embryo forms, there is an increasing level of extracellular material that will direct the cell to the path that it should take and hormonal signals will stimulate its proliferation and further development. These extrinsic cues are important, but there will also be internal changes in the cell chromatin that will serve to limit its expression pattern to that of the desired cell type.

18.10   Structured Nuclei in Differentiated Cells with Limited Expression

As the differentiation program continues, the cell will no longer need to have a wide variety of proteins at its disposal. Instead, differentiated cells require only those proteins within specific pathways, and these programs must be retained even when the cells proliferate as the result of local damage or cell death. To limit the expression of proteins, and thereby confer specialization, heterochromatin regions are preferentially buried in the center of the chromosome. The euchromatin portions are concomitantly moved to the surface of the chromosomes, and are often localized with active regions from other chromosomes. In organs such as the intestine or skin there is constant growth and shedding of differentiated cells. In these cases the stabilization of the differentiated state is manifested in the stem cell regions, which produce the committed cells that then further differentiate and die. This is in contrast to stable epithelia, such as that found in kidney or liver, where there is reason to have cells stably differentiated but ready to grow if their neighbors should die for any reason. Recent studies indicate that in adult organs there are subpopulations of adult stem cells that will preferentially grow in response to loss of their neighbors (Rinkevich et al., 2014). Thus, even in fully differentiated tissues, cells exhibit a range of commitments to that fully differentiated state, but the majority have very stable nuclear expression patterns and extensive heterochromatin.

18.11   Mechanical Aspects of Nuclear Organization in Development

Nuclear organization and the chromatin condensation patterns of differentiated cells are influenced by the mechanical properties of the microenvironment, which may in turn stabilize the cell type. The amount of lamin A, for example, has been correlated with both the rigidity of the matrix to which the cell is attached, and the rigidity of the nucleus. Although it is clear that lamin A expression is determined as a downstream response to the rigidity-sensing pathways, it is not clear why a rigid nucleus benefits cells on a rigid substrate. One hypothesis is that rigid substrates activate an increase in contraction of the cytoskeleton and this in turn creates stress on the nucleus. Growing cells on a glass surface, for example, has been shown to lead to flatter nuclei. Nuclear mechanical rigidity appears to reflect the rigidity of the cellular environment.

Various additional mechanical properties have been shown to influence nuclear dynamics, and thus differentiation programs. For example, the flexibility of the nuclear membrane can aid in the mixing of components. This is important to sustain chromatin dynamics and the movements of mRNA and other diffusing components. There is also some preliminary evidence that the nuclear volume changes during contraction of the cytoplasm and the creation of substrate tension. Small changes in the overall nuclear volume could cause dramatic changes in the volume of the spaces between chromosomes, and that could greatly alter diffusion in those spaces. Distortions of the nucleus would also put considerable strain on the contacts between chromosomes, which typically occur at transcription factories. The mechanical stresses on the nucleus could also be transduced into biochemical signals by the stretching of nuclear proteins. During differentiation, the physical effects of the tissue microenvironment can have major effects on nuclear morphology and function.

18.12   Chemical Changes with Differentiation

The chemical modifications of chromatin that result in the formation of heterochromatin are generally understood. However, a major mystery surrounds how these chemical changes progress in differentiation while changing dynamically during cell division. During prometaphase, the chromatin loops are mechanically consolidated in a parallel fashion, and a condensed chromatid is formed with euchromatin located in the outer portions of the loops. After mitosis, the process of decondensation would naturally proceed from the outer portions of loops inward, leaving heterochromatin internal. This results in the heterochromatin being preferentially condensed in the central portion of the chromosome and the active euchromatin regions on the surface. What remains unclear in this process are the molecular mechanisms that drive condensation, decondensation, and movement of euchromatin to the transcription factories. Actin and myosin 1 are known to localize to the nucleus, but actin filaments, and myosin 1 aggregates bound to filaments, have not been correlated with specific functions. Thus, it is still a mystery how mechanical movements in the nucleus are driven because biased diffusion mechanisms at the scale of chromatin loops appear to be insufficient to account for the dramatic morphological changes. Nevertheless, heterochromatin formation is a gradual process in development and full stabilization of differentiation can take years.

18.13 Repair Involves Dedifferentiation

In cases where there is major damage to a tissue, signals are generated to stimulate the growth of neighboring cells. The primary response will be to activate the growth of adult stem cells in the tissue. With severe trauma or disease, differentiated cells may also be stimulated to dedifferentiate. Recent studies on the formation of induced pluripotent stem cells have shown that removal of chromatin silencing modifications is the major factor in reversing differentiation. Such a process should occur naturally, especially if a tissue loses stem cells and needs to recruit them from neighboring tissues. Dedifferentiation has also been reported in the cases of smooth and cardiac muscles; however, there is a question about which cells are most likely to respond to a trauma signal and whether it could activate otherwise quiescent cancer cells. Thus, trauma can cause the dedifferentiation of some cells and their recruitment to a new environment.

18.14 Aging as a Part of Development

The study of development generally focuses on the early events that go from a single cell to an adult organism. However, additional phases are also programmed that cause the organism to age. Genetic factors play crucial roles in this programming, as evidenced by diseases such as progeria. In this disease, genetic mutations primarily disrupt the anchoring of lamin A to the nuclear membrane, which in turn disrupts nuclear dynamics and genome packaging. The consequence is that young individuals have severe symptoms that are commonly associated with aging. This implies that there is a general program for old age that is not related to wear and tear. Furthermore, experiments with heterochronic mice show that problems with muscle wound-healing in older mice can be corrected by serum factors from

Figure 18.6

Heterochronic mice.

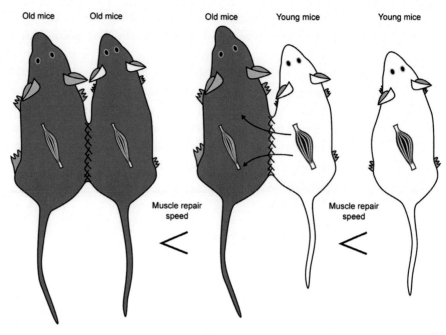

younger mice. Heterochronic mice are formed by sewing together the belly skin of two mice of different ages for several weeks. Because the circulatory systems of the two mice fuse over several weeks, factors from the younger mouse can naturally transfer to the older mouse, where they aid in wound-healing. Muscle repair in old mice is normally much slower than in younger mice, but in heterochronic mice, the muscle repair in the older mouse is greatly improved. This change is linked to the drop in oxytocin that normally occurs in older mice (Elabd et al., 2014). In the heterochronic mice, oxytocin produced in the younger mouse can aid wound repair in the older mouse. Furthermore, the movement of macrophages from younger mice to older mice in the heterochronic pairs has been attributed to improving brain repair in older mice. Changes involved with the normal aging process are clearly not beneficial and hasten the death of older mice, which is consistent with the hypothesis that death is part of the developmental process. Although it is debatable whether programming death for individuals beyond the productive age increases the ability of the species to survive environmental change, there is an increasing acceptance that aging is developmentally controlled.

Analyses of mutations in the *Caenorhabditis elegans* have shown that the lifespan of the nematode worm can be dramatically increased by fivefold when mutations in both the insulin receptor, and TOR pathway, are introduced. This is in contrast to the moderate increase observed when single mutations are introduced individually. Insulin receptor inactivation causes activation of the FOXO transcription factor, which regulates the expression of genes involved in stress response, detoxification, and metabolism. TOR, on the other hand, normally signals through S6K to regulate translation and overlaps with the changes following dietary restriction that include the expansion of mitochondria and increased aerobic metabolism. Wild-type insulin receptor, and proteins within the TOR pathway, are therefore linked to the aging process; however, it remains unclear how these proteins decrease lifespan. It is commonly believed, however, that the insulin receptor, TOR and FOXO are important parts of genetically encoded programs that cause adult organisms to age (reviewed in Gems & Partridge, 2013).

As in any robust biological process, it is hard to block the process of aging with a simple mutation. This is because aging and death are not controlled by a single factor. Instead, many changes occur, including a drop in oxytocin levels, a rise in TGFβ levels, the shortening of telomeres, the inability to repair matrix molecules such as elastin and many other factors. In humans, despite a large number of the population dying from common diseases, many people now live in excess of 80 years. Many of these aged people die from what are described as 'natural causes,' which result from a combination of weakened systems. It has also been suggested that age-related death results from the gradual activity of proteases. In that case, a build up of proteolytic byproducts produce toxic shock symptoms. As we understand factors that contribute to aging, such as the drop in oxytocin levels or the effects of dietary restriction, it is possible to have a better quality of life in old age that can equate with an increased lifespan.

18.15 SUMMARY To move from a single cell to an aging adult organism is a major feat. In early embryogenesis, there are stereotypical changes in the conformation of epithelial cell monolayers that shape the basic body plan. Those involve step-wise contractions of one surface of the monolayer and a mechanosensing of the curvature to trigger the end of the process. Similarly, the formation of multiple tissue layers is stereotypical and follows abrupt transitions from one type of cell behavior to another after checkpoints have been met. The local environment provides many cues, both biomechanical and biochemical, to elicit the proper response from the cells. When cells die during development, either from disease or trauma, their neighbors readily replace them. This ensures that the final organism is normal. As cells move from pluripotency, where they express many different proteins, to a differentiated cell with a more stable expression program, the nucleus changes dramatically. Here, the chromatin becomes modified to block the expression of many unnecessary proteins and there are dramatic mechanical changes in the nucleus that are likely related. In the background, hormones control many growth and/or death programs. Those external signals, along with the condensation of inactive chromatin into heterochromatin, help keep the expression program constant for all cells in a given tissue. Organ size is carefully controlled by YAP/TAZ, TOR and many integrative platforms that sense metabolic, mechanical and other factors to finally control cell size and cell number. Development does include aging, and the evolution of the species is aided by programming death after the individuals pass through reproductive age.

18.16   PROBLEM

A paper for hypothesis generation should be chosen by the Instructor from the past year's literature.

18.17   REFERENCES

Elabd, C., Cousin, W., Upadhyayula, P., et al. 2014. Oxytocin is an age-specific circulating hormone that is necessary for muscle maintenance and regeneration. *Nat Commun* 5: 4082.

Gems, D. & Partridge, L. 2013. Genetics of longevity in model organisms: debates and paradigm shifts. *Annu Rev Physiol* 75: 621–644.

Martin, A.C., Gelbart, M., Fernandez-Gonzalez, R., Kaschube, M. & Wieschaus, E.F. 2010. Integration of contractile forces during tissue invagination. *J Cell Biol* 188 (5): 735–749.

Rinkevich, Y., Montoro, D.T., Contreras-Trujillo, H., et al. 2014. In vivo clonal analysis reveals lineage-restricted progenitor characteristics in mammalian kidney development, maintenance, and regeneration. *Cell Rep* 7(4): 1270–1283.

Teng, X., Qin, L., Le Borgne, R. & Yoyama, Y. 2017. Remodeling of adhesion and modulation of mechanical tensile forces during apoptosis in *Drosophila* epithelium. *Development* 144: 95–105.

18.18   FURTHER READING

Cell–cell adhesion: www.mechanobio.info/topics/mechanosignaling/cell–cell-adhesion/

Cell–matrix adhesion: www.mechanobio.info/topics/mechanosignaling/cell–matrix-adhesion/

Chromatin: www.mechanobio.info/topics/genome-regulation/chromatin/

DNA packaging: www.mechanobio.info/topics/genome-regulation/dna-packaging/

DNA-templated transcription: www.mechanobio.info/topics/genome-regulation/dna-templated-transcription/

Genome regulation in stem cells vs. differentiated cells: www.mechanobio.info/topics/genome-regulation/

Hippo signaling: www.mechanobio.info/topics/mechanosignaling/signaling-pathways/hippo-signaling/

Mechanosignaling: www.mechanobio.info/topics/mechanosignaling/

Microfilament motor activity: www.mechanobio.info/topics/cytoskeleton-dynamics/contractile-fiber/

Physiological relevance of mechanobiology: www.mechanobio.info/physiological-relevance-of-mechanobiology/

19 Cancer versus Regeneration: The Wrong versus Right Response to the Microenvironment

Due to the presence of mechanisms that regenerate portions of tissue after disease or damage, it is possible that excessive or inappropriate growth may occur. In humans, when a portion of tissue grows excessively, it is often described as a cancer; however, there are many forms of excessive growth that are limited and not life-threatening. At the heart of a cancerous growth is the inappropriate response of the cell to its microenvironment. Sensory modules that determine mechanical aspects of the microenvironment contribute to cancerous growth. In particular, the depletion of rigidity-sensing modules enables cells to grow on soft surfaces, much in the way that loss of sensing in self-driving cars causes inappropriate movement. There are common features of inappropriate growth that are considered in the rubric of cancer biology and many books are written on the details. Here we consider growth in the context of cell states and the response of the cell to its microenvironment. By definition, cells in cancers have ignored normal growth controls and have grown excessively. Because tissue growths need vascularization to grow rapidly beyond millimeter dimensions, they are normally self-limiting. Further, telomere shortening will limit the number of divisions of cells. Thus, these two factors act as safeguards to keep most tumors small in size and benign. When cells move to states that continue growth, elongate telomeres and promote vascularization, they can be life-threatening. Even then, surgeons can remove the growths and the normal cells can often heal the area and restore function. What is life-threatening is the ability of inappropriately proliferating cells to disseminate to other parts of the body and to differentiate and grow in many different environments. Then, the surgeon cannot remove all of the tumors and the multiple, differentiated cells will not be killed by a single therapy. Furthermore, the act of killing the tumor cells often elicits a general repair response in the body that will promote further growth of tumors. The lessons of Cancer Biology can tell us a lot about biology in general.

In the previous chapter, we discussed normal development and the process of differentiation. When a loss of cells occurs during normal development, neighboring cells typically divide to fill the gaps. This is important, because with the loss of cells in a tissue, both physical mass and functional performance are lost. In addition, macrophages and tissue-degrading enzymes that catabolize the dead cells produce byproducts that can produce shock or trauma responses in the organism. However, in most cases, loss of cells in a tissue does not pose a problem as the developmental program is robust to the loss of cells. In the embryonic stage of development, whole segments, including portions of fingers, may be regenerated. However, as the embryo matures to a juvenile, this capability is quickly lost. When damage occurs in adults, some cells can revert to a less-differentiated phenotype, move to the damage site, and grow in the damaged area to repair it. However, the repair process becomes slower with age. In cases where the growth pathways are activated inappropriately they can lead to the development of cancer, which even in otherwise healthy individuals can be fatal. Inappropriate tissue growth will occur throughout the life of virtually all individuals. In fact, most readers of this book will have multiple minor growths in their body. Fortunately, the vast majority of these are not life-threatening and are self-limiting. What we will focus on in this chapter are the differences between normal regeneration, self-limiting cancers, and life-threatening cancers.

19.1 Benign vs. Cancerous Tumors

The uncontrolled proliferation of cells that leads to cancerous growth can arise from the activity of many different enzymatic pathways. Due to the vast number of elements involved, and the fact that they will often influence each other, clinicians have so far been unable to cure many cancers. This is a major problem because of the increase in occurrence of cancer with age and the greater longevity of the population. The major consideration here is what factors distinguish a severe malignancy from a benign growth or even a repair process. Seven characteristics of malignancies have been proposed: (1) sustained proliferative signaling, (2) evading growth suppressors, (3) resisting cell death, (4) enabling replicative immortality, (5) inducing angiogenesis, (6) activating invasion, and (7) metastasis (Hanahan & Weinberg, 2011). In most cases mild forms of inappropriate cell growth can be treated, either through surgery or chemotherapy, and will therefore not become life-threatening. However, when the growth displays most or all of these seven characteristics, treatment is difficult. It is then that the cancer is likely to be life-threatening. In most cases these characteristics can be induced by a genetic mutation, even a single point mutation; however, again, and contrary to popular belief, the presence of a causative mutation does not necessarily mean that a cancer will be life-threatening. In fact, if there is a single causative mutation, then the cancer will likely be easily treated by a therapy targeting the activity of the mutated protein. It is the accumulation of multiple abnormalities that creates a persistent growth state even in other environments that can be life-threatening through metastasis.

Figure 19.1

Benign tumors and malignancies. Diagram of the pathways that are altered by elevation of p53 levels following DNA damage or trauma. Adapted from: Araki et al. (2015).

(a)

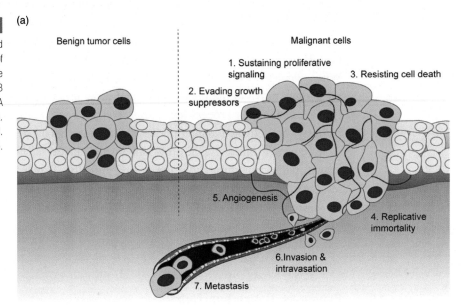

(b)

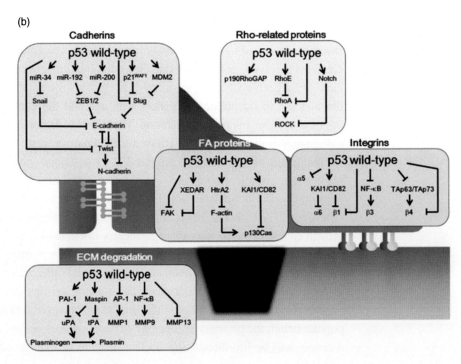

Although underappreciated in cancer generally, the microenvironment can play a major role in cell growth in addition to genetic or epigenetic changes. However, understanding the influence of the microenvironment is complicated by the fact that unknown mechanisms can change what is otherwise a normal environment to a growth-promoting environment. It is believed that tumor-associated fibroblasts can assemble a stiffer matrix that will promote growth of the tumor. However, those fibroblasts are clearly affected by the tumor cells and the metastatic tumor cells can induce cancer fibroblasts at new sites. This raises many questions of which came first, the tumor cell induction of the fibroblast or injury-induced fibroblast changes that favored the growth of the tumor cells? In either case, the critical issue for the growth of a serious tumor is that there is enough time for the tumor to grow and reach sufficient size that it will be able to metastasize.

Severe cancers typically involve cells that not only display many of the above-mentioned characteristics, but also the ability to differentiate in new environments to enable metastatic growth. The presence of mutations or cellular alterations that facilitate cancerous growth does not necessarily mean that they have lost the ability to differentiate in different environments. In fact, the ability to maintain growth without normal environmental growth stimuli allows cells to respond to the stimuli of their new environment, and undergo differentiation. In other words, the metastatic cells that grow in a new environment will often be differentiated from the original tumor cells. Studies of the passage of human breast tumor cells in mice have shown that they can differentiate to target tumor formation in either bone or in lung. This reinforces the concept that tumor cells can evolve, but many of the characteristic changes that enabled the growth of the primary tumor will help metastatic cells to form tumors in new environments even though the growth signal pathways may be different.

19.2 Rigidity-independent Growth from the Loss of Mechanosensory Modules

The rigidity-sensing contractile modules that were described in Chapter 17 constitute mechanosensory machines that tell the cell about its immediate environment. When those machines are depleted in cells, the cells will be transformed to grow on soft surfaces. This can be understood in terms of different cell states and the involvement of different cell machines in each of those states. We suggest that the normal cell state will activate rigidity-sensing contractions upon contacting new matrix molecules and if the surface is soft, those mechanosensing modules will activate growth-inhibitory pathways that can cause anoikis (cell apoptosis on soft surfaces). As noted before, those modules contain many cytoskeletal protein components, such as tropomyosin 2.1, myosin IIA, α-actinin, and filamin. When any of those components are depleted in cells, the mechanosensory pinching modules cannot form and the growth inhibitory signal will not be generated on soft surfaces. Further, the cells appear to be in another state and generate much greater forces on soft surfaces after the loss of rigidity-sensing modules. This all fits well with the idea that cell states are controlled by sensory modules and the loss of a module can cause the growth state.

19.3 Depletion of Mechanosensory Machines is not Reversible by Drugs

Because the rigidity-sensing module generates the signal needed to control growth on soft surfaces, the depletion of modules will remove a normal break on excessive growth. Adding back the depleted proteins will restore rigidity-sensing and inhibition of growth on soft surfaces. However, small-molecule drugs or antibodies will not be able to restore rigidity-sensing and that pathway of growth inhibition. Until the growth state that is activated by loss of rigidity-sensing modules is characterized, the only way to inhibit growth is to target general pathways of growth that will affect normal cell growth as well.

19.4 Many Different Mutations can cause Module Loss and Uncontrolled Growth

If the loss of a modular machine follows the depletion of any one of several different proteins, then many different mutations could cause the depletion of modules. In the case of the rigidity-sensing module, at least four different proteins can be depleted individually to cause the loss of rigidity-sensing and growth on soft surfaces. Because of the complexity of the module, it is hard to imagine how the module could be repaired when one of the components is missing. Thus, many widely different protein mutations could cause growth on soft surfaces, resulting in a disease state. This would cause confusion about the cause of the disease because components of a module can have quite different functions and the loss of binding to other modular components may be sufficient to block module assembly and function. Thus, if cancerous growth is caused by the loss of a sensory module, it could be linked to a wide variety of mutations and virtually impossible to block by drugs without significant side effects.

19.5   Growth States after the Loss of Control Modules

The growth of cells on soft surfaces is not sufficient to cause a serious cancer, but it serves to illustrate how growth states can get out of hand in the absence of proper growth control signaling mechanisms. There are several obvious growth control mechanisms, in addition to soft surfaces, including the absence of ligands, compressive pressures in tissues and the contact with normally inhibitory signals such as matrices or necrotic factors. Nevertheless, growth on soft surfaces is a common feature of many different cancers that have originated from epithelial cells transitioning to a mesenchymal state. In the mesenchymal state, matrices and many receptor tyrosine kinases can stimulate growth. However, one type of alteration will often predominate in a primary tumor, such as a mutation in the EGF receptor due to the tissue environment or other factors. The primary tumor can shed many cells into the bloodstream that will spread throughout the body. Treatment of the primary tumor with a specific inhibitor can often cause it to shrink or disappear, putting the person into remission. However, the tumor cells in other parts of the body will also lack the environmental growth control mechanism and are likely to find other stimulatory signals (matrices or other hormones) in the new regions of the body that are not sensitive to EGFR inhibitors. They would then be able to live through the treatment that killed the primary tumor and proliferate, giving rise to secondary tumors that would need to be treated differently. Because primary tumors can shed billions of cells, many different tissue environments will be sampled and only a few unlucky hits are needed to cause serious problems. Thus, in the absence of proper regulatory environmental sensing systems, the tumor cells can find many different environments for survival outside the primary tumor. Adaptation and eventual growth of the metastasized tumor cells is then possible.

19.6   Repair Mechanisms and Trauma can Stimulate Metastatic Growth

When a tissue is damaged, whether by a normal trauma or the death of cells after cancer treatment, a trauma response is initiated that triggers the recruitment of pro-inflammatory cells. Regulatory T cells often follow and they secrete factors that aid in tissue repair, such as amphiregulin. Growth factor levels within the serum also increase following a bone fracture, brain injury, or side effects of some chemotherapies. The mechanisms that promote hormone release upon trauma are not fully understood. Although these responses are part of the normal repair pathways, they can also stimulate growth of metastatic cells in new environments. This is an increasing concern about many of the general cancer treatments that result in significant damage to normal tissues. Obviously, it is very difficult to trace the cause of a secondary tumor, but one possible source of growth stimulatory factors is trauma or severe treatments as in cancer chemotherapy.

19.7   Fibrosis in Aging and Regeneration can cause Growth

The replacement of a large number of dead or damaged cells following severe trauma or disease often results in an imperfect repair response. Residual debris may build up, or excessive strains from the tissue may distort the matrix, making the repair processes difficult. A common response to a damaged region that cannot be repaired easily is to wall it off in a foreign-body response by collagen deposition. Such fibrosis can also stimulate metastatic cells to proliferate because the fibrous deposits are often rigid and will support large forces at matrix contacts. This is consistent with suggestions that cancers form in regions of constant irritation or damage (e.g. lung cancer is proposed to form from tobacco smoking). A combination of the regenerative response and excess matrix deposition can tip the balance to cancerous growth when a control system is lost or compromised.

19.8   Which Mutations Drive Metastatic Cancers?

Although sustained proliferative signals appear to be important, there are sufficient data on severe metastatic cancers to know which mutations are most commonly present. Therefore, it is logical to consider cancer development from the viewpoint of which proteins and pathways correlate with severe disease. Many cancers have mutations in p53, which is involved in DNA repair pathways and is a general control system that will block growth if DNA damage occurs. p53 is normally degraded rapidly, but when DNA damage or other trauma occurs, degradation is blocked and the higher levels of p53 enhance repair mechanisms. Higher levels of p53 also enhance cadherin synthesis and inhibit migration, which suppresses EMT and migration (Figure 19.1). Thus, p53 has multiple roles in suppressing metastasis and mutations in p53 could act in a variety of ways to support tumor growth. Another common protein mutated in cancer is PTEN, which is a phosphatase that degrades PIP3, a stimulatory lipid. The third common set of alterations in cancer involves the receptor tyrosine kinases (RTKs; mutations in the kinases and/or changes in expression levels). The RTKs feed into only a few growth pathways (most commonly RTK activation causes lipid phosphorylation (PIP3), recruitment of Grb2/SOS with the activation of Ras-ERK). Additional common mutations involve the cancer suppressor proteins that include many of the cytoskeletal components of the rigidity-sensing machinery and force generation (Wolfenson & Sheetz, 2017). From the most common mutations in cancer, it is clear that both growth activation and apoptotic or growth suppression proteins are altered in cancers.

19.9 Sustaining Proliferation by RTKs

As we have discussed, tumor cells are able to sustain growth inappropriately and both genomics and logic indicates that alterations in the levels of EGF/PDGF, or their receptors, are linked to cancerous growth. However, inhibitors of those receptors have not been effective over the long term in curing cancer or blocking cancer relapses. There are several possible explanations for this, including an increase in drug transporter activity, or compensation whereby the tumor cells will switch to other hormones, or other growth pathways. In these latter scenarios, cells from the original cancer have an increased ability to grow irrespective of the growth signal they are receiving. This was recently observed, for example, in lung cancer cells that were initially sensitive to the EGFR inhibitor erlotinib that caused shrinkage of the primary tumor. Unfortunately, tumor cells became resistant to erlotinib over time by responding to a Src-integrin dependent growth pathway (Kanda et al. 2013). *In vitro*, cells that were resistant to erlotinib became sensitive to the drug when Src activity was inhibited. Thus, some tumor cells did not succumb to the initial treatment because an alternative pathway was activated. This involved a change of state to matrix-activated growth. Because RTK inhibitors have generally been much more effective on primary tumors than on metastases, there appears to be a general change in tumor cells that favors growth even with other proliferative signaling pathways.

19.10 Downstream Signaling Pathways Involved in Growth

From a somewhat superficial level, many RTKs activate similar growth signaling pathways and metastatic tumor growth could result from multiple activation signals, such as different growth hormones or growth-stimulating matrices. In many examples such as above, the characteristics of the cell that enabled unregulated growth such as telomere elongation and suppression of growth inhibition enabled tumor expansion in new environments with a new growth signal. At a biochemical level, the activation of RTKs or certain integrins results in the phosphorylation of many proteins in the vicinity that then causes the formation of Grb2-SOS gels through the crosslinking of the many phosphotyrosines on proteins in the region. As a result, Ras is activated by SOS, which starts the cascade through Raf and Mek to cause growth (Figure 19.2). Another pathway that is activated by RTKs involves PI3-kinase that produces high local concentrations of PIP3. In turn, SOS and other proteins in the activating pathways bind to PIP3, which then further favors the formation of local signaling centers. As we have emphasized, the hormone activation of growth is a complex, multistep function, and the RTK activation process not only sets in motion activation of Ras and PIP3 pathways, but also other steps including the reversal of the normally controlled growth state as in the idea of term limits.

19.11 Steps that Could Inhibit Further Growth

In considering the mechanisms of action of the different growth hormones, we need to look beyond the phosphorylation reactions that they catalyze. EGF addition stimulates cell motility, with a rapid extension of the cell edge normally occurring within 1–2 minutes. Following ligand binding, the EGF receptor is endocytosed and cell motility returns to prestimulation levels. After endocytosis, liganded EGFR can be cleaved and the cytoplasmic domain can move to the nucleus. Thus, EGF and other hormones set in motion a number of downstream pathways. This has created confusion in that it is unclear which of these pathways are most important for normal inhibition of growth and are altered in cancerous

Figure 19.2

Proliferating signals.

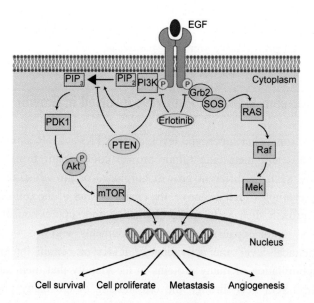

growth. Further, a single stimulus of hormone is not likely to sustain growth and the number of receptor activation-inactivation cycles needed for growth is unknown. The receptor tyrosine kinases are one example of the complex pathways that can inhibit growth. We need much more information about the relative importance of the different possible pathways of growth inhibition to design effective treatments.

19.12 Evading Growth Suppressors

The stimulation of cell growth often occurs in conjunction with the activation of growth suppressors that help to prevent further growth as in the concept of term limits that was discussed earlier. For example, the activation of PI3-kinase (PI3K) (see Figure 19.2) causes the phosphorylation of PIP2 to PIP3, which is needed for activation of several pathways including PDK1 and the Grb2-SOS pathways. Production of PIP3 also causes the recruitment of PTEN phosphatase that will dephosphorylate PIP3 and will turn off the growth pathways. This means that the levels of PIP3 are elevated for only short periods. PTEN is depleted or mutated in many cancer cells, and its restoration slows the growth of the cancer cells. The simplest view is that the loss of PTEN activity results in higher levels of PIP3 for longer periods, which favors increased growth rates. Alternatively, inhibition of PI3K could cause a decrease in PIP3 level, but inactivating PI3K does not cure cancer. At best, drugs that inhibit PI3K will stop cell growth. Because the activation of PI3K is an important part of repair processes, the inhibition of PI3K will generally inhibit repair processes and normal growth pathways. However, once the drug is removed, the cancer cells are able to grow once again. An important aspect of the mutations in PTEN associated with cancers is that they compromise the function of the protein but do not remove it from the cells. There are many other growth-suppression pathways that could control growth in the absence of PTEN; however, if PTEN is present, then the other pathways may not be activated and growth could continue. Complete loss of PTEN may result in cell apoptosis, senescence or movement to another inactive state. However, this is hard to accomplish without damage to normal cells. Thus, the partial inactivation of a growth suppressor can cause excessive growth, whereas total removal of the protein could block cell growth by causing the cell to enter a different cell state.

Another suppressor of growth is tropomyosin, a protein normally involved in the control of muscle contraction. Many cancers are associated with decreased levels of tropomyosin as a result of increased levels of a specific microRNA, MiR21. MicroRNAs bind to mRNAs for specific proteins and cause their degradation, thereby decreasing the level of the protein. MiR21 targets many important proteins, including PTEN, PI3K, and tropomyosin. In studies of cancers with high levels of MiR21, there is no consistent decrease in the levels of PTEN or PI3K, but tropomyosin is consistently decreased. When this protein is decreased in cells grown on soft agar, the cells are transformed for growth; however, restoring tropomyosin to its normal level results in normal cell behavior. In the rigidity-sensing process described in Chapter 16, tropomyosin is a necessary component for the contractile units to form and test matrix rigidity. When those contractile units are absent, the cells will not activate growth-suppressor activities and will grow on soft surfaces. This is all consistent with MiR21's role in severe cancers. Further, there is evidence that some of the anti-cancer effects of the spice, curcumin, are through its inhibition of the production of MiR21 by binding to the promoter of MiR21 transcription. With the loss of the contractile units, cells generate higher forces on soft substrates that then stimulates growth. Thus, the loss of tropomyosin results in the loss of contractile units responsible for suppressing growth on soft surfaces and causes cells to adopt a new state where they generate high forces on soft matrices that will favor growth.

Figure 19.3

Growth suppressors.

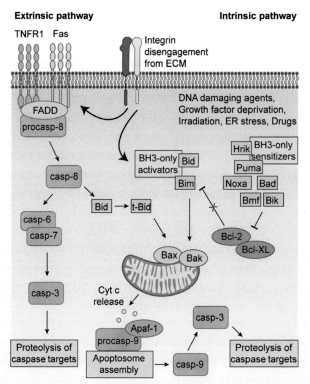

19.13 Resisting Cell Death by Apoptotic Mechanisms

When fully differentiated cells move to a foreign environment for any reason, they should not grow, but rather should commit suicide (i.e. apoptose). There are several mechanisms of apoptosis, but the most common involve the leakage of mitochondrial enzymes into the cytoplasm to set off a cascade of proteolytic and DNA degradation events that cause cell death. Historically, the earliest evidence of apoptotic mechanisms came from the studies of cancer cell growth *in vitro*. Control cells would not grow on soft agarose whereas cancer cells would. The cancer cells were described as transformed but the mechanism of death of the control cells was not understood until much later. As indicated in Figure 19.3, the loss of matrix adhesions activates the pathways for apoptosis, including the Fas-FADD pathway and the more conventional Bid and Bim activation of mitochondrial release of cytochrome C. The connections between integrin disengagement and the apoptotic pathways are not well understood. As in most important cell decisions, the cell typically takes hours to a couple of days to decide whether to apoptose. Cancer cells resist apoptosis, but it is not exactly clear how they do that. One idea is that the growth activation is synonymous with

Figure 19.4

Resisting cell death by apoptotic mechanisms.

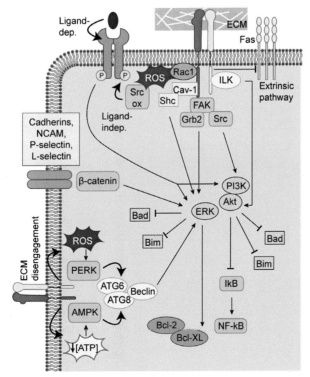

inactivation of apoptosis. Another is that the loss of components like tropomyosin and PTEN make it much more difficult to activate apoptotic pathways. Importantly, apoptosis may be upregulated or induced therapeutically, such as with taxol, which appears to induce apoptosis by acting on a checkpoint of mitosis. An alternative way to stop tumor growth somewhat selectively would be to restore the apoptotic response to matrix or cellular environment signals.

19.14   Enabling Replicative Immortality

Another check on tumor cell growth is provided by telomeres in differentiated cells. Telomeres are specific sequence motifs at the ends of each chromosome and shorten with each replication of the DNA. Upon the shortening of the telomeres, cell growth slows and there is an increased probability that cells will go into a senescent state. Replicative immortality refers to the expansion of the telomeric sequences through telomerase activation that occurs in most cancers to overcome the telomere check on limitless growth. The continued maintenance of telomeres in cancers can be explained as the result of a number of pathways, but is clearly an important component for severe cancers. In terms of possible mechanisms of telomerase activation, it is useful to note that the conversion of normal cells to pluripotent stem cells through the expression of four chromatin-remodeling enzymes will also result in telomerase activation. Further, adult stem cells that are needed for tissue repair should have additional telomerase activity to undergo extra rounds of cell division and the mechanism of activation in stem cells could be adopted for cancer cells. Thus, in thinking about replicative immortality, telomerase activation is needed for continued cancer cell growth, but it is not sufficient for cancer cell growth.

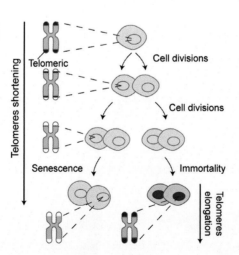

Figure 19.5

Replicative immortality.

19.15   Inducing Angiogenesis

Angiogenesis, or the formation of blood vasculature, is an important factor for the growth of larger tumors. Normally, vascularization is induced in any tissue when cells are placed under hypoxic conditions, and this will typically occur in active tissues when cells are more than ~200 μm from a blood vessel. As tumors become larger than a millimeter, their growth will be severely compromised if vascularization is not induced. This is because of the slow transport rate of nutrients and oxygen. Naturally, tumor cells will generate vascularization signals and the hormones that are common in cancers, EGF and PDGF, will further aid the vascularization of tumors.

Because cell organization in tumors is often unlike that in normal tissues, tumor masses are often not well vascularized in that the blood vessels within them are often leaky and poorly formed. This means that larger objects, including cancer cells from the tumor mass, can move into the bloodstream and travel to another tissue, and vice versa. The poor integrity of tumor vasculature, however, can be exploited to improve the delivery of cancer therapeutics. Such strategies rely upon the fact that most tissue vasculature is not as leaky as in tumors, and therefore nanoparticles (10–100 nm) containing anti-cancer drugs such as taxol will preferentially load into tumors over other tissues. However, the success of this approach has been limited and obviously cannot treat small tumor masses that don't need vascularization. Thus, although vascularization of large tumors is necessary for their growth, this is not a true Achilles heel of tumors in that small tumors do not need vascularization to survive.

19.16 Competition for Space in Confined Organs

In encapsulated organs such as the liver, kidney or brain, tumor cells and normal cells are competing for space as well as nutrients. This not only leads to difficulties in terms of the relative ability of cells to grow in a confined space, but can increase pressures in the organs to the point that blood flow becomes inhibited. As pressure in a tissue increases, the space between cells decreases and nutrient diffusion into the tissue is also inhibited. In many cancers, mucin is overexpressed and this serves to increase the gap between cells, thereby allowing better diffusion of nutrients to the tumor cells, as well as waste products away from the cells. Other factors may also contribute to the ability of the cancer cells to crowd out the normal cells. Naturally, competition between cells in a limited space must occur, yet we know very little about how this is regulated. Recent studies are exploring why some cell types continue to grow in a confined space as other cell types die in the same space. Thus, it appears that factors like increased Mucin expression that increase viability of tumor cells in low oxygen and nutrient environments will also enable tumor cells to grow at the expense of normal cells in the same organ.

19.17 Metastasis and Migration

Once cancer cells spread from the tumor to neighboring tissues or other parts of the body, the prognosis for the patient is considered poor, and death will often result. One major method by which cells spread is tissue migration. This often involves a sheet of cancer cells, and is similar to epiboly in fish embryos where a connected sheet of cells migrates into a new region, typically along a basement membrane. During this migration, the cells at the front of the sheet, or the leading cells, have different behaviors to those that are following. Once the sheet occupies a new environment, however, the cells can consolidate to form a new tumor. Recent studies indicate that tumor-associated fibroblasts are often the leading cells during metastasis. There is interest in understanding why fibroblasts that surround a tumor will adopt a different, more migratory state than normal fibroblasts from the same tissue. Experimentally, tumor fibroblasts keep a distinct phenotype for weeks after culture *in vitro*. Thus, although the migration of tumor cells has been a major focus of research on metastasis, the tumor fibroblasts also appear to have a significant role in the migration of groups of cells from the tumor.

In the classical view of metastasis through dissemination of circulating tumor cells depicted in Figure 19.6, the poor vasculature of the tumor will enable cells to enter the blood stream. Because the tumor cells are larger than red blood cells and most white blood cells, they are often captured by the first capillary bed that they encounter after leaving the tumor. Recent studies that analyzed venous blood from the arms of cancer patients showed that the number of circulating tumor cells can range up to 100–200 cells per ml of blood. This is probably a lower estimate of the number of cells that are exiting the tumor, because the venous blood from the arm will likely have gone through two capillary beds in the transit from the tumor to the arm (the blood from the tumor will normally go through capillary beds in the lung and the hand before reaching the arm). With 5 liters of blood in the body, the number of circulating tumor cells would be greater than 5×10^5 and those should be captured rapidly in capillary beds or the spleen (a long estimate is 30 min). Thus, in a day over 2.4×10^7 tumor cells would be deposited in other parts of the body, possibly dispersing more than 10^9 in a month. Initially these cells would occupy only a couple of milliliters in volume. Once caught in a capillary, the cells will often leave the bloodstream through extravasation and then they can form a new tumor. *In vitro* assays of cell migration and force generation often correlate with the metastatic potential of tumor cells, and this indicates that motility is a critical factor even if the dissemination is through the circulation. Importantly, not all tumor cells will adhere to a new environment and form a new tumor. These cells are particularly difficult to eradicate as they are unlikely to be killed by cancer chemotherapy that targets growing cells. For this reason, there will often be

Figure 19.6

Invasion and metastasis.

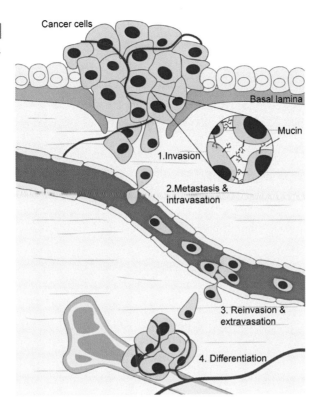

a large number of quiescent but living tumor cells dispersed around the body that can later be activated for growth. Thus, the migration or at least the motility of tumor cells appears to be an important factor in metastasis and the great number of metastatic cells enables a rare activation event to produce a metastatic tumor.

For tumor cells to grow in other regions of the body they will often undergo differentiation, which allows them to adapt to the new environment. In a study of metastasis in nude mice, human breast cancer cells were injected into the bloodstream (nude mice would not reject the human cells), and after a period of several weeks, the mice were sacrificed and the tumors derived from the human breast cancer cells were harvested. Common metastasis sites were lung and bone. By serially passaging the lung tumor cells in multiple mice and isolating lung tumors each time, the group was able to isolate a population of human breast-derived cells that would consistently produce metastatic tumors in mouse lungs. A population of cells that would consistently produce tumors in mouse bone was also prepared. When the group analyzed the expression levels of a number of proteins associated with cancer, they found dramatic changes from the original line, and many differences between each of the metastatic lines. Thus, the original human breast cancer cells were capable of adapting to the lung environment in one path and to the bone environment in another. Differentiation of the metastatic tumor cells may diminish their response to the original chemotherapy protocol, and the presence of tumors in multiple tissues can result in death.

19.18   Therapies

For each type of cancer, and depending on the patient's prognosis, different treatment protocols have been tried. Generally, these treatment regimens will meet with varied levels of success. There are a few cancers that have been cured, but in many cases the primary effect of treatment is to merely delay the course of the disease. Surgery is the most effective primary treatment and targeted radiation therapy is a close second. Drug cocktails typically include microtubule and DNA-altering drugs that inhibit the growth of all cells. The lack of specificity for tumor cells leads to a range of adverse effects, including hair loss.

Targeted inhibitors theoretically provide a rational approach to the treatment of specific cancers because they attack the specific pathways that are needed for tumor growth without causing general inhibition of cell growth. Although there are an increasing number of successful treatments, many cancers are only treatable by more general inhibitors of growth. Gleevec, for example, has been particularly effective in the treatment of BCR-ABL induced chronic myeloid leukemia by targeting the ABL kinase. If, however, additional growth promoting factors occur, then the tumor can avoid being killed by a single drug treatment. In the case of melanoma, which generally has a poor prognosis because of the extensive metastasis and recurrence through alternative pathways, chemotherapy regimens that target two specific growth pathways simultaneously are proving better than protocols that target single pathways sequentially. Thus, it is common to treat with multiple inhibitors, and hopefully, multiple targeted inhibitors. Still, general cytotoxics are used extensively in the treatment of many cancers. A logical problem with the cytotoxic drugs is that they will cause extensive damage to many different tissues and trigger a general trauma response after the treatment is over. The trauma response is important for repair, but can also activate quiescent metastatic cells, causing metastatic tumors. Conventional targeted drug treatments have had several successes, but the general cytotoxic treatments have had many undesired side effects.

Recently, there is considerable excitement about immunotherapies for cancer. There are several possible immunotherapy approaches, but they typically rely upon the ability of the patient's immune system to kill the tumor cells through T-killer cells. Because the tumor will excite an inflammatory response, T cells will collect and some can target the tumor cells. Alternatively, the patient's T cells are themselves transfected with surface proteins that can bind to and kill the tumor cells. The weakness of this approach is that the T-killer cells have a relatively short life span. After injection into the patient, the T-killer cells will disperse throughout the body and kill tumor cells in many different areas, but after several days, the T-killer cells will die. There are efforts now to create memory T cells that can differentiate into T-killer cells with specificity for tumor cells.

Mechanical activity has been shown in mice to suppress tumors, but the basis is not understood. In the experiments in mice, human tumor cells were injected into the mice and they were either left in their cage or put onto a treadmill. After several weeks, the tumors in the exercised mice were significantly smaller. In humans, there are relatively few tumors in muscle and other active tissues and it is probably worthwhile to do controlled studies on the effects of mechanical stimulation on tumor growth. There may be a state of controlled growth with activity as in muscle after exercise that could be induced in other tissues and would suppress tumor growth. In terms of our previous discussions about cell states, there may be ways of mechanically inducing states in tumor cells that would lead to premature cell death or apoptosis.

As was discussed previously, the signals for growth in cells are integrated over time through cyclic pathways, such as the tyrosine phosphorylation–dephosphorylation pathways. The important downstream signals are not understood because metabolism, mechanical activity, microenvironment, and circulating hormones all can contribute to the decision to grow. If the elements were understood at a more quantitative level, then it may be possible to create conditions that would more selectively kill tumor cells. As an example, EGF receptor inhibitors can often kill most of the primary tumor cells when they depend on a known mutation in EGFR. However, the remaining cells will often change state and grow through a different signaling pathway. Thus, we need to not only understand how to kill cells that grow in the primary tumor, but also to understand which alternative growth pathways are likely for the metastasized cells. This is particularly difficult because the metastasized cells are dispersed into many different microenvironments. However, there are limited numbers of pathways that can be activated and it may be possible to identify the normal state transitions so that alternative pathways could also be inhibited to prevent metastasized tumors from growing. Once we understand the different possible growth pathways and how cells can transition from one to another, it may be possible to block both primary and secondary growth pathways. This understanding should include the factors both biochemical (hormones, metabolism and matrix composition) and mechanical (rigidity, oscillations and morphology) to better find the best way to selectively kill cells in a given growth pathway.

One unexpected success for the inhibition of cancer is to target specific ubiquitin-dependent protein degradation pathways. Because the degradation of specific proteins is critical for growth and is often different for cells in different states, a targeted inhibitor of E3 ligase or protein degradation pathways in general can block progression through the cell cycle and cause death. An example is the repurposing of the drug thalidomide that was banned for a long time because of the birth defects that it caused. However, recently, close relatives of thalidomide were found to be effective treatments for multiple myeloma. The killing of the immune cells was caused by the retargeting of an E3 ligase to cause the degradation of a critical protein for the growth of the myeloma cells. Although this seems a fortuitous accident, general inhibition of the proteasome complex has been beneficial in many cancer treatments. In these cases, the benefits are surprising, and it will take much further study to fully understand why some drugs are beneficial.

19.19 SUMMARY Tissue regeneration is critical for the repair of tissues following cell death or the effects of trauma or disease. However, over time, these repair processes can be perverted to produce cancers or periods of inappropriate growth. If the cancer cells can sustain growth, avoid undergoing apoptosis when in the wrong environments, and elongate their telomeres, then they have the chance to become serious cancers that can damage their host. Furthermore, if those cancers can metastasize to other tissue environments and adapt to other growth programs, they can evade treatments targeted to the primary tumor. Again, the signals for growth need to be integrated over time with other factors such as metabolism or mechanics and the normal regulatory mechanisms avoided. An important obstacle for the treatment of cancer is the loss of normal apoptotic pathways as is illustrated by the loss of rigidity-sensing sarcomeres that causes growth on soft surfaces. Several treatments have activated alternative apoptotic pathways through the alteration of ubiquitin-mediated proteolysis or the polymerization of microtubules by taxol. However, many cancer treatments involve killing all growing cells and that produces major trauma to the patient. The normal trauma response is to activate growth and that can activate metastasized tumor cells in new environments where they will differentiate and grow. Thus, it is important to be able to understand the multiple growth states that specific tumor cells can adopt so that the original tumor as well as the secondary tumors can be killed, thereby effectively eliminating all tumor cells from the body. This is the final chapter because the complexity of cancer provides an excellent example of the lessons of robust devices that serve not only to enable the organism to reach reproductive status but also can be perverted to sustain the cancers that typically kill older individuals.

19.20 PROBLEM

Take a recent paper on cancerous growth and develop a hypothesis that can be tested.

19.21 REFERENCES

Araki, K., Ebata, T., Guo, A.K., et al. 2015. p53 regulates cytoskeleton remodeling to suppress tumor progression. *Cell Mol Life Sci* 72(21): 4077–4094.

Hanahan, D. & Weinberg, R.A. 2011. Hallmarks of cancer: the next generation. *Cell* 144(5): 646–674.

Kanda, R., Kawahara, A., Watari, K., et al. 2013. Erlotinib resistance in lung cancer cells mediated by integrin beta1/Src/Akt-driven bypass signaling. *Cancer Res* 73(20): 6243–6253.

Wolfenson, H. & Sheetz, M.P. 2017. *Nature Rev Cell Mol Biol*, in press.

19.22 FURTHER READING

Cell–matrix adhesions: www.mechanobio.info/topics/mechanosignaling/cell–matrix-adhesion/

Clathrin-mediated endocytosis: www.mechanobio.info/topics/cellular-organization/membrane/membrane-trafficking/clathrin-mediated-endocytosis/

Muscle contraction: www.mechanobio.info/topics/cytoskeleton-dynamics/contractile-fiber/

Ras signaling: www.mechanobio.info/topics/mechanosignaling/ras-gtpases/

Src-integrin dependent growth pathway: www.mechanobio.info/topics/mechano signaling/cell–matrix-adhesion/integrin-mediated-signalling-pathway/

Telomere shortening: www.mechanobio.info/topics/genome-regulation/telomere-organization/

Tissue-degrading enzymes: www.mechanobio.info/topics/mechanosignaling/extracel lular-matrix-disassembly/

Index